Food Analysis (Essential Food Science)

Food Analysis (Essential Food Science)

Editor: Samuel Ford

Callisto Reference
www.callistoreference.com

Callisto Reference,
118-35 Queens Blvd., Suite 400,
Forest Hills, NY 11375, USA

Visit us on the World Wide Web at:
www.callistoreference.com

ISBN: 978-1-64116-768-0 (Hardback)

Cataloging-in-Publication Data

Food analysis (essential food science) / edited by Samuel Ford.
p. cm.
Includes bibliographical references and index.
ISBN 978-1-64116-768-0
1. Food--Analysis. 2. Food science. 3. Food. I. Ford, Samuel.
TX541 .F66 2023
664.07--dc23

Table of Contents

Preface

It is often said that books are a boon to mankind. They document every progress and pass on the knowledge from one generation to the other. They play a crucial role in our lives. Thus I was both excited and nervous while editing this book. I was pleased by the thought of being able to make a mark but I was also nervous to do it right because the future of students depends upon it. Hence, I took a few months to research further into the discipline, revise my knowledge and also explore some more aspects. Post this process, I begun with the editing of this book.

Food analysis is a discipline that is concerned with the development, application and study of analytical procedures for determining the properties of foods in terms of their composition, structure, physicochemical properties and sensory attributes. The structure of the food can be analyzed at molecular, physical and chemical levels. The analysis of the physiochemical properties of foods involves examining the rheological properties, optical properties, stability, and flavor of the food. The analysis of the sensory attributes of food helps in determining its quality and desirability. There are a range of methods used in food analysis including magnetic resonance (NMR) spectroscopy, gas chromatography (GC), atomic absorption spectroscopy (AAS), and high-performance liquid chromatography (HPLC). The technique of polymerase chain reaction (PCR) is also used for food analysis. PCR testing analyses specific DNA sequences and is primarily employed for identifying animal species in meat products. It is also used for detecting harmful pathogens and other microorganisms. This book traces the most recent advancements and trends related to food analysis. It is an invaluable source for reference for undergraduate students and professionals associated with the analysis of food.

I thank my publisher with all my heart for considering me worthy of this unparalleled opportunity and for showing unwavering faith in my skills. I would also like to thank the editorial team who worked closely with me at every step and contributed immensely towards the successful completion of this book. Last but not the least, I wish to thank my friends and colleagues for their support.

Editor

1

Analysis of 27 β-Blockers and Metabolites in Milk Powder by High Performance Liquid Chromatography Coupled to Quadrupole Orbitrap High-Resolution Mass Spectrometry

Jian-Qiao Cheng [1,2,†], Tong Liu [1,†], Xue-Mei Nie [1], Feng-Ming Chen [1], Chuan-Sheng Wang [2] and Feng Zhang [1,*]

1 Institute of Food Safety, Chinese Academy of Inspection & Quarantine, Beijing 100176, China; 13352479167@163.com (J.-Q.C.); liutongyes@163.com (T.L.); niexuemei_00@163.com (X.-M.N.); chenfengmingok@163.com (F.-M.C.)

2 College of Applied Chemistry, Shenyang University of Chemical Technology, Shenyang 110142, China; wchsh18@163.com

* Correspondence: fengzhang@126.com

† These authors contributed equally to this work.

Academic Editors: Alessandra Gentili and Chiara Fanali

Abstract: This paper presents an application of high performance liquid chromatography coupled with quadrupole orbitrap high-resolution mass spectrometry (HPLC-Q-Orbitrap HRMS) for the analysis of 27 β-blockers and metabolites in milk powder. Homogenized milk power samples were extracted by acetonitrile and purified by using Oasis PRiME HLB solid-phase extraction cartridges. The Ascentis® C8 chromatographic column was used to separate the analytes. The quantification was achieved by using matrix-matched standard calibration curves with carazolol-d_7 and propranolol-d_7 as the internal standards. The results show an exceptional linear relationship with the concentrations of analytes over wide concentration ranges (0.5–500 μg kg^{-1}) as all the fitting coefficients of determination r^2 are > 0.995. All the limits of detection (LODs) and quantitation (LOQs) values were within the respective range of 0.2–1.5 μg kg^{-1} and 0.5–5.0 μg kg^{-1}. Overall average recoveries were able to reach 66.1–100.4% with the intra- and inter-day variability under 10%. This method has been successfully applied to the screening of β-blockers and metabolites in commercial milk powders. At the same time, the corresponding characteristic fragmentation behavior of the 27 compounds was explored. The characteristic product ions were determined and applied to the actual samples screening.

Keywords: β-blockers; metabolites; milk powder; Q-Orbitrap

1. Introduction

β-blockers (BBS) are structurally analogous to the catecholamines, which can act as non-specific β-adrenergic receptor blocking agents. They play an extremely important role in the treatment of cardiovascular diseases such as coronary heart disease, hypertension, arrhythmia and cardiac insufficiency. However, improper use of β-blockers can cause an increase of myocardial oxygen consumption, vascular resistance, oxygen free radicals and myocardial cell apoptosis, etc. [1]. β-blockers are usually used in animals to reduce morbidity and mortality during transportation (to the slaughterhouse or livestock farm), mating, childbirth and in other stressful situations. Such stress usually results in a poor quality of meat, or even in the premature death of the animal [2–4]. The illegal use of β-blockers gives rise to drug residues in edible animal tissue, which can be metabolized in the body. Several metabolites of β-blockers are pharmacologically active and also harmful to the body. For example, 4-hydroxyphenyl carvedilol (the metabolite of carvedilol) exhibits an approximately thirteen-fold higher-adrenoreceptor blocking potency compared to carvedilol itself [5]. In order to protect public health, many countries and organizations began to establish regulations. For example, carazolol has maximum residue limits (MRLs) in animal-based foods. The European Union and the International Codex Alimentarius Commission have asked for MRLs of carazolol in edible animal tissues, with MRLs of 25 $\mu g\ kg^{-1}$ for porcine kidney and 15 $\mu g\ kg^{-1}$ for bovine kidney. The European Union has also asked for a maximum residue limit of 1.0 $\mu g\ kg^{-1}$ of carazolol in the milk powder [1,6]. However, no restrictions have been placed upon their metabolites (4-hydroxyphenyl carvedilol, etc.). Due to the fact that eating food containing high levels of carazolol and other β-blockers can be harmful to consumer health (especially to infants and children), control of β-blockers is required [3]. Therefore, it is necessary to establish a high-throughput analytical method for β-blockers and their metabolites.

Many approaches for the detection of β-blockers have been reported, such as enzyme-linked immunosorbent assay (ELISA) [7], gas chromatography coupled with mass spectrometry (GC-MS) [8–11], liquid chromatography with fluorescence detection (LC-UV) [12], and liquid chromatography mass spectrometry (LC-MS) [13–17]. Although these methods play important roles in the detection of β-blockers, they also have some drawbacks. For example, the quantification of the ELISA method is not accurate and the operation is troublesome, and GC-MS involves derivative steps before chromatographic separation, which are time-consuming and which increase the possibility of contamination [18]. And the LC-UV method is limited by poor sensitivity, so it cannot meet the requirements for ultra-trace analyses. LC-MS was the most widely used method for the qualitative and quantitative detection of β-blockers multiple residues. Liquid chromatography is generally used, coupled with a low resolution mass spectrometry (LRMS) analyzer such as triple-quadrupole (QqQ). Orbitrap is the newest HRMS analyzer. Most identification and determination studies of β-blockers were undertaken using the LTQ Orbitrap (linear ion trap quadrupole Orbitrap high resolution mass spectrometry), achieving LODs below 2 $\mu g\ kg^{-1}$ and 5 $ng\ mL^{-1}$ [19,20]. The Q-Orbitrap (Q-Exactive^{TM}, hybrid quadrupole-orbitrap mass spectrometer, Thermo Fisher Scientific, Bremen, Germany) combines high-performance quadrupole precursor selection with high resolution and accurate mass (HR/AM) Orbitrap detection, which has great potential to avoid both false positive and negative results in residue analyses. Compared with the LTQ Orbitrap, the Q-Orbitrap has higher sensitivity, and its use has become widespread in the confirmation and quantification of drugs residues in food [21,22]. In addition, it can realize real-time positive and negative switches; therefore, the time spent on the preparation process and method optimization is significantly reduced. However, analysis of β-blockers and their metabolites using the Q-Orbitrap has not been reported.

In this study, a high-throughput, high performance liquid chromatography coupled to quadrupole Orbitrap high-resolution mass spectrometry (HPLC-Q-Orbitrap HRMS) has been developed for the screening of 27 analytes, including 21 β-blockers and 6 metabolites in milk powder samples. In addition to this, the corresponding characteristic fragmentation behavior and the product ions of the 27 compounds are described in detail. They will provide a basis for the target-free screening of these drugs and the identification markers of the newly-emerging β-blockers residues.

2. Results and Discussion

2.1. LC Parameters Optimization

Chromatographic conditions were studied in order to obtain the best separation and retention for the compounds. Four HPLC columns (Waters ACQUITY UPLC® BEH C18 (1.7 μm, 50 × 2.1 mm, Milford, MA, USA), Thermo Accucore aQ (2.6 μm, 150 × 2.1 mm, Bellefonte, PA, USA), ALDRICH Ascentis® C8 (3 μm, 10 cm × 4.6 mm, Bellefonte, PA, USA), Waters ACQUITY UPLC^{TM} BEH Phenyl (1.7 μm, 50 × 2.1 mm, Milford, MA, USA) were evaluated in (0.1% FA) H_2O-MeCN in their appropriate gradient elution at 0.5 $mL\ min^{-1}$. Similar separation performances for total analytes were observed by the first three columns; however, the Ascentis® C8 column provided better shape and retention for hydroxyatenolol, as shown in Figure 1. Therefore, the Ascentis® C8 column was chosen.

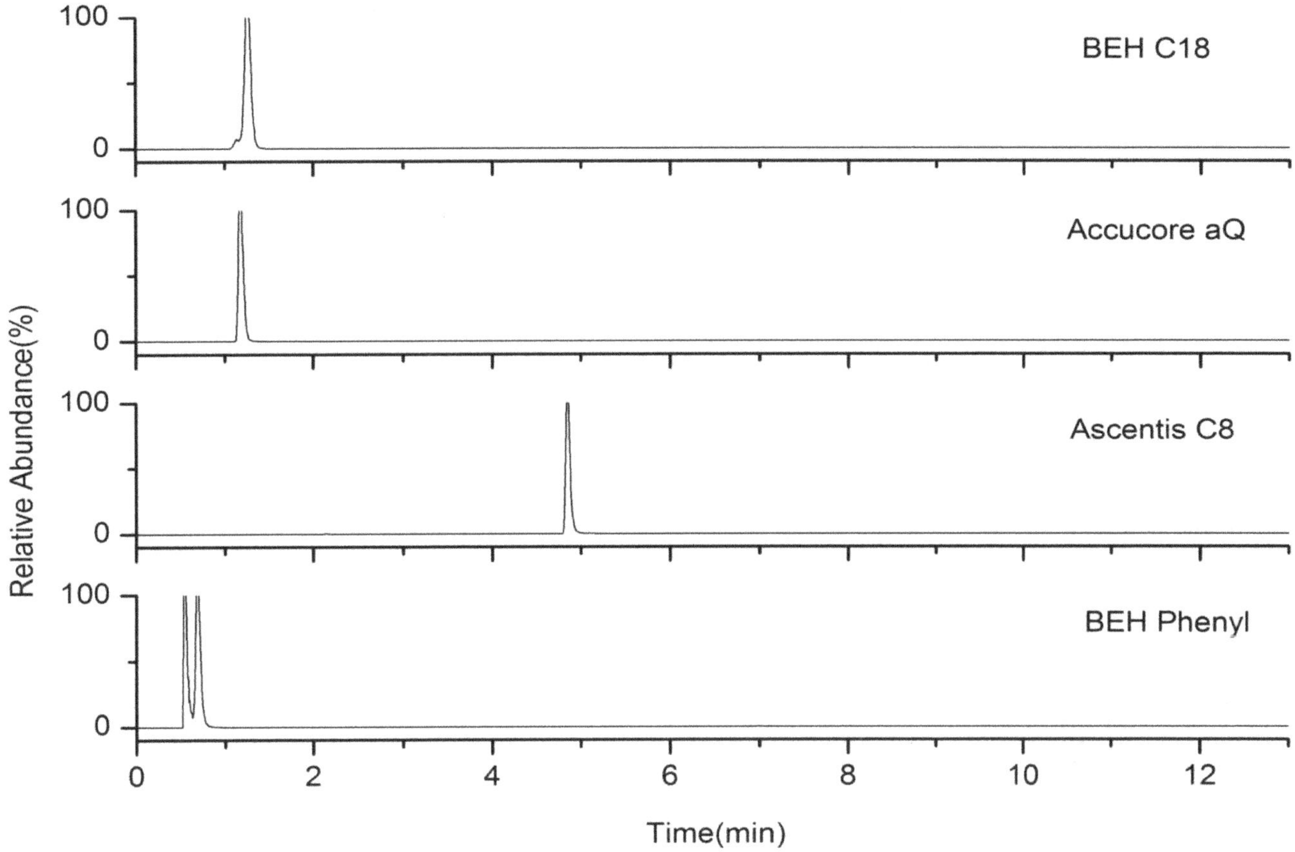

Figure 1. The effect of the columns on the chromatographic separation of hydroxyatenolol.

Several mobile phases were tested using MeCN or MeOH as an organic solvent and water as a polar solvent with FA addition (from 0% to 0.5%). As shown in Figure 2A, the mixture of MeOH and water showed higher responses and better separation for isomers of practolol and atenolol. The proton donor tendency of MeOH contributed to the formation of positive adducts. The better volatility and lower surface tension of MeOH can also improve desolvation of the droplets. Therefore, MeOH was chosen as the organic solvent. On the other hand, the addition of FA could improve the phenomenon of peak tailing and the response, which is probably because the excess silanols in the stationary phase combine with the acid rather than with their targets. The retention time of isomers of atenolol and practolol were greatly affected by pH, so a good chromatographic separation could not obtained until 0.1% FA was added (mobile phase pH = 2.58) (Figure 2B). Therefore, 0.1% FA-H_2O was chosen as the water phase.

In addition, injection volumes of 2 μL to 5 μL were evaluated using the aforementioned conditions. Taking the isomers of atenolol and practolol as examples, the experimental results showed that the separation factor decreased from 1.41 to 0.77 with increasing injection volume. So, 2 μL was chosen as the experimental injection volume.

Furthermore, several elution gradient profiles were also optimized to obtain better chromatographic separation and less analysis time (within 15 min). Other parameters (flow rate, column temperature) were also characterized to achieve better target separation and peak shapes. Under these conditions (see Section 3.2), the retention times (RT) of these 27 analytes were constant, and ranged ranging from 3.85 min to 9.01 min.

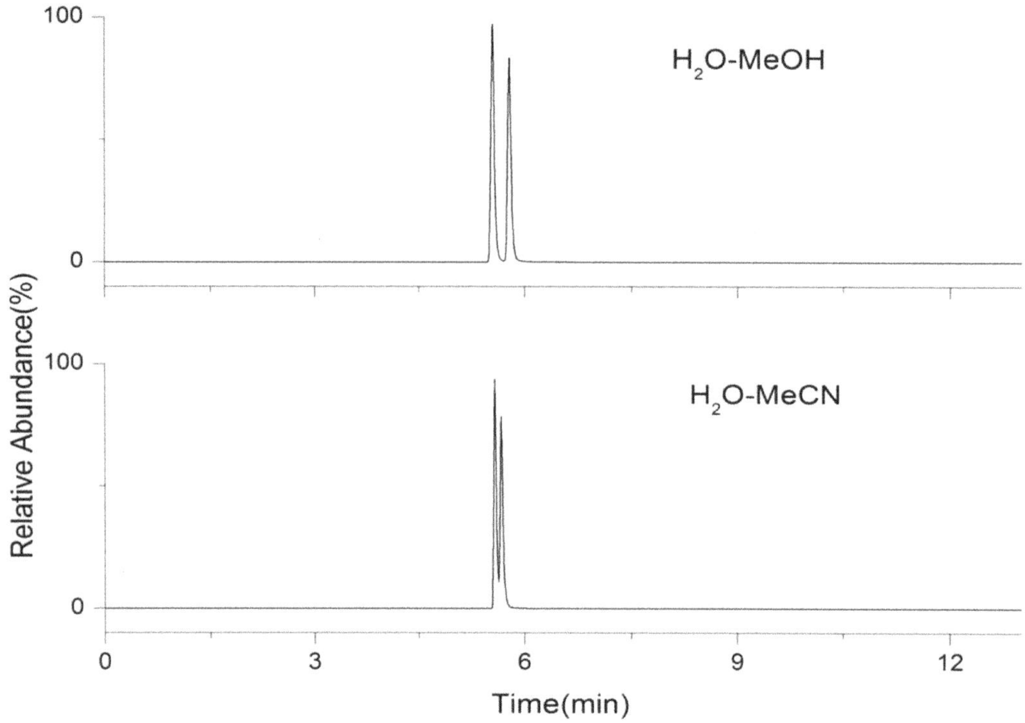

(A)

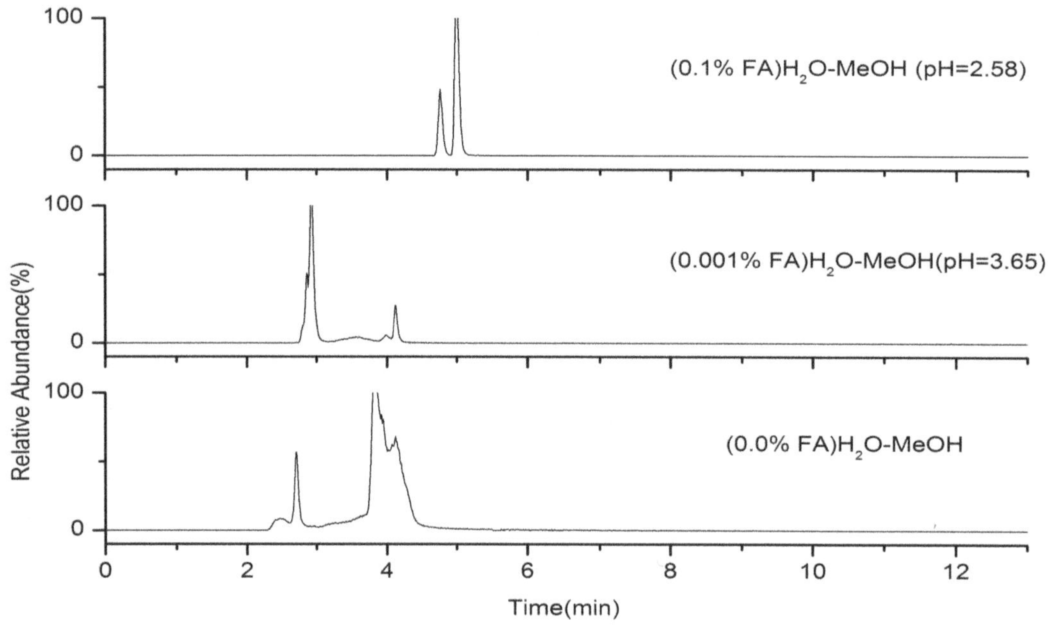

(B)

Figure 2. The effect of different mobile phases on the chromatographic separation of the isomers of atenolol and practolol. (**A**) Different organic solvents (atenolol:practolol = 2:1); (**B**) Different concentrations of FA in water phase (atenolol:practolol = 1:1).

2.2. Optimization of the Mass Spectrometric Parameters

The optimization of MS parameters was performed by infusing a standard solution of 100 μg L^{-1} of each β-blockers in methanol-water (50:50, v/v) as the mobile phase under Full scan mode (Full MS). The precursor ions were selected in both positive and negative modes. Consistent with previous report [2], β-blockers and metabolites tend to form $[M + H]^+$ adduct ions in positive mode. The Full

MS/ddMS^2 scan mode, which can achieve the non-target list qualitative and quantitative detection in a single run, was used for screening all samples. All the MS parameters were optimized to provide the best responses of the analytes. The optimized parameters values are summarized in Section 3.2.

2.3. *The Proposed Fragmentation Pathways for 27 β-Blockers*

In this study, secondary mass spectrometry data of all the substances were extracted, and the fragment ions occurring many times were selected to analyze the fragmentation pathway. β-blockers are mainly classified into three kinds of structures. Labetalol and sotalol have the structure of phenylethanolamine, and others have the structure of aryloxypropanolamine except for timolol and hydroxytimolol, which have special chemical structures. The principal structures of phenylethanolamine and aryloxypropanolamine compounds are shown in Figure 3.

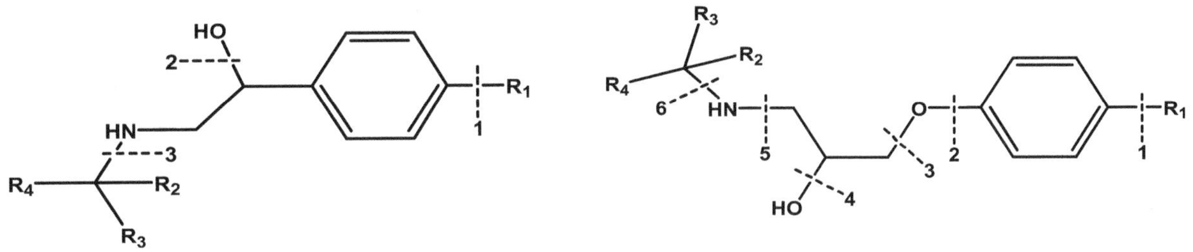

(**A**) (**B**)

Figure 3. Principal structure of (**A**) phenylethanolamine compounds; (**B**) aryloxypropanolamine compounds.

2.3.1. Phenylethanolamines Structure

For phenylethanolamine structure, there is a characteristic loss of one molecule of water at position 2 at first, and then a characteristic cleavage at positions 1 and 3, as shown in Figure 3A. This phenomenon was consistent with the fragmentation of β-agonists compounds with the phenylethanolamine structure that were studied in our laboratory [23].

2.3.2. Aryloxypropanolamines Structure

For the aryloxypropanolamine structure (Figure 3B), the proposed fragmentation pathways can be divided into three types, depending on the different substitution groups.

Type I (R_6 was H, R_7 and R_8 were methyl)

For type I, the bonds between carbon and oxygen were preferentially broken at position 2. At first, the phenyl structure (C_6H_4-R_5) was lost to form the fragment ion $[C_6H_{14}NO_2 + H]^+$ at *m/z* 133.06412. Then, the potential loss of $[C_3NH_9]^+$ or $[OH]^+$ gave *m/z* 74.06063 (formula $C_3H_6O_2$) or 116.10702 (formula $C_6H_{14}NO$) moieties. And the loss of one molecule of water and a series of fragmentations occurred. The suggested fragmentation pathway of type I is shown in Figure 4.

It is worth mentioning that the ϒ-H of the amine structure with 1-propene ($C_6H_{12}N$ at *m/z* 98.09663) rearranged to an unsaturated group, accompanied with the cleavage of the β-bond in the amine structure to produce the McLafferty Rearrangement. Imine structure of the fragment C_3H_7N (at *m/z* 57.07101) was produced by this reaction. The mechanism of the McLafferty Rearrangement reaction is shown in Figure 5.

Type II (R_6, R_7 and R_8 were methyl group)

For type II, the R_6 position of the aryloxypropanolamine structure was substituted with a methyl group. The methyl group can easily get lost to produce a type I structure, and then have a similar fragmentation pathway to the type I mentioned above.

Type III (R_6 or R_7 was an isophthalic ether structure, others were H)

For type III, the R_6 or R_7 position of the aryloxypropanolamine structure was substituted with an isophthalic ether structure, which has strong electronegativity. The bond between carbon and nitrogen at position 8 was easily broken, and it was difficult to form the fragment at *m/z* 116.10702 (formula $C_6H_{14}NO$).

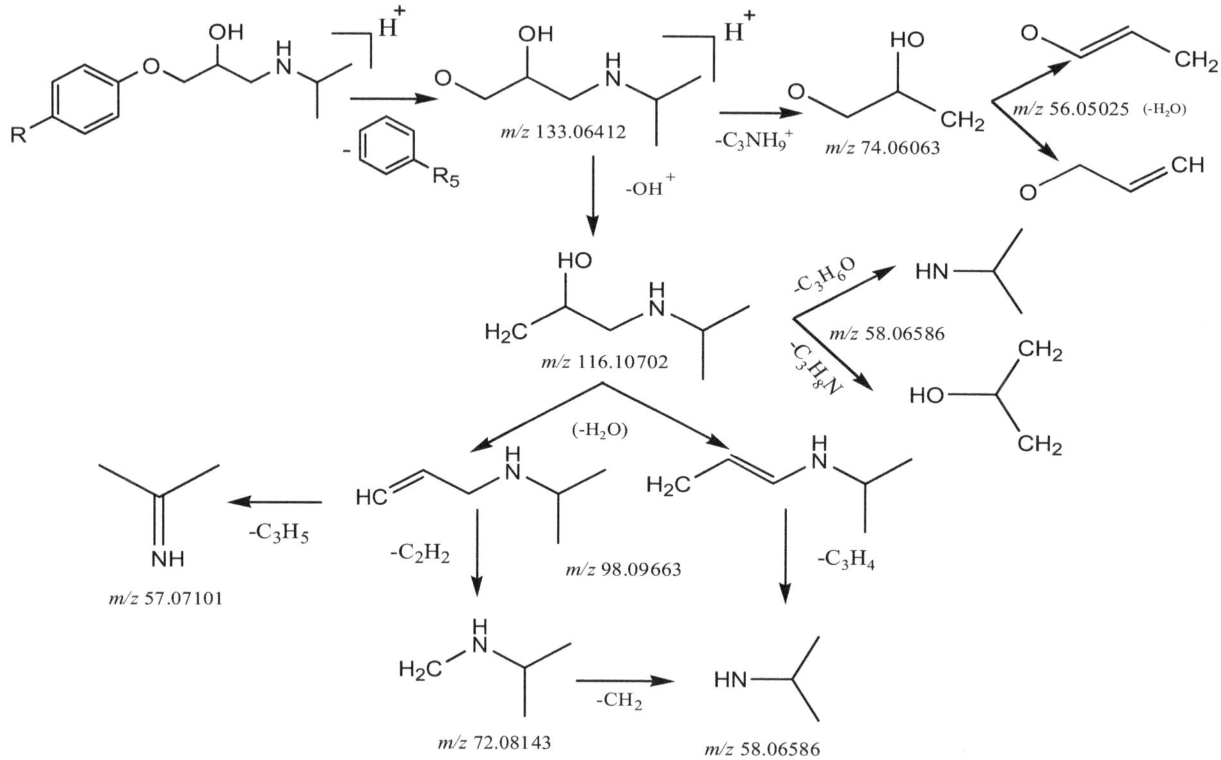

Figure 4. The suggested fragmentation pathways of aryloxypropanolamine compounds.

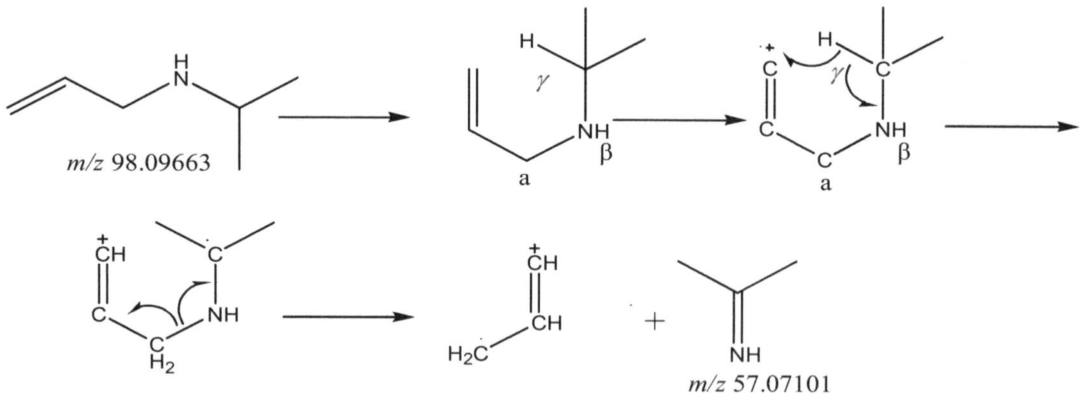

Figure 5. The McLafferty Rearrangement of *m/z* 98.09663.

2.3.3. Special Structures

For timolol and hydroxytimolol (structures see Figure 6.), they can be considered as β-blockers for newly-emerging aryloxypropanolamine compounds, which differ in structure from the others. These two compounds have characteristic fragments at *m/z* 74.06063, *m/z* 57.07101 and *m/z* 56.05025.

The structure types of the 21 β-blockers and 6 metabolites are listed in Table 1. The possible structures of the corresponding characteristic fragments are described and summarized in Table 2. By exploring the exact mass of these identification markers, it is possible to find newly-emerging β-blockers residues in a complex food matrix.

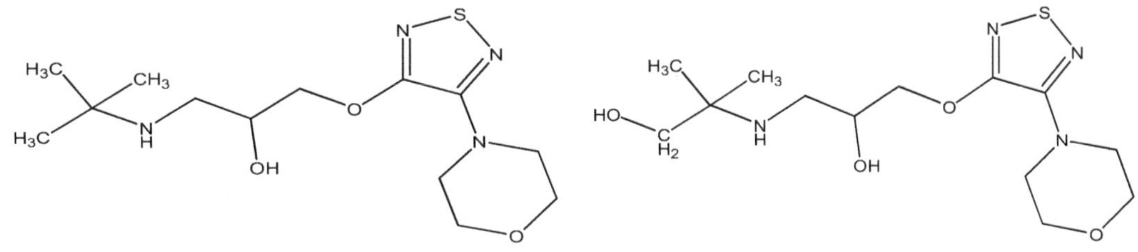

(A) (B)

Figure 6. Chemical structures of timolol and hydroxytimolol, (**A**) Timolol, Theoretical m/z 317.16419 (**B**) Hydroxytimolol, Theoretical m/z 333.15910.

Table 1. Formula, ionization mode, theoretical mass, measured mass, mass accuracy and MS^2 data for 27 β-blockers.

Analytes	Formula	Theoretical Mass [m/z]	Measured Mass [m/z]	Accuracy $^{\alpha}$ [ppm]	MS^2	Structure Type
Carazolol	$C_{18}H_{22}N_2O_2$	299.17540	299.17484	1.87	222.09090 116.10712 98.09675 72.08148 56.05032	Type I
Oxprenolol	$C_{15}H_{23}NO_3$	266.17507	266.17496	0.41	133.06451 116.10696 98.09662 72.08138 56.05027	Type I
Propranolol	$C_{16}H_{21}NO_2$	260.16451	260.16373	3.00	183.07979 116.10689 98.09659 72.08135 58.06586	Type I
Alprenolol	$C_{15}H_{23}NO_2$	250.18016	250.17953	2.52	173.09550 116.10687 98.09660 72.08134 56.05024	Type I
Bisoprolol	$C_{18}H_{31}NO_4$	326.23258	326.23169	3.00	133.06441 116.10689 98.09659 74.06059 56.05026	Type I
Betaxolol	$C_{18}H_{29}NO_3$	308.22202	308.22174	0.91	133.06445 116.10691 98.09661 72.08135 56.05004	Type I
Sotalol	$C_{12}H_{20}N_2O_3S$	273.12674	273.12680	0.46	255.11484 213.06822 176.12991 133.07552 198.05713	phenylethanolamine
Pindolol	$C_{14}H_{20}N_2O_2$	249.15975	249.15961	0.56	172.07532 116.10711 98.09679 72.08147 58.06597	Type I
Nadolol	$C_{17}H_{27}NO_4$	310.20128	310.20084	1.42	354.13795 236.12750 201.09059 74.06068 56.05030	Type II
Timolol	$C_{13}H_{24}N_4O_3S$	317.16419	317.16367	1.64	261.10089 244.07440 188.04840 74.06068 57.07074	Special structure
Acebutolol	$C_{18}H_{28}N_2O_4$	337.21218	337.21310	2.73	218.11726 116.10712 98.09680 72.08150 56.05036	Type I
Celiprolol	$C_{20}H_{33}N_3O_4$	380.25438	380.25299	3.67	324.19070 307.16397 251.10155 74.06061 56.05026	Type II

Table 1. *Cont.*

Analytes	Formula	Theoretical Mass [*m*/*z*]	Measured Mass [*m*/*z*]	Accuracy $^{\alpha}$ [ppm]	MS^2	Structure Type
Labetalol	$C_{19}H_{24}N_2O_3$	329.18597	329.18613	0.49	311.17395 294.14755 207.11201 179.08063 162.05423	phenylethanolamine
Cloranolol	$C_{13}H_{19}Cl_2NO_2$	292.08656	292.08658	0.07	236.02318 218.01273 174.97054 74.06063 56.05020	Type II
Penbutolol	$C_{18}H_{29}NO_2$	292.22711	292.22672	1.33	236.16374 201.12683 133.06451 74.06063 57.07070	Type II
Practolol	$C_{14}H_{22}N_2O_3$	267.17032	267.16965	2.51	190.08589 116.10711 98.09682 72.08146 56.05036	Type I
Carvedilol	$C_{24}H_{26}N_2O_4$	407.19653	407.19565	2.16	283.14340 224.12755 100.07599 74.06063 56.05036	Type III
Bupranolol	$C_{14}H_{22}ClNO_2$	272.14118	272.14020	3.60	216.07790 198.06741 181.04089 74.06061 56.05027	Type II
Atenolol	$C_{14}H_{22}N_2O_3$	267.17032	267.16983	1.83	133.06412 116.10690 98.09663 74.06060 56.05026	Type I
Esmolol	$C_{16}H_{25}NO_4$	296.18563	296.18558	0.17	133.06467 116.10737 98.09705 72.08168 56.05050	Type I
Metoprolol	$C_{15}H_{25}NO_3$	268.19072	268.19028	1.64	133.06435 116.10693 98.09660 74.06057 56.05026	Type I
Diacetolol	$C_{16}H_{24}N_2O_4$	308.18088	308.18088	0.00	291.16943 116.10702 98.09670 72.08143 56.05031	Type I
α-hydroxymetoprolol	$C_{15}H_{25}NO_4$	284.18563	284.18472	3.20	133.06435 116.10691 98.09663 74.06059 56.05026	Type I
α-hydroxyatenolol	$C_{14}H_{22}N2O_4$	283.16523	283.16507	0.57	133.08632 116.10760 89.06059 74.06103 57.07010	Type I
(*S*)-Hydroxytimolol	$C_{13}H_{24}N_4O_4S$	333.15910	333.15823	2.61	261.10059 188.04814 146.11705 74.06059 56.05025	Special structure
7-Hydroxyproprenolol	$C_{16}H_{21}NO_3$	276.15942	276.15930	0.43	199.07463 116.1067 98.09663 74.06057 58.06586	Type I
4-Hydroxyphenylcarvedilol	$C_{24}H_{26}N_2O_5$	423.19145	423.19141	0.09	283.14267 240.12180 100.07578 74.06049 56.05022	Type III
1	56.05025	C_3H_4O	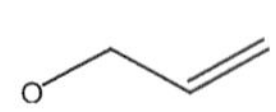 or 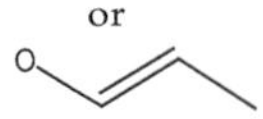			

$$^{\alpha}ppm = \frac{|m_{measured} - m_{theoretical}|}{m_{theoretical}} \times 10^6$$

Table 2. The possible structure of corresponding characteristic fragments.

No.	*m/z*	The Molecular Formula	The Possible Structure
2	57.07101	C_3H_7N	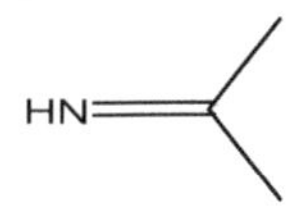
3	58.06586	C_3H_6O or C3H8N	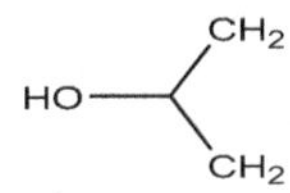 or 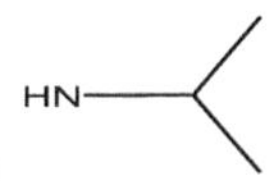
4	72.08143	$C_4H_{10}N$	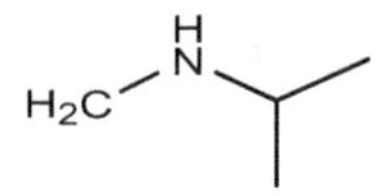
5	74.06063	$C_3H_6O_2$	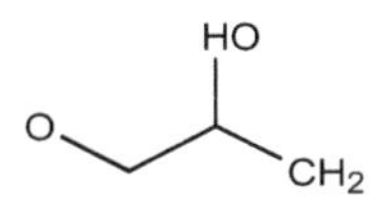
6	98.09663	$C_6H_{12}N$	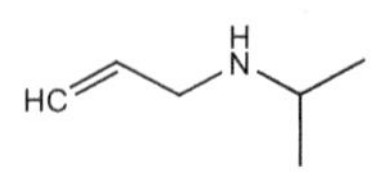 or 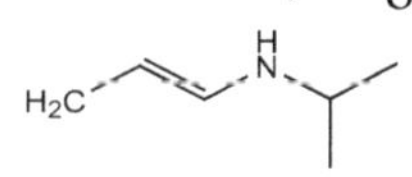
7	116.10702	$C_6H_{14}NO$	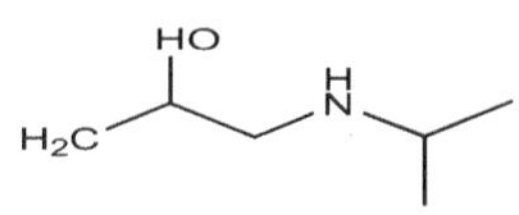
8	133.06412	$C_6H_{15}NO_2^+$	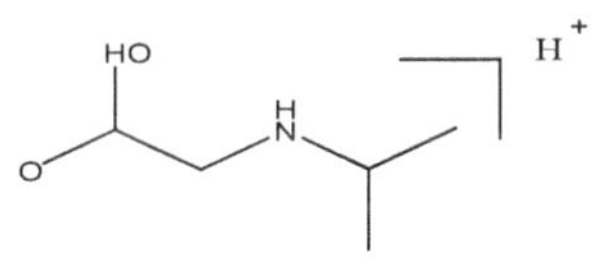

2.4. *The Optimization of the Sample Preparation Procedure*

A rapid enzymolysis method was chosen to ensure the processing flux and dissociate the possible bound residual drug. Neutral enzymatic environment (closing to pH 5.2) can significantly reduce matrix co-extraction. Na_2EDTA was added to the buffer to reduce the chelation between metal ions and strongly polar targets. The alkalized aqueous phase and salting out after the enzymolysis facilitate the target extraction into the organic solvent. pH at 9.0, 9.5, 10.0, 10.5, 11.0 and 12.0 were evaluated, and the best extraction efficiency was obtained at pH 10.0, which was consistent with the literature [1].

For food samples, MeCN was commonly used as the extraction solvent due to its protein precipitation ability. Since the acetic-buffer could be used to increase the recoveries of pH-dependent compounds, pure MeCN and different contents of acids (0.1% HOAc, 1% HOAc, 0.1% FA, 1% FA, v/v) in MeCN were compared for extraction efficiency in this study. For extraction solvents containing HOAc or FA, the recoveries of some analytes (such as sotalol, hydroxymetoprolol, labetalol, epractolol and hydroxytimoloven) were lower than 60%, as can be seen in Figure 7A. Pure MeCN provided better extraction efficiency with all analyte recoveries being higher than 65%; therefore, pure MeCN was found to be the most suitable extract solvent. Then, the solvent volume was investigated for optimization of the recoveries of the targets. It can be observed in Figure 7B that the recoveries of analytes increased with the solvent volume. When the solvent volume reached 15 mL, recoveries began to be stable. In order to ensure the stability of the recoveries, 20 mL extract solvent was chosen to extract all analytes.

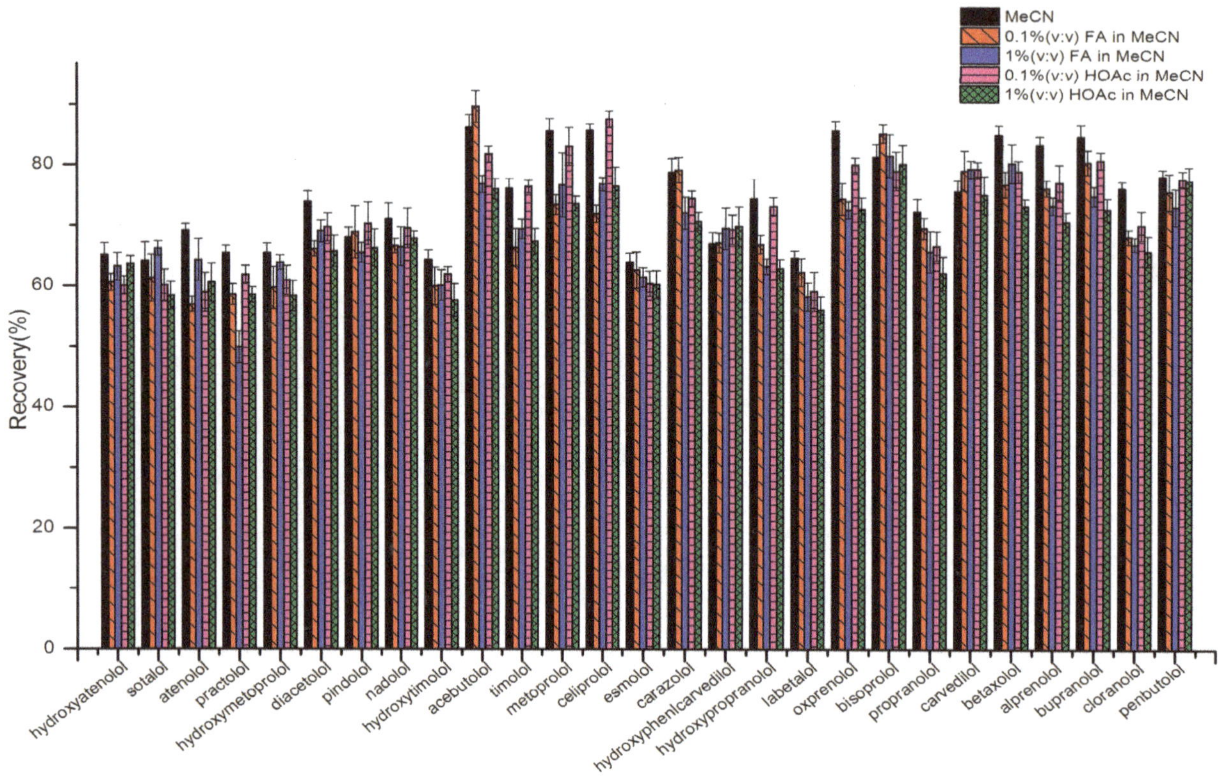

(A)

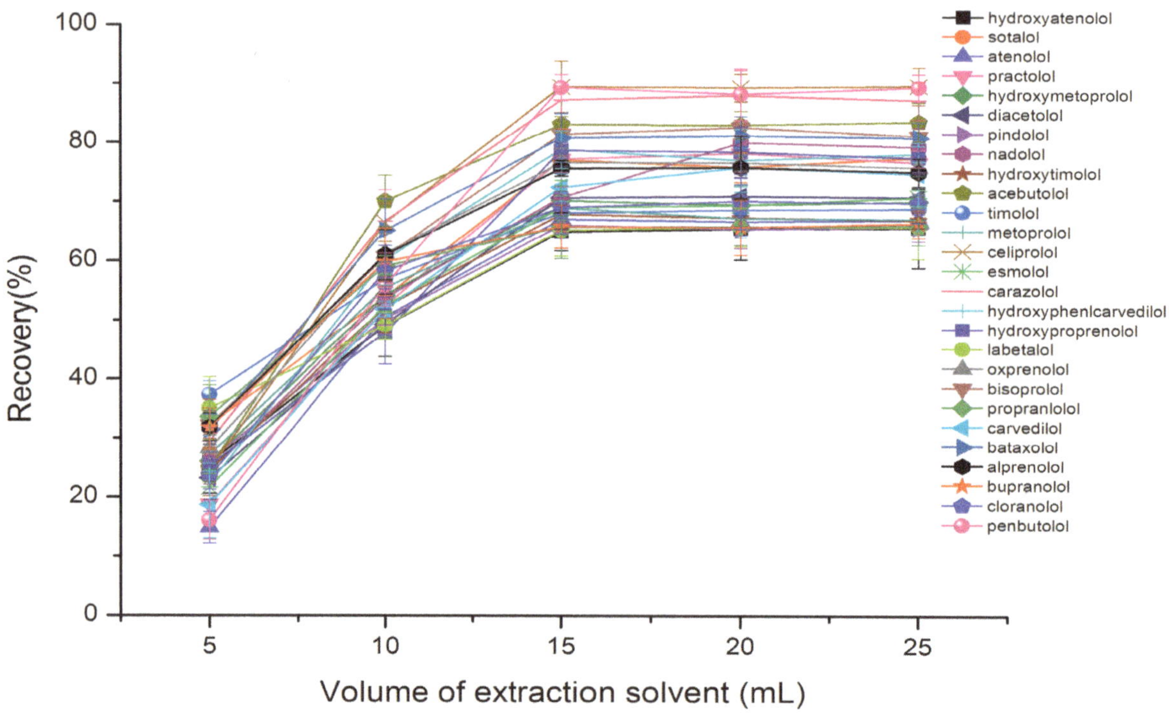

(B)

Figure 7. The effect of extracting solvents on the recovery of 27 analytes in milk powder. (**A**) The type of extract solvent, (**B**) the volume of extraction solvent (n = 3).

Taking into account the characteristics of the β-blockers and matrix interferences in milk powder samples, low temperature high-speed centrifugation, low temperature high-speed centrifugation + SPE (PRiME HLB column), and low temperature high-speed centrifugation + QuEChERS (quick, easy, cheap, effective, rugged and safe) methods were tested as purification steps. For the low temperature, high-speed centrifugation + QuEChERS method, the water removal step (using anhydrous magnesium sulfate) might take some water-soluble targets (such as sotalol) away, resulting in a low recovery rate. Compared with low temperature, high-speed centrifugation, PRiME HLB column used in SPE step could adsorb nonpolar interferences (some fats and phospholipids) in milk powder samples, which results in a smaller matrix effect and better target recoveries, as shown in Table 3. Therefore, the solid phase extraction PRiME HLB cartridge was selected for purification step.

2.5. *Validation of the Proposed Method*

2.5.1. Linearity and Sensitivity

The linearity of the proposed method was evaluated using matrix-matched spiked samples over the range of 0.5–500 μg kg^{-1}. Calibration curves resulted from the ratios of the peak area of the target compounds to the peak area of the isotope-labeled internal standards. The results showed a good linearity relationship with correlation coefficients (r^2) higher than 0.995 (Table 4). Limits of detection (LODs) and quantification (LOQs) are fundamental parameters used to evaluate the sensitivity of instructions and methods. The LODs were determined by the injection of a series of diluted standard solutions corresponding to a signal-to-noise (S/N) ratio of 3. The LOQs were determined by the injection of a series of spiked samples corresponding to a signal-to-noise (S/N) ratio of 10. Under the optimum condition, the LODs and LOQs were in the range of 0.2–1.5 μg kg^{-1} and 0.5–5.0 μg kg^{-1}, respectively, which allows the quantification of analytes presented at low content, indicating that good sensitivity was obtained.

Table 3. Validation parameters of the developed method.

NO.	Analyte	Matrix Effect C (%)		QC Concentration ($\mu g\ kg^{-1}$)	Average Recovery (%)	Intra-Day Precision (%) (n = 5)	Inter-Day Precision (%) (n = 5)
		PRiME HLB	Centrifugation				
1	Atenolol	115.7	118.5	2	72.5	2	3.1
				4	76.3	3.1	2.2
				8	74.8	4.1	5
2	Sotalol	85.6	86.8	2	83.6	3.2	3.2
				4	87	5.6	1.9
				8	81.1	2.7	7.9
3	Pindolol	101.7	108.2	1	89.2	1.6	5.2
				2	100.4	4.8	2.7
				4	83.7	3.6	6.4
4	Nadolol	102.2	112.8	0.5	83.6	7.9	5.5
				1	93.2	2.1	3.1
				2	78.5	4.4	5.3
5	Metoprolol	120.9	140	1	80.4	3.9	5
				2	90	6.2	3.5
				4	84.2	2.1	7.7
6	Timolol	116.9	133.7	1	76.5	2.1	7.4
				2	83.8	5.2	3.9
				4	78.8	3.8	3.2
7	Acebutolol	129.2	155.3	0.5	95.6	2.2	3.2
				1	89.3	5.7	5
				2	92.1	2.4	7.5
8	Oxprenolol	109.9	123.3	1	69.6	1.9	4.4
				2	89.7	6.7	7.5
				4	84.8	3.3	3.9
9	Celiprolol	165.1	181.5	1.5	98.5	7.3	5.6
				3	87.4	3.5	5.4
				6	93.3	2.1	7.7
10	Bisoprolol	134.6	156.3	0.5	93.8	7.1	7.2
				1	90.4	4.4	3.6
				2	84.6	2.5	5.6
11	Labetalol	91.2	102.1	0.5	91.7	2.1	2.8
				1	86.6	7.5	6.4
				2	83.4	3.3	6.5

Table 3. *Cont.*

NO.	Analyte	Matrix Effect C (%)		QC Concentration (μg kg^{-1})	Average Recovery (%)	Intra-Day Precision (%) (n = 5)	Inter-Day Precision (%) (n = 5)
		PRiME HLB	Centrifugation				
12	Alprenolol	102.1	117	0.5	74.8	5.5	4.2
				1	81.3	7.2	8.9
				2	81.1	4.1	5.4
13	Propranolol	98.5	120.7	0.5	80.2	8.7	2.3
				1	83.5	3.5	1.8
				2	80.4	6.4	5.6
14	Betaxolol	117.6	146.5	2	79.8	7.1	4.5
				4	91.5	2.7	3.7
				8	85.2	5.6	2.5
15	Cloranolol	109.2	126.1	2	72.1	8.1	3.5
				4	75.5	3.5	5.4
				8	76.6	3.1	7.2
16	Penbutolol	109.4	134.3	1	85.5	3.4	6.5
				2	97.6	2.6	4.4
				4	76.4	1.7	2.5
17	Practolol	115.7	120.6	0.5	71.9	4.4	2.6
				1	73.6	3.4	7.5
				2	75.2	5	5.8
18	Carazolol	80.1	96.1	0.5	99.3	3.1	2.8
				1	85.3	4	5.7
				2	85.9	7.9	4.9
19	Carvedilol	82.7	105.6	2	79.5	5.7	2.5
				4	78	2.5	7.5
				8	84.6	3.4	5.6
20	Esmolol	101.8	106.9	3	72.5	5.1	3.6
				6	83.2	8.1	2.3
				12	73.4	4.3	7.5
21	Bupranolol	112.6	122.9	0.5	73.4	3.2	5.9
				1	79.4	5.2	3.5
				2	80	1.7	5.4
22	Diacetolol	134	152.2	1	81.7	3.8	3.5
				2	87.7	2.3	5.9
				4	82.9	6.9	2.5

Table 3. *Cont.*

NO.	Analyte	Matrix Effect C (%)		QC Concentration (μg kg^{-1})	Average Recovery (%)	Intra-Day Precision (%) (n = 5)	Inter-Day Precision (%) (n = 5)
		PRiME HLB	Centrifugation				
				1	85.8	2.4	7.1
23	α-Hydroxymetoprolol	107.5	113.8	2	89.8	4.7	4.6
				4	84.5	5.6	5.6
				5	67.7	4.3	3.2
24	α-Hydroxyatenolol	80.1	72.8	10	66.1	3.4	5.4
				20	68.6	7.7	6.9
				1	78.5	2.2	5.8
25	(s)-Hydroxytimolol	93.4	103.6	2	91.3	5.5	5.7
				4	85.7	6.3	4.6
				1	69.8	5.6	3.5
26	7-Hydroxypropranolol	84.7	99.9	2	73.6	2.2	5.3
				4	78.8	3.2	7.2
				2	73.5	3.7	1.8
27	4-Hydroxyphenlcarvedilol	85.6	103.1	4	66	7	4.5
				8	67.4	6.3	6.9

Table 4. Regression data, Precision, LODs, and LOQs for the investigated compounds.

Analytes	Linear Equation	Linear Range (μg kg^{-1})	Correlation Coefficient (r^2)	LOD (μg kg^{-1})	LOQ (μg kg^{-1})
Atenolol	Y = −0.0196094 + 0.0399798X	2–200	0.9994	0.6	2
Sotalol	Y = −0.0302291 + 0.0364729X	2–200	0.9995	0.6	2
Pindolol	Y = 0.374297 + 0.109209X	1–200	0.9967	0.3	1
Nadolol	Y = 0.00696148 + 0.0330124X	0.5–50	0.9987	0.2	0.5
Metoprolol	Y = 0.250919 + 0.0925317X	0.5–50	0.9975	0.3	1
Timolol	Y = −0.0667935 + 0.0862132X	1–100	0.9997	0.3	1
Acebutolol	Y = 0.0484461 + 0.0478361X	0.5–50	0.9977	0.2	0.5
Oxprenolol	Y = −0.157711 + 0.0272438X	1–100	0.9990	0.3	1
Celiprolol	Y = −0.00691133 + 0.068192X	2–200	0.9966	0.5	1.5
Bisoprolol	Y = 0.130461 + 0.111734X	0.5–50	0.9969	0.2	0.5
Labetalol	Y = −0.0396352 + 0.0389041X	0.5–50	0.9997	0.2	0.5
Alprenolol	Y = −0.164222 + 0.588251X	0.5–50	0.9997	0.2	0.5
Propranolol	Y = 0.0870785 + 0.148381X	0.5–50	0.9996	0.2	0.5
Betaxolol	Y = 0.132601 + 0.130952X	2–200	0.9987	0.6	2
Cloranolol	Y = 0.214191 + 0.103506X	2–200	0.9991	0.6	2

Table 4. *Cont.*

Analytes	Linear Equation	Linear Range (μg kg^{-1})	Correlation Coefficient (r^2)	LOD (μg kg^{-1})	LOQ (μg kg^{-1})
Penbutolol	Y = −0.0422939 + 0.063028X	1–100	0.9989	0.3	1
Practolol	Y = −0.0153134 + 0.114317X	0.5–50	0.9990	0.2	0.5
Carazolol	Y = −0.015998 + 0.0807354X	0.5–50	0.9998	0.2	0.5
Carvedilol	Y = −0.00662109 + 0.0671112X	2–200	0.9998	0.6	2
Esmolol	Y = −0.142646 + 0.159519X	5–500	0.9994	1	3
Bupranolol	Y = −0.126666 + 0.325598X	0.5–50	0.9995	0.3	0.5
Diacetolol	Y = 0.204209 + 0.0797893X	1–100	0.9973	0.3	1
α-Hydroxymetoprolol	Y = 0.0134667 + 0.105121X	1–100	0.9992	0.3	1
α-Hydroxyatenolol	Y = −0.0738747 + 0.0396782X	5–500	0.9993	1.5	5
(S)-Hydroxytimolol	Y = 0.168441 + 0.119248X	1–100	0.9989	0.3	1
7-Hydroxypropranolol	Y = 0.0104 + 0.167864X	1–100	0.9999	0.3	1
4-Hydroxyphenlcarvedilol	Y = 0.0993695 + 0.159881X	2–200	0.9991	0.6	2

Y: The ratio of the peak area of the target to the area of the isotope peak, X: Corresponding concentration (μg kg^{-1}).

2.5.2. Matrix effect

Suppression or enhancement of the target signal usually occurs in the HESI source, especially for complicated food matrices. With a matrix effect (ME) value was between 80% and 120%, signal suppression or enhancement effect can be considered tolerable. As shown in Table 3, many of the analytes did not significantly express the matrix effect, except metoprolol, acebutolol, celiprolol, bisoprolol, diacetolol and α-hydroxyatenolol, which showed a significant matrix effect (ME < 80%, or ME > 120%). In order to accurately quantify the compounds, the assay was quantified with matrix-matched internal standard calibration.

2.5.3. Trueness and Precision

Recovery experiments were performed to evaluate the trueness of the method due to the lack of certified reference materials (CRM). As shown in Table 3, recoveries at three spiking levels (LOQ, 2 × LOQ, 4 × LOQ) ranged from 66.1% to 100.4%. The precision was calculated in terms of intra-day repeatability and inter-day reproducibility, which were expressed as relative standard deviations (RSDs). The results of intra-day and inter-day analyses performed at three spiking levels are presented in Table 3. Repeatability and reproducibility were in the range of 1.6–8.7% and 1.8–8.9%, respectively. Consequently, these results indicated that the developed method in this study is quite reliable, accurate and reproducible for determining β-blockers and their metabolites in milk powder samples.

2.6. Real Samples Analysis

In order to estimate the reliability and practicability of the developed method, 30 samples of infant formula milk powder purchased at local markets were analyzed in this study. The samples were regarded as representative, since they were ranged from phase 1 to stage 4 produced by reputable manufactures. High accuracy parent ions and product ions were used for qualitative analysis simultaneously. Full MS data of this mode were used for quantitative analysis. None of the 27 targeted analytes were detected by the developed Q-Orbitrap high resolution mass spectrometry method. However, an unknown compound which has the same molecule mass (m/z 309.18005) but different retention time (7.15 min vs. 5.75 min) with diacetolol has been screened in one sample. As shown in Figure 8, under the MS^2 conditions, the detected unknown compound had high-accuracy product ions 98.09692 (m/z) and 72.08157 (m/z), which was similar to diacetolol, and also the characteristic fragment ion m/z 57.07080. So, it is reasonable to speculate that this unknown substance may be an isomer of diacetolol. Considering the possible structure of product ions, the detected diacetolol's isomer could have two possible chemical structures ($C_{16}H_{25}O_4N_2$), as shown in Figure 9. For the first one, the acetyl group on the phenyl ring is in the presence of an acetyl amino group, which is more likely. The reason for this is that intermediate isomers are produced during the synthesis process, resulting in the isomerism of the target compounds. The second possibility may be diastereoisomers. Imino NH, phenolic hydroxyl oxygen and carbonyl oxygen through hydrogen bonding make the nitrogen atom form a relatively stable chiral center, as shown in Figure 9B. The specific structure needs to be further confirmed by NMR or other techniques.

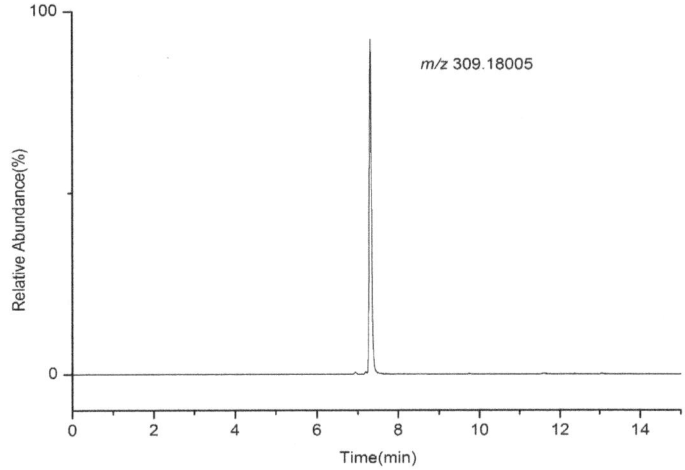

(A)

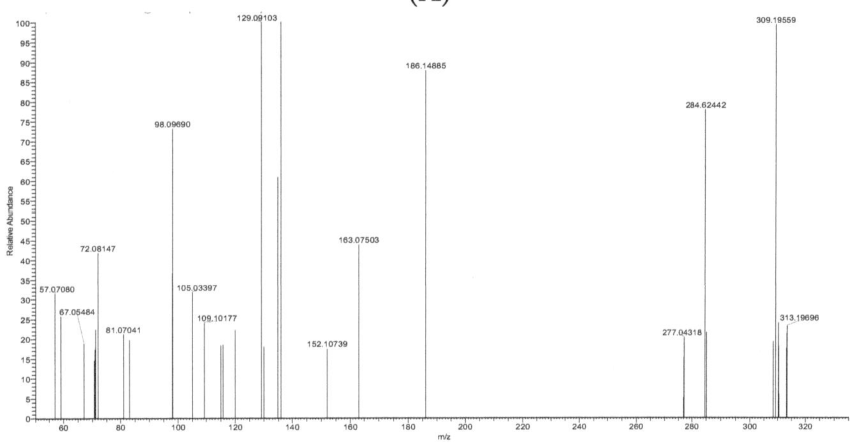

(B)

Figure 8. Extracted ion chromatogram and product ion spectrum of the suspected sample. (**A**) The extracted ion chromatogram of the suspected sample. (**B**) The product ion spectrum of the suspected sample.

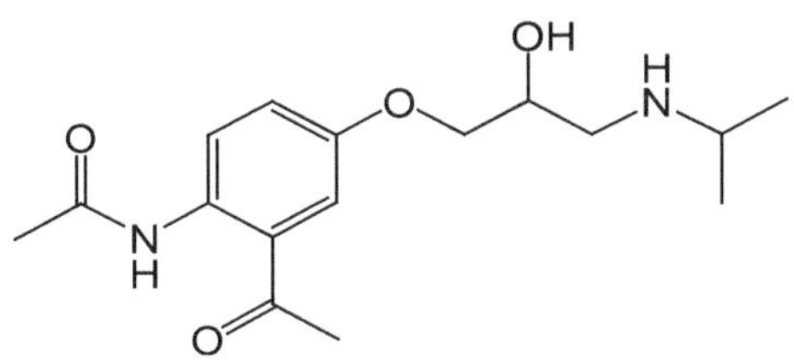

(A)

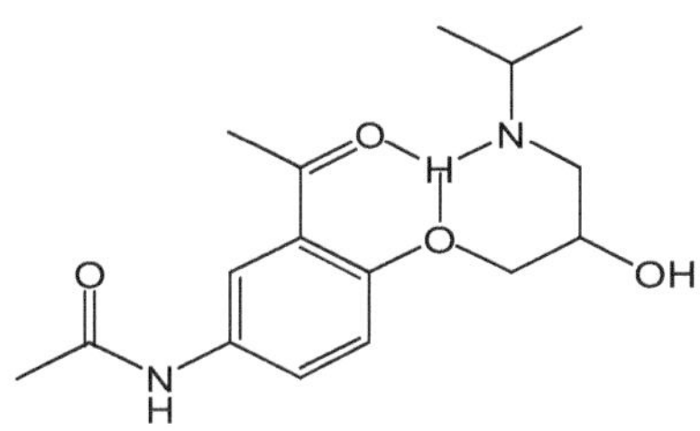

(B)

Figure 9. The possible chemical structure of diacetolol isomer detected in sample. (**A**) The first possible chemical structure. (**B**) The second possible chemical structure.

3. Materials and Methods

3.1. Chemicals and Reagents

Formic acid (FA), acetic acid (HOAc) were purchased from Sigma-Aldrich (Steinheim, Germany). Sodium chloride, ammonium acetate, ethylenediaminetetraacetic acid disodium salt (Na_2EDTA) were obtained from Beijing Chemical Company (Beijing, China). HPLC-grade methanol (MeOH) and acetonitrile (MeCN) were supplied by Fisher Scientific (Lough borough, UK). Ultra-pure water (H_2O) was obtained using a Milli-Q Ultrapure system (Millipore, Brussels, Belgium). β-glucuronidase/arylsulfatase was supplied by Merck (Darmstadt, Germany). Solid phase extraction cartridges Oasis PRiME HLB (500 mg, 6 cm^3) were obtained from Waters (Milford, MA, USA).

Standards of carazolol, timolol maleate, nadolol, sotalol hydrochloride, pindolol, atenolol, metoprolol, carazolol-d_7 (internal standard, ISTD), propranolol-d_7 (ISTD) (purity > 96%) were purchased from Dr. Ehrenstorfer (Augsburg, Germany). Acebutolol, carvedilol, penbutolol sulfate, propranolol, betaxolol hydrochloride, alprenolol, oxprenolol, celiprolol, bisoprolol fumarate, labetalol hydrochloride (purity > 96%) were purchased from U.S. Pharma-copoeia (Rockville, MD, USA). Diacetolol, cloranolol, esmolol, bupranolol, practolol, 7-hydroxypropranolol, hydroxytimolol, 4-hydroxyphenylcarvedilol, α-hydroxyatenolol (purity > 96%) were obtained from Toronto Research Chemicals (North York, Canada). α-hydroxymetoprolol (100 μg mL^{-1}, methanol) were obtained from AccuStandard (Chiron, Norway).

All standard stock solutions were prepared in MeOH at 100 μg mL^{-1}. The mixed working standard solutions were prepared daily via proportional dilution of the stock solutions. All of the standard solutions were stored at −20 °C in a dark amber bottle.

Matrix-matched standard working solutions were prepared in blank sample extracts, which were obtained from a commercial product purchased from a local market and affirmed in advance not to contain any of the tested analytes. All of the standard solutions were stored at −20 °C in a dark amber bottle.

Extraction solution: 37.5 g Na_2EDTA was dissolved in an ammonium acetate buffer produced by dissolving 15.4 g ammonium acetate in 1 L deionized water and then using acetic acid to adjust the pH to 5.2.

3.2. Instrument and Analytical Conditions

The UHPLC/HESI Q-Orbitrap system consisted of a Thermo UltiMate 3000 $UHPLC^+$ system coupled with a Q Exactive mass spectrometer (Thermo Fisher Scientific, Bremen, Germany). The system was controlled by Exactive Tune 1.1 and Xcalibur 2.2 software (Thermo Fisher Scientific, San Jose, CA, USA).

Chromatographic separation was achieved on an Ascentis® C8 chromatographic column (100 × 4.6 mm, 3 μm) (SUPELCO® Analytical, Bellefonte, PA, USA). The autosampler tray temperature, column oven temperature, flow rate and injection volume were set at 10 °C, 30 °C, 0.5 mL min^{-1} and 2 μL, respectively. The mobile phase consisted of water containing 0.1% FA (A) and MeOH (B). The gradient used for eluting analytes with mobile phase is as follows: 0–0.5 min, 5% B; 0.5–9 min, 5–95% B; 9–12.5 min, 95% B; 12.5–14 min, 95–5% B; 14–15 min, 5% B.

The Q-Orbitrap HRMS was equipped with a heated electrospray ionization (HESI, Waltham, MA, USA) source and the analysis was operated in the Full MS/dd-MS^2 (data-dependent MS^2) scanning mode, which includes a Full scan followed by MS/MS scan of precursors in the inclusion list. All analytes were measured in positive mode and precursor ion selected was $[M + H]^+$ in all cases. To keep a balance between the selectivity and the sensitivity with Full MS, a mass resolution of 70,000 FWHM was selected, and this turned out to be optimal for the majority of the analytes. For the dd-MS^2 scan, 35,000 FWHM was used for time-saving and to ensure sufficient scan points of the Full MS. The stepped normalized collision energy (NCE) was set to 15%, 25% and 35%. The spray voltage, capillary temperature, aux gas heater temperature were set to 3.0 kV, 350 °C and 350 °C, respectively. The sheath gas, auxiliary gas, sweep gas and S-lens RF level were set to 40, 10, 0 (arbitrary units) and 50 V, respectively. Main MS acquisitions parameters are listed in Table 1. All the extracted mass traces were based on a 2 ppm mass window (accuracy).

3.3. Sample Preparation

Two grams of each sample were precisely weighed in polypropylene centrifuge tube (50 mL). Then, 100 μL mixed ISTD solution (1μg mL^{-1}), 40 μL β-glucuronidase/arylsulfatase and 5 mL EDTA extract solution were added to the sample after being fully dissolved in 5 mL H_2O. The mixture was placed in a water bath shaker at 50 °C for 60 min after vortex-mixing for 1 min. After cooling to room temperature, the pH of the extract was adjusted to 10.0 with 3 mol L^{-1} NaOH solution. MeCN (20 mL) and NaCl (2.5 g) were added to the mixture and then shaken for 30 min. After that, the extract was centrifuged at 10,000 rpm at 4 °C for 10 min. The supernatant was decanted to another polypropylene centrifuge tube. The above procedure was repeated and combined with the supernatant. Next, 2 mL of the supernatant directly passed through the Oasis PRiME HLB (500 mg, 6 cm^3) cartridge. After sample loading, the cartridge was washed with 2 mL H_2O/MeOH (95:5, *v*/*v*) and 2 mL MeOH/MeCN (1:9, *v*/*v*), and all of the effluent was collected. The mixture was evaporated with a gentle N_2 stream at 40 °C, and redissolved in 1 mL of H_2O/MeOH (1:1). The sample extract was vortexed for 0.5 min and filtered through a 0.22 μm nylon membrane, and was ready for Q-Orbitrap HRMS analysis.

3.4. Method Validation

Linearity, precision and recovery were carried out to validate the method. An internal standard method which using carazolol-d_7 and propranolol-d_7 as ISTD was utilized for quantification. A matrix-matched calibration cure was constructed by linear regression of the ratios of chromatographic peak areas of the standards and the ISTD. The linearity was discussed by the coefficient of determination (r^2).

Blank milk sample powders spiked at three concentration levels (LOQ, 2 × LOQ and 4 × LOQ) which were tested for the recovery experiments. Each level was analyzed in five replicates. Intra-day precision was performed by spiking blank milk at three concentration levels (LOQ, 2 × LOQ and 4 × LOQ) with five replicates in one day. To evaluate inter-day precision, the same concentration levels were performed during over consecutive days.

The matrix effect (ME) was calculated by comparing the response of analytes prepared in solvent and in extracted blank matrix at the same concentration, respectively. The value of matrix effect can be calculated as (Equation (1)):

$$\text{ME } (\%) = \text{B}/\text{A} \times 100 \qquad (1)$$

A refers to the peak areas obtained from neat solution standards, while B refers to the corresponding peak areas of standards spiked after extraction from matrix [24,25].

4. Conclusions

In this study, a rapid HPLC-Q-Orbitrap HRMS method for simultaneous analyses of 27 compounds (21 β-blockers and 6 metabolites) in milk powder has been developed. Simultaneous qualitative and quantitative analysis of analytes were achieved using Full MS/dd-MS^2 acquisition mode of the Q-Orbitrap mass analyzer and the preparation procedure comprised a simple acetonitrile step, followed by a cleanup using cartridges. The method has been well validated, and is particularly effective and valuable for the routine screening of β-blockers and metabolites in infant formula milk powder. At the same time, the corresponding characteristic fragmentation behavior of the 27 compounds were explored, the characteristic product ions were determined and applied to the actual sample screening.

Author Contributions: J.-Q.C. and T.L. conceived and designed the experiments, performed the experiments, analyzed the data, wrote the paper; X.-M.N. and F.-M.C. prepared the sample, analyzed the data; C.-S.W. analyzed the data; F.Z. designed the experiment, wrote and revised the paper.

References

1. Xu, H.; Zhang, H.W.; Wang, F.M. Determination of 9 β-blockers residues in dairy products by liquid chromatography-tandem mass spectrometry. *J. Food Safety Qual.* **2014**, *12*, 3884–3890.
2. Zhang, J.; Shao, B.; Yin, J. Simultaneous detection of residues of β-adrenergic receptor blockers and sedatives in animal tissues by high-performance liquid chromatography/tandem mass spectrometry. *J. Chromatogr. B Anal. Technol. Biomed. Life Sci.* **2009**, *877*, 1915–1922. [CrossRef] [PubMed]
3. Mitrowska, K.; Posyniak, A.; Zmudzki, J. Rapid method for the determination of tranquilizers and a β-blocker in porcine and bovine kidney by liquid chromatography with tandem mass spectrometry. *Anal. Chim. Acta* **2009**, *637*, 185–192. [CrossRef] [PubMed]
4. Hao, J.; Jiang, J.; Shao, R.T.; Ding, X.Y.; Shi, N.; Lu, Y. Determination of Eleven β-Blocker residues in Animal Derived Foods by Ultra High Performance Liquid Chromatography-Tandem Mass Spectrometry with Molecularly Imprinted Solid Phase Extraction. *J. Instrum. Anal.* **2016**, *10*, 1278–1282.
5. Gehr, T.W.B.; Tenero, D.M.; Boyle, D.A.; Qian, Y.; Sica, D.A.; Shusterman, N.H. The pharmacokinetics of carvedilol and its metabolites after single and multiple dose oral administration in patients with hypertension and renal insufficiency. *Eur. J. Clin. Pharmaco.* **1999**, *55*, 269–277. [CrossRef]
6. The Council of the European Communities. Council Regulation 2377/90/EEC. *Off. J. Eur. Commun.* **1990**, *224*, 1.
7. Cooper, J.; Delahaut, P.; Fodey, T.L. Development of a rapid screening test for veterinary sedatives and the β-blocker carazolol in porcine kidney by ELISA. *Analyst* **2004**, *129*, 169–174. [CrossRef] [PubMed]
8. Delbeke, F.T.; Debackere, M.; Desme, T.N. Comparative study of extraction methods for the GC and GC-MS screening of urine for β-blocker abuse. *J. Pharmaceu. Biomed. Anal.* **1988**, *6*, 827–835. [CrossRef]
9. Ternes, T.A.; Hirsch, R.; Mueller, J. Methods for the determination of neutral drugs as well as β-blockers and β_2-sympathomimetics in aqueous matrices using GC/MS and LC/MS/MS. *Anal. Bioanal. Chem.* **1998**, *362*, 329–340.
10. Magiera, S.; Uhlschmied, C.; Rainer, M. GC–MS method for the simultaneous determination of β-blockers, flavonoids, isoflavones and their metabolites in human urine. *J. Pharmaceu. Biomed. Anal.* **2011**, *56*, 93–102. [CrossRef] [PubMed]
11. Amendola, L.; Molaioni, F.; Botrè, F. Detection of β-blockers in human urine by GC-MS-MS-EI: Perspectives for the antidoping control. *J. Pharmaceu. Biomed. Anal.* **2000**, *23*, 211–221. [CrossRef]
12. Delamoye, M.; Duverneuil, C.; Paraire, F. Simultaneous determination of thirteen β-blockers and one metabolite by gradient high-performance liquid chromatography with photodiode-array UV detection. *Forensic Sci. Int.* **2004**, *141*, 23–31. [CrossRef] [PubMed]
13. Zuppa, A.F.; Shi, H.; Adamson, P.C. Adamson, Liquid chromatography-electrospray mass spectrometry (LC-MS) method for determination of esmolol concentration in human plasma. *J. Chromatogr. B* **2003**, *796*, 293–301. [CrossRef]
14. Delahaut, P.; Levaux, C.; Eloy, P. Validation of a method for detecting and quantifying tranquillisers and a β-blocker in pig tissues by liquid chromatography–tandem mass spectrometry. *Anal. Chim. Acta* **2003**, *483*, 335–340. [CrossRef]
15. Lee, H.B.; Sarafin, K.; Peart, T.E. Determination of β-blockers and β_2-agonists in sewage by solid-phase extraction and liquid chromatography–tandem mass spectrometry. *J. Chromatogr. A* **2007**, *1148*, 158–167. [CrossRef] [PubMed]
16. Umezawa, H.; Lee, X.P.; Arima, Y. Simultaneous determination of β-blockers in human plasma using liquid chromatography–tandem mass spectrometry. *Biomed. Chromatogr.* **2008**, *22*, 702–711. [CrossRef] [PubMed]
17. Farré, M.; Gros, M.; Hernández, B. Analysis of biologically active compounds in water by ultra-performance liquid chromatography quadrupole time-of-flight mass spectrometry. *Rapid Commun. Mass Spectrom.* **2008**, *22*, 41–51. [CrossRef] [PubMed]
18. Cheng, Y.; Nie, X.M.; Wu, H.Q. A high-throughput screening method of bisphenols, bisphenols digycidyl ethers and their derivatives in dairy products by ultra-high performance liquid chromatography-tandem mass spectrometry. *Anal. Chim. Acta* **2017**, *950*, 98–107. [CrossRef] [PubMed]
19. Choi, J.H.; Lamshöft, M.; Zühlke, S. Determination of sedatives and adrenergic blockers in blood meal using accelerated solvent extraction and Orbitrap mass spectrometry. *J. Chromatogr. A* **2012**, *1260*, 111–119. [CrossRef] [PubMed]

20. Li, X.; Shen, B.; Jiang, Z.; Huang, Y.; Zhuo, X. Rapid screening of drugs of abuse in human urine by high-performance liquid chromatography coupled with high resolution and high mass accuracy hybrid linear ion trap-Orbitrap mass spectrometry. *J. Chromatogr. A* **2013**, *1302*, 95–104. [CrossRef] [PubMed]
21. Guo, C.; Shi, F.; Gong, L. Ultra-trace analysis of 12 β-agonists in pork, beef, mutton and chicken by ultrahigh-performance liquid-chromatography-quadrupole-orbitrap tandem mass spectrometry. *J. Pharmaceu. Biomed. Anal.* **2015**, *107*, 526–534. [CrossRef] [PubMed]
22. Hong, Y.H.; Xu, X.L.; Li, W.Q. A high-accuracy screening method of 44 cephalosporins in meat using liquid chromatography quadrupole-orbitrap hybrid mass spectrometry. *Anal. Methods* **2017**, *10*. [CrossRef]
23. Wang, X.J.; Zhang, F.; Li, W.Q. Simultaneous determination of 12 β-agonists in feeds by ultra-high-performance liquid chromatography-quadrupole-time-of-flight mass spectrometry. *J. Chromatogr. A* **2013**, *1278*, 82–88.
24. Matuszewski, B.K.; Constanzer, M.L.; Chavez-Eng, C.M. Strategies for the assessment of matrix effect in quantitative bioanalytical methods based on HPLC–MS/MS. *Anal. Chem.* **2003**, *75*, 3019–3030. [CrossRef] [PubMed]
25. Cappiello, A.; Famiglini, G.; Palma, P.; Pierini, E.; Termopoli, V.; Trufelli, Y. Overcoming matrix effects in liquid chromatography–mass spectrometry. *Anal. Chem.* **2008**, *80*, 9343–9348. [CrossRef] [PubMed]

2

Ultrasensitive Electrochemical Sensor based on Polyelectrolyte Composite Film Decorated Glassy Carbon Electrode for Detection of Nitrite in Curing Food at Sub-Micromolar Level

Jingheng Ning, Xin Luo, Min Wang, Jiaojiao Li, Donglin Liu, Hou Rong, Donger Chen and Jianhui Wang *

School of Chemistry and Biological Engineering, Changsha University of Science & Technology, Changsha 410110, China; jinghengning@csust.edu.cn (J.N.); luoxin_gl@126.com (X.L.); wang_min1993@126.com (M.W.); lijiaojiao_2013@126.com (J.L.); dong993@163.com (D.L.); hourong0406@163.com (H.R.); chendonger9@163.com (D.C.)

* Correspondence: wangjh0909@csust.edu.cn

Academic Editors: Alessandra Gentili and Chiara Fanali

Abstract: To ensure food quality and safety, developing cost-effective, rapid and precision analytical techniques for quantitative detection of nitrite is highly desirable. Herein, a novel electrochemical sensor based on the sodium cellulose sulfate/poly (dimethyl diallyl ammonium chloride) (NaCS/PDMDAAC) composite film modified glass carbon electrode (NaCS/PDMDAAC/GCE) was proposed toward the detection of nitrite at sub-micromolar level, aiming to make full use of the inherent properties of individual component (biocompatible, low cost, good electrical conductivity for PDMDAAC; non-toxic, abundant raw materials, good film forming ability for NaCS) and synergistic enhancement effect. The NaCS/PDMDAAC/GCE was fabricated by a simple drop-casting method. Electrochemical behaviors of nitrite at NaCS/PDMDAAC/GCE were investigated by cyclic voltammetry (CV) and differential pulse voltammetry (DPV). Under optimum conditions, the NaCS/PDMDAAC/GCE exhibits a wide linear response region of 4.0×10^{-8} mol·L^{-1}~1.5×10^{-4} mol·L^{-1} and a low detection 1imit of 43 nmol·L^{-1}. The NaCS/PDMDAAC shows a synergetic enhancement effect toward the oxidation of nitrite, and the sensing performance is much better than the previous reports. Moreover, the NaCS/PDMDAAC also shows good stability and reproducibility. The NaCS/PDMDAAC/GCE was successfully applied to the determination of nitrite in ham sausage with satisfactory results.

Keywords: polyelectrolyte composite film; nitrite detection; differential pulse voltammetry; cyclic voltammetry

---

1. Introduction

Nitrite is widely used as a preservative in food industry, especially in the production and processing of cured meat and fishery products [1]. Additionally, it is also utilized as a fertilizing agent in agriculture and an inhibitor in corrosion science [2]. Furthermore, nitrite has become one of the widespread inorganic contaminants present in soil, food, ground water, and even physiological systems [3]. Nitrite has been proven to be of potential risk to human health and leads to many serious diseases, such as methemoglobinemia, esophageal cancer, spontaneous abortion, and birth defects of the central nervous system [4–7]. Therefore, the content of nitrite demands strict control. The World

Health Organization (WHO) has set a fixed maximum limit in drinking water for nitrite as 65 μM [8]. In China, according to national standard GB 2760-2014, the maximum permitted level of nitrite in meat food is 0.15 g/kg [9]. To ensure food quality and safety, developing cost-effective, rapid, and precision analytical techniques for quantitative detection of nitrite is highly desirable.

To date, many techniques have been developed for the detection of nitrite, including chromatography [10,11], UV spectrophotometry [12,13], fluorometry [14,15], and flow injection analysis [16,17]. Although these analytical methods are well-recognized, they also suffered from limits, such as high cost, being time-consuming, cumbersome, and requiring complicated pretreatment. Recently, electrochemical analysis has become one of most preferred techniques to detection small biomolecules, food additives, and environmental contaminants, owing to its overwhelming merits including cost- and time-effective, facile operation, and high sensitivity and specificity [18–23]. In the last few years, electrochemical nitrite sensors have received increasing attention. Various modified materials have been proposed for nitrite sensors, including graphene [24,25], carbon nanotubes [26], metal or metal oxide nanoparticles [27–29], metal organic frameworks [30,31], and conducting polymers [32,33]. However, these modified electrodes have several drawbacks, such as low detection limits (almost at the mmol/L level, far lower than sub-micromolar level for HPLC and fluorometry), narrow linear range, and susceptibility to interference. Thus, it is urgent to develop novel electrochemical sensors for the detection of nitrite at sub-micromolar level.

Poly (dimethyldiallylammonium chloride) (PDMDAAC) is a ubiquitous linear cationic polymer that has unique properties, such as biocompatibility, low cost, high charge density, as well as good water solubility [34]. It has been reported that PDMDAAC film was successfully immobilized on electrochemical sensors [35–37]. These previous studies demonstrated that PDMDAAC film has excellent conductivity due to the presence of electron conductive networks, and good ion-exchange performances that can effectively increase the concentration of anions round the film as compared with the bulk solution. For these reasons, PDMDAAC film was chosen as a modified material to construct electrochemical sensors, aiming to facilitate preconcentration and enhance selectivity properties of nitrite sensors, and eventually enhance the electrochemical response. However, PDMDAAC was found to exhibit relatively poor performances in film formation in our preliminary experiments, which could influence the electrode stability. Additionally, the sensitivity of PDMDAAC modified electrode is very limited. To overcome these problems, PDMDAAC was combined with sodium cellulose sulfate (NaCS), which has good film-formation ability. NaCS/PDMDAAC is an emerging composite film that has unique properties such as excellent mechanical strength, transparency and small pore size [38]. NaCS/PDMDAAC has been widely used in the immobilization of bacteria, fungus and microalgae, drug delivery systems, and membrane systems [38–40]. However, to our best knowledge, NaCS/PDMDAAC has rarely been applied as a modified material in electrochemical sensors.

Herein, NaCS/PDMDAAC composite film as prepared to construct a novel electrochemical sensor for the detection of nitrite at the sub-micromolar level, aiming to make full use of the merits of individual components (biocompatible, low cost, good electrical conductivity for PDMDAAC; non-toxic, abundant raw materials, and the good film forming ability of NaCS) and synergistic enhancement effect (Scheme 1). NaCS/PDMDAAC/GCE was fabricated by coating the PDMDAAC and NaCS sequentially on the surface of GCE. The electrochemical performance were investigated by cyclic voltammetry (CV) and electrochemical impedance spectroscopy (EIS) using $[Fe(CN)_6]^{3-/4-}$ as redox probe solution. The analytical parameters (including pH) were systemically explored. Finally, the content of nitrite in ham sausage was determined by differential pulse voltammetry (DPV) with satisfactory results.

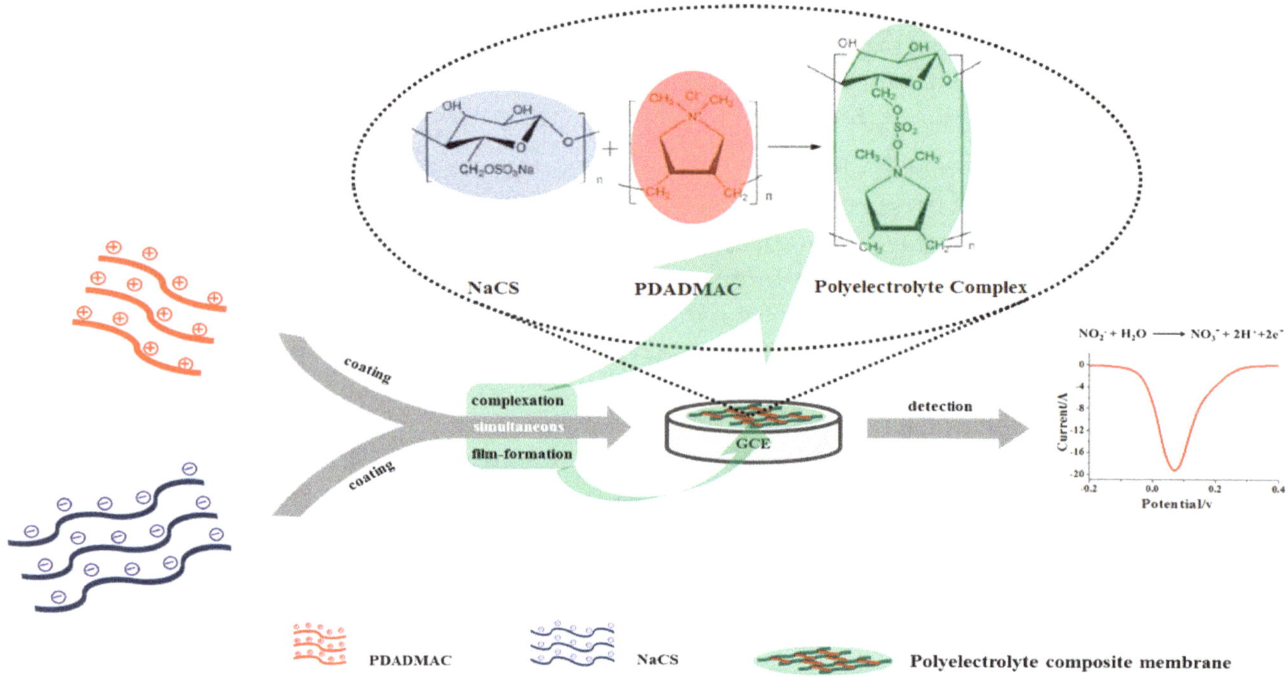

Scheme 1. Schematic of NaCS/PDMDAAC/GCE for the detection of nitrite detection.

2. Results and Discussion

2.1. Characterization of Surface Morphology

The surface morphologies of NaCS film and NaCS/PDMDAAC composite film were characterized by scanning electron microscope (SEM, Figure 1). As shown in Figure 1A, the NaCS film consists of acicular crystal structure. After compositing with NaCS/PDMDAAC, the surface is changed from acicular crystal to smooth film (Figure 2a), which is favorable for the electron transfer between the surface of modified electrode and analytes, and would eventually enhance electrocatalytic activity toward nitrite.

2.2. Electrochemical Characterization of NaCS/PDMDAAC/GCE

The electrochemical performance of bare GCE, NaCS/GCE, PDMDAAC/GCE, and NaCS/PDMDAAC/GCE were investigated using cyclic voltammetry (CV) recorded in 5×10^{-3} mol L^{-1} $K_3[Fe(CN)_6]$ solution. The cyclic voltamograms recorded on various electrodes are shown in Figure 2. A pair of redox peaks appears on all the electrodes, suggesting the electrochemical process is quasi-reversible. A pair of weak and broad peaks was observed on the bare GCE (curve a), with redox peak separation (ΔE_p) of 137 mV. The anodic peak current (i_{pa}) and cathodic peak current (i_{pc}) is 43 μA, and 58 μA, respectively. On the NaCS modified electrode (curve b), the ΔE_p increases to 186 mV. Meanwhile, the peak currents also increase slightly, with i_{pa} of μA and i_{pc} of 69 μA, respectively. When modified with PDMDAAC alone (curve c), a pair of obvious and well-shaped redox peaks appears with a significant decrease on ΔE_p (101 mV). The corresponding redox peaks increase greatly, with i_{pa} of 81 μA and i_{pc} of 80 μA, respectively. When modified with NaCS/PDMDAAC composite film (curve d), the ΔE_p (143 mV) has a slight increase compared to that on the NaCS/GCE. The i_{pa} and i_{pc} is 72 μA and 89 μA, indicating that the NaCS/PDMDAAC film facilitates the electron transfer between the electrode and analyses, and eventually improves the electrochemical response.

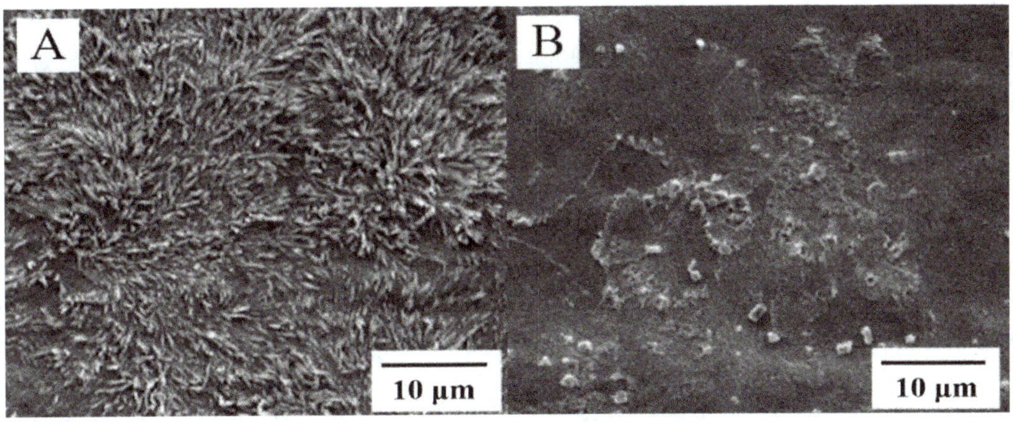

Figure 1. SEM images of NaCS film (**A**) and NaCS/PDMDAAC composite film (**B**).

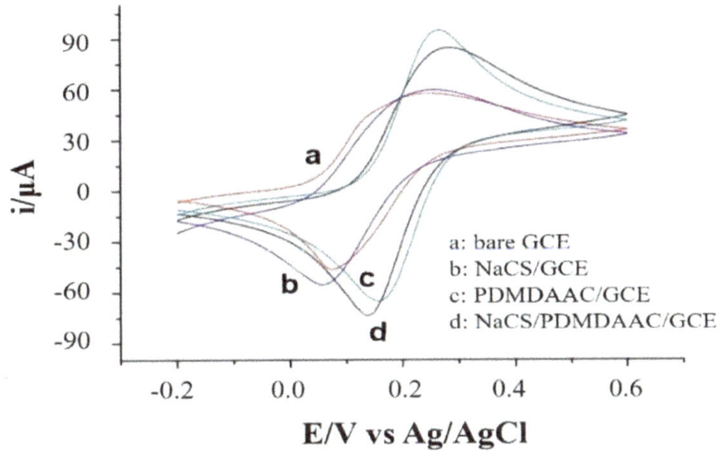

Figure 2. Cyclic voltammograms of bare GCE (a), NaCS/GCE (b), PDMDAAC/GCE (c), and NaCS/PDMDAAC/GCE (d) recorded in 5×10^{-3} mol·L^{-1} $K_3[Fe(CN)_6]$ solution.

The influence of layers of NaCS/PDMDAAC composite film was also investigated in 5×10^{-3} mol L^{-1} $K_3[Fe(CN)_6]$ solution by CV. As illustrated in Figure 3, the response peak currents decrease with the increase of layers of NaCS/PDMDAAC composite film. It is probably related to the charge transfer resistance that increases with the thickness increasing. Hence, monolayer NaCS/PDMDAAC composite film was employed for subsequent experiments.

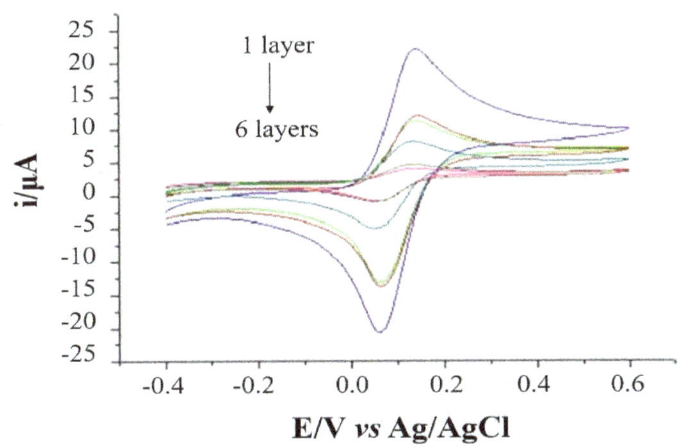

Figure 3. Cyclic voltammograms of different layer of PDMDAAC/NaCS modified GCE recorded in 5×10^{-3} mol·L^{-1} $K_3[Fe(CN)_6]$ solution.

Electrochemical impedance spectroscopy (EIS) is a useful tool to acquire abundant information about electrode interface, which have been widely used in the electrocatalysis [41,42], chemosensors, and biosensors [43,44]. The EIS of bare GCE, NaCS/GCE, PDMDAAC/GCE, and NaCS/PDMDAAC/GCE were also measured in the 5×10^{-3} mol L^{-1} $K_3[Fe(CN)_6]$ solution. The Nyquist plots of different electrodes were shown in Figure 4. The charge transfer resistance (R_{ct}) were estimated by EIS fitting with the equivalent circuit. The R_{ct} of bare GCE, NaCS/GCE, PDMDAAC/GCE, and NaCS/PDMDAAC/GCE are 37,830 Ω, 104,800 Ω, 36,810 Ω, and 52,190 Ω, respectively, i.e., the conductivity ($1/R_{ct}$) of them are 2.6434×10^{-5} S, 9.5420×10^{-6} S, 2.7167×10^{-5} S, and 1.9161×10^{-5} S, respectively. Apparently, the conductivity of the bare GCE (2.6434×10^{-5} S) is slightly lower than that of PDMDAAC/GCE (2.7167×10^{-5} S) but much higher than that of NaCS/GCE (9.5420×10^{-6} S), indicating that when each of them is independently modified onto the bare electrode, the polycation PDMDAAC can improve the conductivity while the polyanion NaCS is not favorable for electron transfer. As is well known, both PDMDAAC and NaCS are polyelectrolytes and, theoretically, should have good conductivities. However, these polyelectrolytes are quite different in charge density, solubility, and counter-ion effects in solution (ions will naturally attract the opposite-charged ions to form counter-ion layers), all of which are thought to be able to influence the conductivity, and the former two factors are thought to be favorable for conductivity while the last one is unfavorable. In fact, the molecular weight of PDMDAAC used in our experiment is less than 1.0×10^5, while NaCS has a much bigger average molecular weight more than 1.9×10^6, and so it is obvious that PDMDAAC contains much shorter molecular chains, which reinvests it with a higher charge density, a better solubility, as well as a weaker counter-ion effect due to narrower interfaces to attract less opposite-charged ions. Thus, although both being polyelectrolytes, PDMDAAC can improve the conductivity of the modified GCE while NaCS cannot, as the above results show. However, it is worth noting that when using the composite NaCS/PDMDAAC modifies the bare GCE, the R_{ct} of NaCS/PDMDAAC/GCE (52,190 Ω) is only about half of that on the NaCS/GCE (104,800 Ω), suggesting the unfavorable counter-ion effect of NaCS is significantly weakened after the introduction of PDMDAAC, which actually improves the charge transfer efficiency. Obviously, the conductivity of NaCS/PDMDAAC/GCE (1.9161×10^{-5} S) is better than that of NaCS/GCE (9.5420×10^{-6} S). Thus, by considering the outstanding conductivity of PDMDAAC as well as the good performances of NaCS for film formation, the composite NaCS/PDMDAAC is the most suitable material for the modification of GCE to construct the new and effective sensor for nitrite detection, herein with a moderate R_{ct}/conductivity value, which is consistent with the above research result conducted by the CV method.

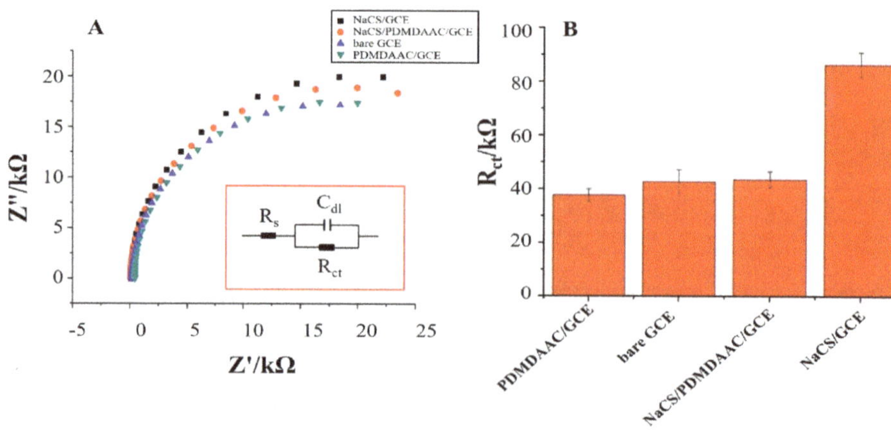

Figure 4. (**A**) Nyquist plots of different electrodes in 5×10^{-3} mol·L^{-1} $K_3[Fe(CN)_6]$ solution obtained from electrochemical impedance spectroscopy (EIS). The frequency investigated was between 10,000 and 0.1 Hz with a pulse amplitude of 5 mV. The inset is the equivalent circuit of electrode/electrolyte interface. (**B**) The R_{ct} of different electrodes by EIS fitting.

2.3. *Electrochemical Behavior of Nitrite on NaCS/PDMDAAC/GCE*

The electrochemical behaviors of nitrite on bare GCE and NaCS/PDMDAAC/GCE were investigated by CV in 0.1 mol·L^{-1} PBS solution containing 4.0 × 10^{-3} mol·L^{-1} nitrite (Figure 5). The anodic peak current on the bare GCE is 59 μA. When modified with NaCS/PDMDAAC composite film, the anodic peak current increases to 97 μA, suggesting the NaCS/PDMDAAC composite film improve the electrochemical responses of nitrite.

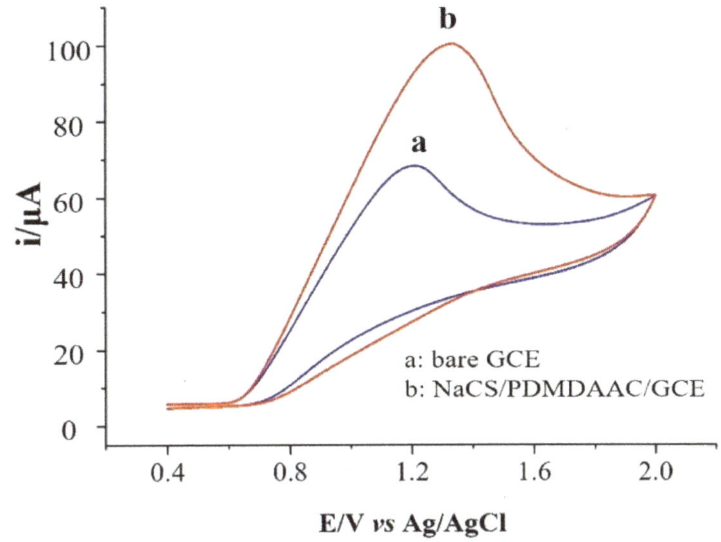

Figure 5. Cyclic voltammograms of bare GCE (a) and NaCS/PDMDAAC/GCE (b) in 0.1 mol·L^{-1} PBS solution (pH 7.0) containing 4.0 × 10^{-3} mol·L^{-1} nitrite.

2.4. *Optimization of Analytical Conditions*

2.4.1. Effect of the Modifier Loading Amount

The effect of the modifier loading amount is one of important factors that affect the electrochemical response of modified electrode. The bare GCEs were modified with various dosage of NaCS/PDMDAAC dispersion (v/v = 1:1) firstly. Then the response peak currents of 1.0 × 10^{-4} mol·L^{-1} nitrite on these electrodes were recorded and compared. The effect of modifier loading on the response peak current was shown in Figure 6A. The response peak currents of nitrite vary with the loading amount of NaCS/PDMDAAC composite. The response peak current increases gradually with the loading amount increasing from 2–6 μL, afterwards the response peak currents decrease with the further increase on loading amount. The maximum response peak current is obtained when the loading amount is 6 μL. Namely, the loading amounts of PDMDAAC and NaCS are 3 μL. Hence, the 6 μL loading amount was used in the subsequent experiments. The trend of response peak current to loading amount is directly related to the amount of reactive active sites and the thickness of modifier films. With loading amount increasing, the amount of available reactive active sites increases, which causes the increase on the response peak current. However, the thickness of the modified film increases when the loading amount beyond 6 μL, which hinders the electron transfer between the electrode and nitrite. As a result, the response peak current decreases with the further increase on the loading amount.

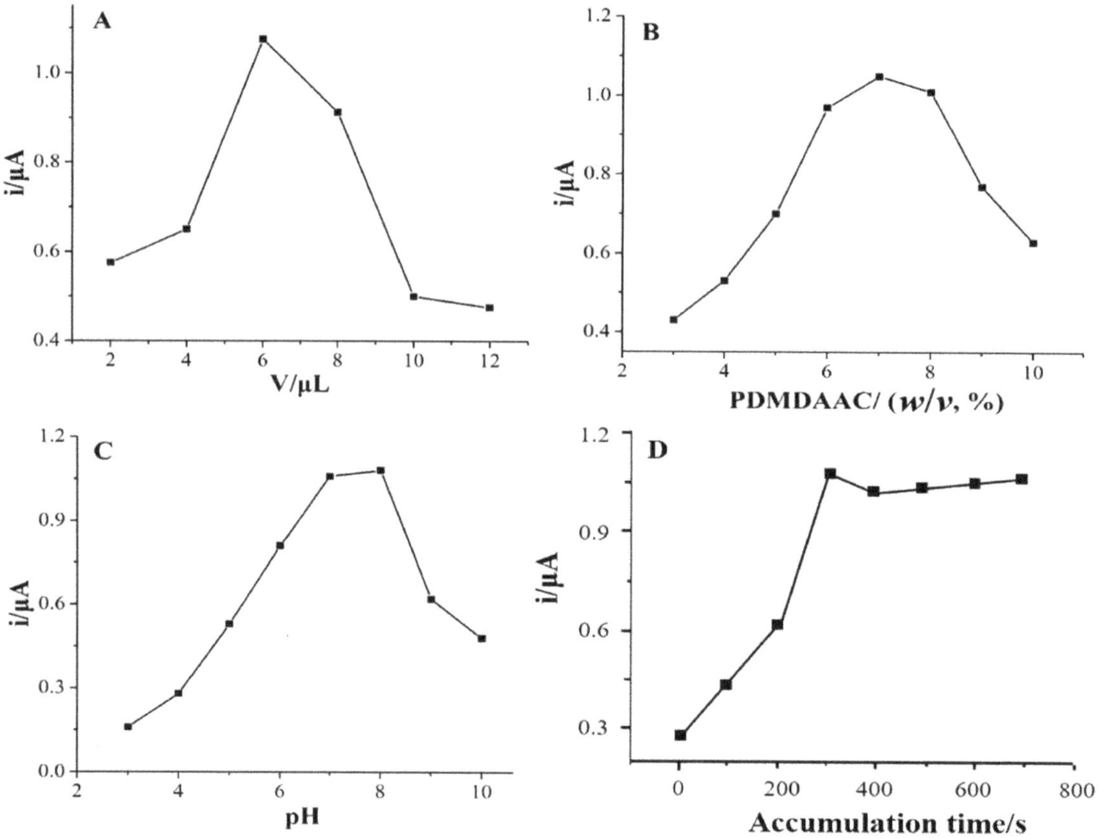

Figure 6. Effect of loading amount of NaCS/PDMDAAC (**A**), concentration of PDMDAAC(*w*/*v*) (**B**), pH (**C**) and accumulation time (**D**) on the response peak current of nitrite (1.0×10^{-4} mol·L^{-1}) at the NaCS/PDMDAAC/GCE.

2.4.2. Effect of PDMDAAC Concentration

The effect of the PDMDAAC concentration on the response peak current of nitrite was also explored. The results are shown in Figure 6B. It observed that the response peak current increases as the increase of PDMDAAC concentration. When the concentration (*w*/*v*) is 7%, the peak current reaches the maximum. Afterwards the response peak current decreases with further increase on PDMDAAC concentration. Therefore, 7% was selected to be the suitable PDMDAAC concentration for the following measurements.

2.4.3. Effect of pH

It is well known that pH plays a crucial role on the oxidation of nitrite, so it's well worth investigating the influence of pH. The response peak currents of 1.0×10^{-4} mol·L^{-1} nitrite were measured in various pH (3.0–10.0), and the result was presented in Figure 6C. When the electrolyte solution closes to neutral (pH 7.0 or 8.0), the response peak current is largest. Therefore, pH 7.0 was recommended as the optimum pH. Under acidic conditions, NO_2^- is very unstable and easily reacted with H^+ to produce NO and NO_2 gas. As a result, the response peak current declines as the pH decreasing. In addition, most amount of the NO_2^- is protonated in such acidic environment ($pK_a(HNO_2) = 3.3$). Under alkaline pH ranging of 8.0–10.0, the increase on pH leads to the decline of the response peak current due to the lack of sufficient H^+.

2.4.4. Effect of Accumulation Time

As we all know, accumulation time directly influences the adsorption of analyses at the surface of modified electrode. The NaCS/PDMDDA/GCE was accumulated for various times firstly, then corresponding response peak currents were record and compared (Figure 6D). During the first 300 s, the response peak currents increase with the time prolonging. It is probably due to the increasing amount of nitrite adsorbed on the reaction active sites. The highest response peak current was obtained for 300 s accumulation. Afterwards the response peak currents remain stable, suggesting the nitrite adsorption has achieved saturation. Thus, 300 s was selected as the optimal accumulation time.

2.5. Standard Curves, Linear Ranges, and Detection Limit

Under the optimized analytical conditions, the response peak currents of various concentrations of nitrite were measured by NaCS/PDMDAAC/GCE using differential pulse voltammetry (DPV). The response DPV curves of various concentrations of nitrite are shown in Figure 7. There is a good relationship between the response peak current and the nitrite concentration ranging from 4.0×10^{-8} mol·L^{-1} to 1.5×10^{-4} mol·L^{-1}. The corresponding linear regression equation can expressed as i (μA) = 0.01027c (μmol·L^{-1}) + 0.02771 (r = 0.9984, n = 8). The limit of detection (LOD, S/N = 3) is estimated to 4.3×10^{-8} mol·L^{-1}. A comparison on sensing performances between reported electrodes and NaCS/PDMDAAC/GCE is summarized on the Table 1. The sensing performances (in terms of linear response range and limit of detection) of the NaCS/PDMDAAC/GCE are at least comparable to, and even better than, the most reported modified electrodes [45–50].

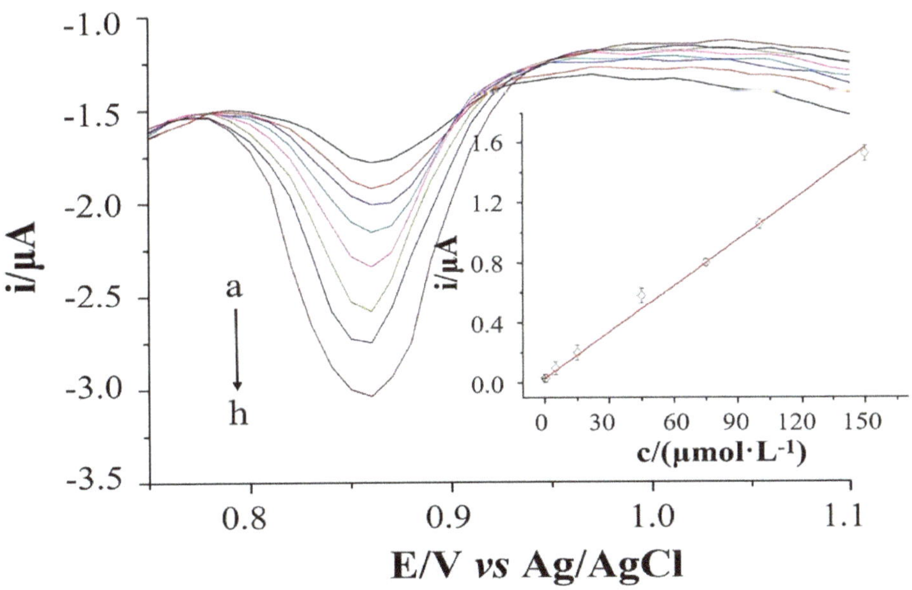

Figure 7. Differential pulse voltammograms of NaCS/PDMDAAC/GCE in the nitrite solution of different concentrations with 0.1 V pulse amplitude, 0.01 s pulse width and 0.1 V s^{-1} scan rate. (a) 1.5 $\times 10^{-4}$ mol·L^{-1}; (b) 1.0×10^{-4} mol·L^{-1}; (c) 7.5×10^{-5} mol·L^{-1}; (d) 4.5×10^{-5} mol·L^{-1}; (e) 1.5×10^{-5} mol·L^{-1}; (f) 5×10^{-6} mol·L^{-1}; (g) 5×10^{-7} mol·L^{-1}; (h) 4×10^{-8} mol·L^{-1}.

Table 1. A comparison on sensing performances between reported electrodes and NaCS/PDMDAAC/GCE.

Electrodes	Methods	Linear Ranges (μmol·L^{-1})	LOD (μmol·L^{-1})	References
POSS/rGO/GCE	CA	0.5–120	0.08	[45]
PNB/GCE	DPV	0.5–100	0.1	[46]
β-MnO_2 NRs/GCE	CA	0.29–26,090	0.29	[47]
Ag/HNT/MoS_2/CPE	CA	2–425	0.7	[48]
MWCNTs-TiN/GCE	CA	1–2000	0.0014	[49]
hemin/TNT/GCE	CA	0.6–130	0.084	[50]
NaCS/PDMDAAC/GCE	DPV	150–0.04	0.043	This work

2.6. Selectivity, Reproducibility, and Stability of NaCS/PDMDAAC/GCE

Before detection nitrite in real samples, the practicability (including selectivity, reproducibility, and stability) of the proposed NaCS/PDMDAAC/GCE were checked carefully. Firstly, the interference investigation was performed in 1×10^{-5} mol·L^{-1} sodium nitrite mixed with the equal concentration of potential interfering substances including sodium chloride, glucose, L-cysteine, and ascorbic acid. The response peak current is hardly changed with these interfering compounds (peak current variation within 5%), indicating the nitrite is bonded specifically to the NaCS/PDMDAAC composite film and interfering substances cannot interfere with the nitrite detection. In other words, the proposed sensor has good specificity and selectivity toward nitrite.

Then, the reproducibility of NaCS/PDMDAAC/GCE was validated by repeatedly measuring the response peak currents for 9 times in 8.0×10^{-5} mol·L^{-1} and 8.0×10^{-6} mol·L^{-1} nitrite solution alternately. The result is shown in Table 1. Obviously, the response peak current remains stable after repeated measurements for 9 times, with relative standard deviation (RSD) of 1.05% and 4.89%, respectively (Table 2). It demonstrates that the proposed NaCS/PDMDAAC exhibits good reproducibility for nitrite detection.

Table 2. The reproducibility of NaCS/PDMDAAC/GCE for nitrite determination.

No.	Current Response (μA)	
	8.0 × 10−5 mol·L^{-1}	**8 × 10−6 mol·L^{-1}**
1	0.896	0.115
2	0.874	0.123
3	0.887	0.112
4	0.888	0.121
5	0.878	0.114
6	0.890	0.117
7	0.898	0.112
8	0.894	0.129
9	0.903	0.122
Average	0.889	0.118
SD	0.009	0.006
RSD (%)	1.05%	4.89%

Finally, the stability of the proposed sensor was investigated by continuously monitoring the response peak current for 30 days. The NaCS/PDMDAAC/GCE was always immersed in 0.1 mol·L^{-1} PBS solution (pH 7.0) and stored at 4 °C for 30 days. The cyclic voltammetric response of nitrite at the NaCS/PDMDAAC/GCE was measured every 3 days. As shown in Figure 8, after 24 days, the current response of nitrite remains unchanged nearly. After 30 days, the response peak current value still retains 73.6% of the initial value. It can be inferred that the NaCS/PDMDAAC/GCE have good stability, which is very especially important for real sample detection.

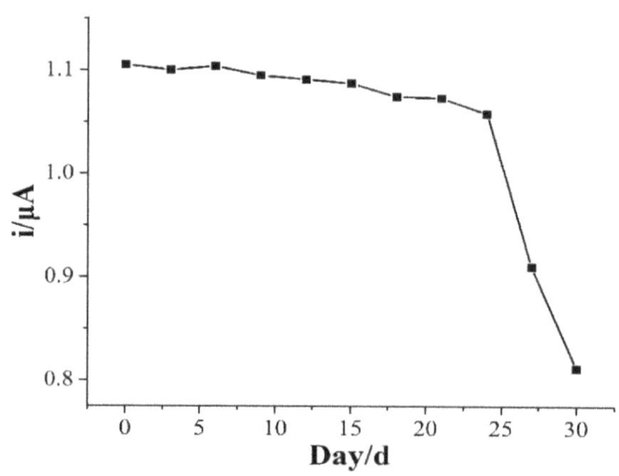

Figure 8. Stability for NaCS/PDMDAAC/GCE monitored in the 1.0×10^{-4} mol·L^{-1} nitrite for 30 days.

2.7. *Determination of Actual Samples*

Under the optimal analytical conditions, the pretreated ham sausage filtrate was used as the actual sample. The concentration of nitrite was detected by DPV method in 0.1 mol·L^{-1} PBS (pH 7.0) using the proposed NaCS/PDMDAAC/GCE, and determination results are listed in Table 3. The unspiked content of nitrite are 26.24−27.32 mg/kg, which does not exceed the maximum permitted content (0.15 g/kg) [9]. Then these samples were spiked with known concentration nitrite, the recovery is estimated to 98.2–104.0%, with relative standard deviation (RSD) lower than 3.3%, suggesting the satisfactory results are obtained. Together with outstanding advantages including low cost, sensitivity, and selectivity, the proposed NaCS/PDMDAAC/GCE shows great prospects on the detection of nitrite in diverse food samples.

Table 3. Determination of nitrite in ham sausage samples using the NaCS/PDMDAAC/GCE.

Samples	Content (mg/kg)	Added (mg/kg)	Founded (mg/kg)	Recovery (%)	RSD (%)
1	27.32	30	58.52	104.0	2.3
2	26.87	30	57.32	102.6	1.4
3	26.24	30	55.71	98.2	3.3

3. Conclusions

In this work, a novel electrochemical sensor based on NaCS/PDMDAAC composite film was proposed for the detection of nitrite at sub-micromolar level. The NaCS/PDMDAAC/GCE was fabricated by a facile drop-casting method. The surface morphology of NaCS film and NaCS/PDMDAAC composite was characterized by SEM. The electrochemical behavior of nitrite at the NaCS/PDMDAAC was investigated by CV and DPV. The NaCS/PDMDAAC not only shows the inherent properties from the individual component, but also exhibits synergetic enchantment effect toward sensing nitrite. Under optimal detection conditions, the proposed NaCS/PDMDAAC/GCE exhibits a wide linear response region of 4.0×10^{-8} mol·L^{-1}~1.5×10^{-4} mol·L^{-1}, and low detection limit of 43 nmol·L^{-1} (S/N = 3). Moreover, the proposed NaCS/PDMDAAC/GCE also has good stability and reproducibility. Finally, the NaCS/PDMDAAC/GCE was successfully used to detect the nitrite content in the ham sausage sample with satisfactory results. The proposed NaCS/PDMDAAC/GCE provided a promising sensing platform on the detection nitrite in diverse real samples at the sub-micromolar level.

4. Materials and Methods

4.1. *Chemicals*

Poly(dimethyldiallylammonium chloride) (PDMDAAC) was supplied by Aldrich Co., Ltd (Shanghai, China). Cellulose sulfate sodium salt was purchased from J&K SCIENTIFIC LTD (Shanghai, China). Borax and sodium nitrite were purchased from Hunan General Chemical Reagent Works (Changsha, China). Potassium ferricyanide, potassium hexacyanoferrate, sodium dihydrogen phosphate and disodium hydrogen phosphate were all bought from Sinopharm Chemical Reagent Co., Ltd (Shanghai, China). All the chemicals were of analytical reagent grade and were used as received without any further purification. Both PDMDAAC and NaCS were also synthesized in our experiments (details can be seen in the Supporting Information).

4.2. *Fabrication of NaCS/PDMDAAC/GCE*

The synthesis of NaCS and PDMDAAC has been described in the Supporting Information. The infrared spectra of cellulose sodium sulfate, ^{1}H NMR spectrum of DMDAAC, infrared spectra of (a) PDMDAAC & (b) DMDAAC have been also plotted in the Supporting Information (Figure S1,

Figure S2 & Figure S3). GCE was polished to mirror-like surface with 0.3 μm and 0.05 μm alumina powder, rinsed twice with ethanol and distilled water, and dried under natural condition. Firstly, 3 μL PDMDAAC (7% (*w*/*v*)) solution was drop-casted on the surface of GCE, dried naturally to obtain PDMDAAC/GCE. Secondly, 3 μL NaCS (3.5% (*w*/*v*)) solution was dropped-casted to the surface of PDMDAAC/GCE, dried naturally to obtain NaCS/PDMDAAC/GCE.

4.3. Electrochemical Measurements

All electrochemical experiments were performed on the CHI-760B electrochemical workstation (Shanghai Chenhua Instrument Co., Ltd., Shanghai, China). A standard three-electrode system is used for all electrochemical experiments, consisting of bare GCE or modified electrodes as the working electrode, platinum wire as the counter electrode, and an Ag/AgCl electrode (saturated potassium chloride) as the reference electrode. The electrochemical performance was measured by cyclic voltammetry (CV) and electrochemical impedance spectroscopy (EIS) in 5×10^{-3} mol·L^{-1} $K_3[Fe(CN)_6]$ solution. The electrochemical behavior of nitrite at the NaCS/PDMDAAC/GCE was investigated by CV. The content of nitrite was detected by differential pulse voltammetry (DPV). The supporting electrolyte for all electrochemical measurements is 0.1 mol·L^{-1} Na_2HPO_4-NaH_2PO_4 buffer solution (PBS, pH 7.0, containing 0.1 M NaCl) unless otherwise stated.

4.4. Pretreatment of Ham Sausage Samples

The ham sausage samples were purchased from a local supermarket. 12.5 mL of borax saturation solution was added into 5 g ground ham sausage, and stirred to mix well. Then the ham sausage sample was washed into 50 mL volumetric flask by a small amount of hot water, and heated in boiling water bath for 15 min. Afterwards the volumetric flask was cooled to room temperature. Subsequently, 2.5 mL PBS solution was added into the volumetric flask, and then dilute with distilled water to 50 mL. After 30 min rest, the upper layer of fat or oil was removed, the spare liquor was extracted by suction filtration to obtain sample solution for quantitative analysis.

Author Contributions: J.N. and J.W. conceived and designed the experiments; J.N., X.L., M.W. and J.L. performed the experiments; D.L., R.H., D.C., J.N., and J.W. analyzed the data; J.N., and J.W. contributed reagents/materials/analysis tools; J.N. and J.W. wrote the paper; all authors read and approved the final manuscript.

References

1. Noor, N.S.; Tan, L.L.; Heng, L.Y.; Chong, K.F.; Tajuddin, S.N. Acrylic microspheres-based optosensor for visual detection of nitrite. *Food Chem.* **2016**, *207*, 132–138. [CrossRef] [PubMed]
2. Kumar, J.V.; Karthik, R.; Chen, S.M.; Balasubramanian, P.; Muthuraj, V.; Selvam, V. A novel cerium tungstate nanosheets modified electrode for the effective electrochemical detection of carcinogenic nitrite ions. *Electroanalysis* **2017**, *29*, 2385–2394. [CrossRef]
3. Rastogi, P.; Ganesan, V.; Krishnamoorthi, S. A promising electrochemical sensing platform based on a silver nanoparticles decorated copolymer for sensitive nitrite determination. *J. Mater. Chem. A* **2013**, *2*, 933–943. [CrossRef]
4. Manassaram, D.M.; Backer, L.C.; Moll, D.M. A review of nitrites in drinking water: Maternal exposure and adverse reproductive and developmental outcomes. *Environ. Health Perspect.* **2006**, *114*, 320–327. [CrossRef] [PubMed]
5. Bukowski, J.; Somers, G.; Bryanton, J. Agricultural contamination of groundwater as a possible risk factor for growth restriction or prematurity. *J. Occup. Environ. Med.* **2001**, *43*, 377–383. [CrossRef] [PubMed]
6. Brender, J.D.; Olive, J.M.; Felkner, M.; Suarez, L.; Marckwardt, W.; Hendricks, K.A. Dietary nitrites and nitrites, nitrosatable drugs, and neural tube defects. *Epidemiology* **2004**, *15*, 330–336. [CrossRef] [PubMed]

7. Kikura-Hanajiri, R.; Martin, R.S.; Lunte, S.M. Indirect measurement of nitric oxide production by monitoring nitrite and nitrite using microchip electrophoresis with electrochemical detection. *Anal. Chem.* **2002**, *74*, 6370–6377. [CrossRef] [PubMed]
8. WHO. *Guidelines for Drinking-Water Quality*, 3rd ed.; World Health Organization: Geneva, Switzerland, 2008; ISBN 9789241547611.
9. Mao, Y.; Bao, Y.; Han, D.; Zhao, B. Research progress on nitrite electrochemical sensor. *Chin. J. Anal. Chem.* **2018**, *46*, 147–156. [CrossRef]
10. Li, Y.; Whitaker, J.S.; McCarty, C.L. Reversed-phase liquid chromatography/electrospray ionization/mass spectrometry with isotope dilution for the analysis of nitrite and nitrite in water. *J. Chromatogr.* **2011**, *1218*, 476–483. [CrossRef] [PubMed]
11. He, L.; Zhang, K.; Wang, C.; Luo, X.; Zhang, S. Effective indirect enrichment and determination of nitrite ion in water and biological samples using ionic liquid-dispersive liquid-liquid microextraction combined with high-performance liquid chromatography. *J. Chromatogr.* **2011**, *1218*, 3595–3600. [CrossRef] [PubMed]
12. Manzoori, J.L.; Sorouraddin, M.H.; Hajishabani, A.M. Spectrophotometric determination of nitrite based on its catalytic effect on the oxidation of carminic acid by bromate. *Talanta* **1998**, *46*, 1379–1386. [CrossRef]
13. Pourreza, N.; Fat'Hi, M.R.; Hatami, A. Indirect cloud point extraction and spectrophotometric determination of nitrite in water and meat products. *Microchem. J.* **2012**, *104*, 22–25. [CrossRef]
14. Peng, Q.; Xie, M.G.; Koon-Gee, N.; Kang, E.T. A new nitrite-selective fluorescent sensor fabricated from surface-initiated atom-transfer radical polymerization. *Chem. Lett.* **2005**, *34*, 1628–1629. [CrossRef]
15. Jiao, C.X.; Niu, C.G.; Huan, S.Y.; Shen, Q.; Yang, Y.; Shen, G.L.; Yu, R.Q. A reversible chemosensor for nitrite based on the fluorescence quenching of a carbazole derivative. *Talanta* **2004**, *64*, 637–643. [CrossRef] [PubMed]
16. Lin, K.; Ma, J.; Yuan, D.; Feng, S.; Su, H.; Huang, Y.; Shangguan, Q. Sequential determination of multi-nutrient elements in natural water samples with a reverse flow injection system. *Talanta* **2017**, *167*, 166–171. [CrossRef] [PubMed]
17. Feng, S.; Zhang, M.; Huang, Y.; Yuan, D.; Zhu, Y. Simultaneous determination of nanomolar nitrite and nitrite in seawater using reverse flow injection analysis coupled with a long path length liquid waveguide capillary cell. *Talanta* **2013**, *117*, 456–462. [CrossRef] [PubMed]
18. Ning, J.; He, Q.; Luo, X.; Wang, M.; Liu, D.; Wang, J.; Liu, J.; Li, G. Rapid and sensitive determination of vanillin based on a glassy carbon electrode modified with Cu_2O-electrochemically reduced graphene oxide nanocomposite film. *Sensors* **2018**, *18*, 2762. [CrossRef] [PubMed]
19. He, Q.; Liu, J.; Liang, J.; Liu, X.; Li, W.; Liu, Z.; Ding, Z.; Tuo, D. Towards Improvements for Penetrating the Blood–Brain Barrier—Recent Progress from a Material and Pharmaceutical Perspective. *Cells* **2018**, *7*, 24. [CrossRef] [PubMed]
20. He, Q.; Liu, J.; Liu, X.; Li, G.; Deng, P.; Liang, J. Preparation of Cu_2O-reduced graphene nanocomposite modified electrodes towards ultrasensitive dopamine detection. *Sensors* **2018**, *18*, 199. [CrossRef] [PubMed]
21. He, Q.; Liu, J.; Liu, X.; Li, G.; Chen, D.; Deng, P.; Liang, J. Fabrication of amine-modified magnetite-electrochemically reduced graphene oxide nanocomposite modified glassy carbon electrode for sensitive dopamine determination. *Nanomaterials* **2018**, *8*, 194. [CrossRef] [PubMed]
22. He, Q.; Liu, J.; Liang, J.; Liu, X.; Tuo, D.; Li, W. Chemically Surface Tunable Solubility Parameter for Controllable Drug Delivery—An Example and Perspective from Hollow PAA-Coated Magnetite Nanoparticles with R6G Model Drug. *Materials* **2018**, *11*, 247. [CrossRef] [PubMed]
23. He, Q.; Liu, J.; Liu, X.; Li, G.; Deng, P.; Liang, J. Manganese dioxide Nanorods/electrochemically reduced graphene oxide nanocomposites modified electrodes for cost-effective and ultrasensitive detection of Amaranth. *Colloid Surf. B* **2018**, *172*, 565–572. [CrossRef] [PubMed]
24. Ma, Y.; Song, X.; Ge, X.; Zhang, H.; Wang, G.; Zhang, Y.; Zhao, H. In situ growth of α-Fe_2O_3 nanorod arrays on 3d carbon foam as an efficient binder-free electrode for highly sensitive and specific determination of nitrite. *J. Mater. Chem. A* **2017**, *5*, 4726–4736. [CrossRef]
25. Mehmeti, E.; Stanković, D.M.; Hajrizi, A.; Kalcher, K. The use of graphene nanoribbons as efficient electrochemical sensing material for nitrite determination. *Talanta* **2016**, *159*, 34–39. [CrossRef] [PubMed]
26. Rabti, A.; Aoun, S.B.; Raouafi, N. A sensitive nitrite sensor using an electrode consisting of reduced graphene oxide functionalized with ferrocene. *Microchim. Acta* **2016**, *183*, 3111–3117. [CrossRef]
27. Dai, J.; Deng, D.; Yuan, Y.; Zhang, J.; Deng, F.; He, S. Amperometric nitrite sensor based on a glassy carbon electrode

modified with multi-walled carbon nanotubes and poly(toluidine blue). *Microchim. Acta* **2016**, *183*, 1553–1561. [CrossRef]

28. Shivakumar, M.; Nagashree, K.L.; Manjappa, S.; Dharmaprakash, M.S. Electrochemical detection of nitrite using glassy carbon electrode modified with silver nanospheres (AgNS) obtained by green synthesis using pre-hydrolysed liquor. *Electroanalysis* **2017**, *29*, 1434–1442. [CrossRef]
29. Seo, Y.; Manivannan, S.; Kang, I.; Lee, S.W.; Kim, K. Gold dendrites co-deposited with m13 virus as a biosensor platform for nitrite ions. *Biosens. Bioelectron.* **2017**, *94*, 87–93. [CrossRef] [PubMed]
30. Kung, C.W.; Li, Y.S.; Lee, M.H.; Wang, S.Y.; Chiang, W.H.; Ho, K.C. In-situ growth of porphyrinic metal-organic framework nanocrystals on graphene nanoribbons for electrocatalytic oxidation of nitrite. *J. Mater. Chem. A* **2016**, *4*, 10673–10682. [CrossRef]
31. Kung, C.W.; Chang, T.H.; Chou, L.Y.; Hupp, J.T.; Farha, O.K.; Ho, K.C. Porphyrin-based metal–organic framework thin films for electrochemical nitrite detection. *Electrochem. Commun.* **2015**, *58*, 51–56. [CrossRef]
32. Xu, G.; Liang, S.; Fan, J.; Sheng, G.; Luo, X. Amperometric sensing of nitrite using a glassy carbon electrode modified with a multilayer consisting of carboxylated nanocrystalline cellulose and poly(diallyldimethyl ammonium) ions in a pedot host. *Microchim. Acta* **2016**, *183*, 2031–2037. [CrossRef]
33. Chen, L.; Liu, X.; Wang, C.; Lv, S.; Chen, C. Amperometric nitrite sensor based on a glassy carbon electrode modified with electrodeposited poly(3,4-ethylenedioxythiophene) doped with a polyacenic semiconductor. *Microchim. Acta* **2017**, *184*, 2073–2079. [CrossRef]
34. Masumi, T.; Matsushita, Y.; Dan, A.; Takama, R.; Saito, K.; Kuroda, K.; Fukushima, K. Adsorption behavior of poly(dimethyl-diallylammonium chloride) on pulp fiber studied by cryo-time-of-flight secondary ion mass spectrometry and cryo-scanning electron microscopy. *Appl. Surf. Sci.* **2014**, *289*, 155–159. [CrossRef]
35. Wang, B.; Okoth, O.K.; Yan, K.; Zhang, J. A highly selective electrochemical sensor for 4-chlorophenol determination based on molecularly imprinted polymer and PDDA-functionalized graphene. *Sens. Actuators B Chem.* **2016**, *236*, 294–303. [CrossRef]
36. Yang, J.; Lin, Q.; Yin, W.; Jiang, T.; Zhao, D.; Jiang, L. A novel nonenzymatic glucose sensor based on functionalized pdda-graphene/cuo nanocomposites. *Sens. Actuators B Chem.* **2017**, *253*, 1087–1095. [CrossRef]
37. Yang, T.; Zhang, W.; Du, M.; Jiao, K. A PDDA/poly(2,6-pyridinedicarboxylic acid)-CNTs composite film DNA electrochemical sensor and its application for the detection of specific sequences related to PAT gene and NOS gene. *Talanta* **2008**, *75*, 987–994. [CrossRef] [PubMed]
38. Chen, G.; Yao, S.J.; Guan, Y.X.; Lin, D.Q. Preparation and characterization of NaCS-cmc/PDMDAAC capsules. *Colloids Surf. B Biointerfaces* **2005**, *45*, 136–143. [CrossRef] [PubMed]
39. Zeng, X.H.; Danquah, M.K.; Zheng, C.; Potumarthi, R.; Chen, X.D.; Lu, Y.H. NaCS-PDMDAAC immobilized autotrophic cultivation of chlorella sp. For wastewater nitrogen and phosphate removal. *Chem. Eng. J.* **2004**, *99*, 185–192. [CrossRef]
40. Zhang, J.; Guan, Y.X.; Ji, Z.; Yao, S.J. Effects on membrane properties of NaCS–PDMDAAC capsules by adding inorganic salts. *J. Membr. Sci.* **2006**, *277*, 270–276. [CrossRef]
41. Wang, M.; Chen, K.; Liu, J.; He, Q.; Li, G.; Li, F. Efficiently enhancing electrocatalytic activity of α-MnO_2 nanorods/N-doped ketjenblack carbon for oxygen reduction reaction and oxygen evolution reaction using facile regulated hydrothermal treatment. *Catalysts* **2018**, *8*, 138. [CrossRef]
42. Chen, K.; Wang, M.; Li, G.; He, Q.; Liu, J.; Li, F. Spherical α-MnO_2 supported on N-KB as efficient electrocatalyst for oxygen reduction in al–air battery. *Materials* **2018**, *11*, 601. [CrossRef] [PubMed]
43. Li, G.; Wang, S.; Duan, Y.Y. Towards gel-free electrodes: A systematic study of electrode-skin impedance. *Sens. Actuators B Chem.* **2017**, *241*, 1244–1255. [CrossRef]
44. Li, G.; Zhang, D.; Wang, S.; Duan, Y.Y. Novel passive ceramic based semi-dry electrodes for recording electroencephalography signals from the hairy scalp. *Sens. Actuators B Chem.* **2016**, *237*, 167–178. [CrossRef]
45. Bai, W.; Sheng, Q.; Zheng, J. Hydrophobic interface controlled electrochemical sensing of nitrite based on one step synthesis of polyhedral oligomeric silsesquioxane/reduced graphene oxide nanocomposite. *Talanta* **2016**, *150*, 302–309. [CrossRef] [PubMed]
46. Chen, X.; Wang, F.; Chen, Z. An electropolymerized nile blue sensing film-based nitrite sensor and application in food analysis. *Anal. Chim. Acta* **2008**, *623*, 213–220. [CrossRef] [PubMed]
47. Feng, J.J.; Zhang, P.P.; Wang, A.J.; Zhang, Y.; Dong, W.J.; Chen, J.R. One-pot hydrothermal synthesis of uniform β-MnO_2 nanorods for nitrite sensing. *J. Colloid Interface Sci.* **2011**, *359*, 1–8. [CrossRef] [PubMed]

48. Ghanei-Motlagh, M.; Taher, M.A. A novel electrochemical sensor based on silver/halloysite nanotube/molybdenum disulfide nanocomposite for efficient nitrite sensing. *Biosens. Bioelectron.* **2018**, *109*, 279–285. [CrossRef] [PubMed]

49. Haldorai, Y.; Hwang, S.K.; Gopalan, A.I.; Huh, Y.S.; Han, Y.K.; Voit, W.; Saianand, G.; Lee, K.P. Direct electrochemistry of cytochrome c immobilized on titanium nitride/multi-walled carbon nanotube composite for amperometric nitrite biosensor. *Biosens. Bioelectron.* **2016**, *79*, 543–552. [CrossRef] [PubMed]

50. Ranjani, B.; Kalaiyarasi, J.; Pavithra, L.; Devasena, T.; Pandian, K.; Gopinath, S. Amperometric determination of nitrite using natural fibers as template for titanium dioxide nanotubes with immobilized hemin as electron transfer mediator. *Microchim. Acta* **2018**, *185*, 194. [CrossRef] [PubMed]

3

Development of New Analytical Microwave-Assisted Extraction Methods for Bioactive Compounds from Myrtle (*Myrtus communis* L.)

Ana V. González de Peredo [1], Mercedes Vázquez-Espinosa [1], Estrella Espada-Bellido [1], Ana Jiménez-Cantizano [2], Marta Ferreiro-González [1], Antonio Amores-Arrocha [2], Miguel Palma [1], Carmelo G. Barroso [1] and Gerardo F. Barbero [1,*]

1 Department of Analytical Chemistry, Faculty of Sciences, University of Cadiz, Agrifood Campus of International Excellence (ceiA3), IVAGRO, Puerto Real, 11510 Cadiz, Spain; ana.velascogope@uca.es (A.V.G.d.P.); mercedes.vazquez@uca.es (M.V.-E.); estrella.espada@uca.es (E.E.-B.); marta.ferreiro@uca.es (M.F.-G.); miguel.palma@uca.es (M.P.); carmelo.garcia@uca.es (C.G.B.)

2 Department of Chemical Engineering and Food Technology, Faculty of Sciences, University of Cadiz, Agrifood Campus of International Excellence (ceiA3), IVAGRO, Puerto Real, 11510 Cadiz, Spain; ana.jimenezcantizano@uca.es (A.J.-C.); antonio.amores@uca.es (A.A.-A.)

* Correspondence: gerardo.fernandez@uca.es

Abstract: The phenolic compounds and anthocyanins present in myrtle berries are responsible for its beneficial health properties. In the present study, a new, microwave-assisted extraction for the analysis of both phenolic compounds and anthocyanins from myrtle pulp has been developed. Different extraction variables, including methanol composition, pH, temperature, and sample–solvent ratio were optimized by applying a Box–Behnken design and response surface methodology. Methanol composition and pH were the most influential variables for the total phenolic compounds (58.20% of the solvent in water at pH 2), and methanol composition and temperature for anthocyanins (50.4% of solvent at 50 °C). The methods developed showed high repeatability and intermediate precision (RSD < 5%). Both methods were applied to myrtle berries collected in two different areas of the province of Cadiz (Spain). Hierarchical clustering analysis results show that the concentration of bioactive compounds in myrtle is related to their geographical origin.

Keywords: anthocyanins; bioactive compounds; Box–Behnken design; microwave-assisted extraction; myrtle; *Myrtus communis*; phenolic compounds

1. Introduction

Myrtus communis L., the common myrtle, is an evergreen shrub that grows spontaneously in the Mediterranean area and in the Middle East. Myrtle berries have a maximum ripening period from October to February. These berries have multiple shapes and colors [1], but are mainly dark blue in color. The ancient Mediterranean populations already used myrtle mainly for ornamental and aromatic purposes [2]. Recent developments in the fields of health and food have markedly increased their interest in natural compounds with antioxidant potential [3]. The extraction of natural antioxidants in fruits is very useful when substituting synthetic antioxidants, which are being restricted because of their potential health risks and side-effects, and their safety has been questioned for a long time [4]. Therefore, nowadays, myrtle has gained greater recognition in the food and medicinal industries due to its potential beneficial effects [5]. Myrtle presents anti-diabetic, anti-inflammatory, anticancer, antioxidant, antihyperglycaemic, antimycotic, and antiseptic properties [6,7]. For example, myrtle oil is recommended for the treatment of respiratory diseases [8], and it is normally taken as an infusion. In addition, essential oils are used in the perfume and cosmetics industries [9]. Despite the aforementioned uses, myrtle is still mainly known for the production of an aromatic liqueur called "Mirto", which is obtained by alcoholic maceration of its leaves and fruit. This liquor is very popular and traditional in Sardinia, where it is usually served very cold after meals due to its digestive powers [10].

The phenolic compounds and anthocyanins present in myrtle berries are the main contributors to these beneficial health properties. The quantities of phenolic compounds and anthocyanins present in myrtle berries are extraordinarily high [11,12]. The major phenolic compounds identified in myrtle are quercetin 3-*O*-galactoside, quercetin 3-*O*-rhamnoside, myricetin 3-*O*-rhamnoside, quercetin 3-*O*-glucoside, ellagic acid, and myricetin [13,14]. The major anthocyanins identified are delphinidin 3,5-*O*-diglucoside, delphinidin 3-*O*-glucoside, cyanidin 3-*O*-galactoside, cyanidin 3-*O*-glucoside, cyanidin 3-*O*-arabinoside, petunidin 3-*O*-glucoside, delphinidin 3-*O*-arabinoside, peonidin 3-*O*-glucoside, malvidin 3-*O*-glucoside, petunidin 3-*O*-arabinoside, and malvidin 3-*O*-arabinoside [15,16].

For the extraction of bioactive compounds, specifically for the extraction of phenolic compounds and anthocyanins in vegetable matrices, solid–liquid extraction is usually carried out. Microwave-assisted extraction (MAE) is one of the more advanced extraction methods. MAE is widely used as it is a promising green-extraction method that has the advantage of reducing both extraction time and solvent consumption [17,18]. The microwave technique is based on the application of electromagnetic radiation, with a frequency from 0.3 to 300 GHz. This radiation, which leads to rapid and localized heating of the solvent and sample, is based on a direct effect on the molecules through ionic conduction and dipole rotation [19]. The localized heating leads to a pressure build-up within the cells of the sample, resulting in a rapid transfer of the compound of interest from the cells to the extraction solvent [20]. This extraction method has been widely employed for the extraction of phenolic compounds and anthocyanins from a wide variety of vegetable matrices, such as grapes [21], tomatoes [22], and blackberries [23]. In a recent study, MAE has been applied to extract polyphenols, tannins, and flavonoids from myrtle leaves [24]. Based on the previous results obtained from this study and the high content of bioactive compounds in myrtle leaves, the development of new methods for myrtle analyses to improve the quality of final products, such as liqueurs, is required. Besides, this extraction technique has not yet been completely developed and optimized for the extraction of phenolic compounds and anthocyanins from myrtle berries. As in myrtle leaves, its berries are expected to contain an important amount of the same bioactive compounds besides the anthocyanins due to its dark blue color. This approach is of great interest due to the importance of developing techniques for analysis of the raw material to improve the liqueur quality, which is the main use of myrtle.

The efficiency of MAE can be affected by several variables, such as the solvent (volume, composition, pH), the temperature, the time of application, and the power level [25]. For this reason, experimental designs are usually applied in order to study the effects of the different variables and their interactions, and to determine the optimal conditions. In the present study, a Box–Behnken design (BBD) with a response surface methodology (RSM) was chosen [26,27]. It was employed because the number of experiments necessary to provide sufficient information for statistically acceptable results in a BBD is lower than other statistical designs, and it also ensures that each experiment is in the region of interest, avoiding extreme conditions [28].

The aim of the present study was to develop and optimize MAE methods for the extraction of bioactive compounds (phenolic compounds and anthocyanins) in myrtle in order to evaluate the quality of myrtle berries and to study the possible effect of the geographical origin in the total amount of bioactive compounds.

2. Results and Discussion

2.1. Development of the MAE Methods

A Box–Behnken design was carried out for the development and optimization of the microwave-assisted extraction of both total phenolic compounds and total anthocyanins (as the sum of individual components) in the myrtle berries. Analysis of variance (ANOVA) was carried out to evaluate the effects of the factors and the possible interactions between them. The factors studied in this work were: solvent composition (% methanol in water), solvent pH, extraction temperature, and sample–solvent ratio. The results of this analysis are shown in Tables 1 and 2 for the total phenolic compounds and total anthocyanins, respectively. The coefficients for the different parameters of the quadratic polynomial equation and their significance (*p*-values) are presented. The factors and/or interactions that showed a *p*-value lower than 0.05 were considered to be significant factors that influenced the response at the selected level of significance (95%).

Table 1. Analysis of variance of the quadratic model adjusted to the extraction of total phenolic compounds. The studied ranges for each parameter were: methanol (50–100%), pH (2–7), temperature (50–100 °C), and sample–solvent ratio (0.5 g/10 mL–0.5 g/20 mL).

Variable	Source	Coefficient	Sum of Squares	Degrees of Freedom	Mean Square	*F*-Value	*p*-Value
	Model		1016.28	14	72.59	2.92	0.0352
Methanol	X_1	−4.39257	231.54	1	231.54	9.31	0.0101
pH	X_2	−3.62017	157.27	1	157.27	6.32	0.0272
Temperature	X_3	0.915742	10.06	1	10.06	0.4047	0.5366
Ratio	X_4	1.11842	15.01	1	15.01	0.6036	0.4522
Methanol-pH	X_1X_2	−5.57764	3.80	1	3.80	0.1528	0.7027
Methanol-Temperature	X_1X_3	0.974675	47.75	1	47.75	1.92	0.1911
Methanol-Ratio	X_1X_4	−3.45505	3.43	1	3.43	0.1380	0.7167
pH-Temperature	X_2X_3	0.926275	51.01	1	51.01	2.05	0.1776
pH-Ratio	X_2X_4	4.93415	25.11	1	25.11	1.01	0.3348
Temperature-Ratio	X_3X_4	−3.571	8.33	1	8.33	0.3350	0.5734
Methanol-Methanol	X_1^2	−2.50565	165.92	1	165.92	6.67	0.0240
pH-pH	X_2^2	1.45055	129.84	1	129.84	5.22	0.0413
Temperature-Temperature	X_3^2	−1.44313	11.22	1	11.22	0.4513	0.5145
Ratio-Ratio	X_4^2	0.889012	4.22	1	4.22	0.1695	0.6878
Residual		42.6799	298.41	12	24.87		
Lack of Fit			266.57	10	26.66	1.67	0.4311
Pure Error			31.83	2	15.92		
Total			1314.68	26			

With regard to the total phenolic compounds (Table 1), the *p*-values for solvent composition and pH were less than 0.05, meaning that these factors had significant effects. The quadratic interactions of solvent composition (X_1^2) and pH (X_2^2) had a significant effect on the extraction of phenolic compounds. The interactions between factors were not significant (p-value > 0.05). Among the linear terms, the most significant factor was the solvent composition (p-value < 0.01), and this had a negative effect ($b_1 = -4.39257$), which means that the phenolic compounds were extracted more efficiently when the solvent had a low methanol content in water in this range. The pH also had a negative effect ($b_2 = -3.62017$), which means that the extraction of phenolic compounds is more favorable at a low pH. Among the quadratic effects, the methanol effect was more significant than the pH effect, and methanol had a negative effect, whereas pH had a positive effect.

In the case of anthocyanins (Table 2), only the linear term temperature (X_3) had an influence on the response, with a p-value < 0.01. With regard to quadratic effects, X_1^2 (solvent composition) once again had a significant effect on the extraction. The interaction between the factors' pH and temperature (X_2X_3) had a significant effect. The temperature had a negative effect ($b_2 = -2.56032$), which indicates that a decrease in its value led to a higher recovery of anthocyanins. This should be due to the degradation of anthocyanins when high temperatures were used [29]. The quadratic effect for methanol and the interaction between pH-temperature also had positive coefficients.

Table 2. Analysis of variance of the quadratic model, adjusted to the extraction of total anthocyanins. The ranges studied for each parameter were: methanol (50–100%), pH (2–7), temperature (50–100 °C), and sample–solvent ratio (0.5 g/10 mL–0.5 g/20 mL).

Variable	Source	Coefficient	Sum of Squares	Degrees of Freedom	Mean Square	*F*-Value	*p*-Value
	Model		445.11	14	31.79	4.05	0.0100
Methanol	X_1	0.249033	0.7442	1	0.7442	0.0949	0.7633
pH	X_2	−0.207067	0.5145	1	0.5145	0.0656	0.8022
Temperature	X_3	−2.56032	78.66	1	78.66	10.03	0.0081
Ratio	X_4	−0.92215	10.20	1	10.20	1.30	0.2762
Methanol-pH	X_1X_2	5.98483	15.51	1	15.51	1.98	0.1850
Methanol-Temperature	X_1X_3	−1.9692	21.20	1	21.20	2.70	0.1260
Methanol-Ratio	X_1X_4	2.3022	0.6398	1	0.6398	0.0816	0.7800
pH-Temperature	X_2X_3	0.39995	41.29	1	41.29	5.27	0.0406
pH-Ratio	X_2X_4	−0.758496	0.5627	1	0.5627	0.0718	0.7933
Temperature-Ratio	X_3X_4	3.21277	9.76	1	9.76	1.24	0.2864
Methanol-Methanol	$X_1{}^2$	0.375075	191.03	1	191.03	24.36	0.0003
pH-pH	$X_2{}^2$	−0.830096	3.07	1	3.07	0.3913	0.5433
Temperature-Temperature	$X_3{}^2$	−1.56213	3.67	1	3.67	0.4687	0.5066
Ratio-Ratio	$X_4{}^2$	1.0868	6.30	1	6.30	0.8034	0.3877
Residual		18.6496	94.10	12	7.84		
Lack of Fit			87.43	10	8.74	2.62	0.3076
Pure Error			6.67	2	3.33		
Total			539.21	26			

The standardized Pareto chart, which allows for knowledge of the influencing variables and their order of influence from a graphical point of view, is presented in Figure 1. As mentioned earlier, for phenolic compounds (Figure 1a), it can be seen that the significant factors in decreasing order of influence on the response are: methanol percentage, the quadratic interaction of methanol percentage, and the quadratic interaction of pH. For anthocyanins (Figure 1b), the significant factors in the same order are: the quadratic interaction of methanol percentage, the temperature, and the interaction pH temperature.

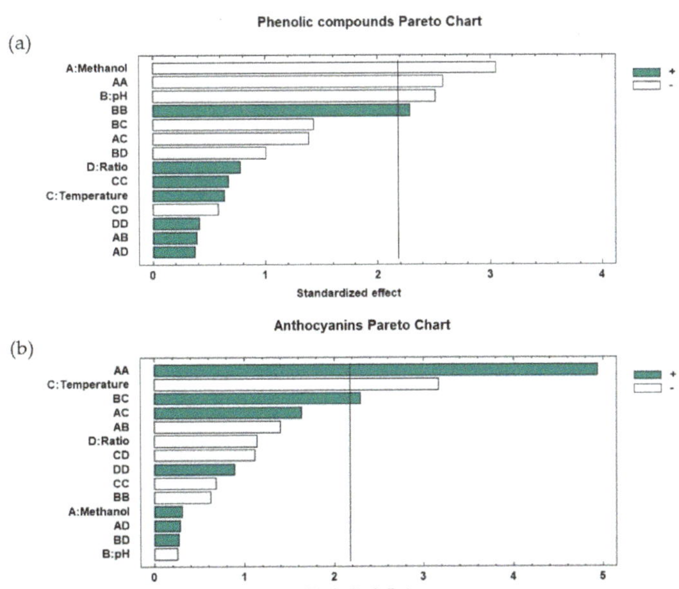

Figure 1. Standardized Pareto charts for: (**a**) total phenolic compounds; (**b**) anthocyanins.

The complete second-order polynomial model correlates the relationship between independent variables and responses. The correlation was evaluated using the squared correlation coefficients (R^2). The coefficients obtained for the total phenolic compounds, R^2 = 77.30%, and the anthocyanins, R^2 = 82.55%, indicate a statistically significant agreement between the measured and estimated responses. More specifically, the lack-of-fit test showed a p-value higher than 0.05 (not significant) for both phenolic compounds and anthocyanins, which means that the models fit well.

The reduced Equations (1) and (2), which show acceptable agreement between the experimental data and the estimated values, are expressed as follows:

$$Y_{TP}\ (\mu g\ g^{-1}) = 42.6799 - 4.3927X_1 - 3.62017X_2 - 5.57764X_{12} + 4.93415X_{22} \quad (1)$$

$$Y_{TA}\ (\mu g\ g^{-1}) = 18.6496 - 2.56032X_3 + 5.98483X_{12} + 2.3022X_1X_3 \quad (2)$$

The trends outlined above were recorded in three-dimensional surface plots obtained by using the polynomial equations. Solvent composition and pH were selected as the most significant factors for phenolic compounds (Figure 2a), and temperature and solvent composition as the most significant factors for anthocyanins (Figure 2b) according to the previous results mentioned above. The plots illustrate the combined effects of the most significant variables on: (**a**) total anthocyanins; and (**b**) total phenolic compound recovery, respectively.

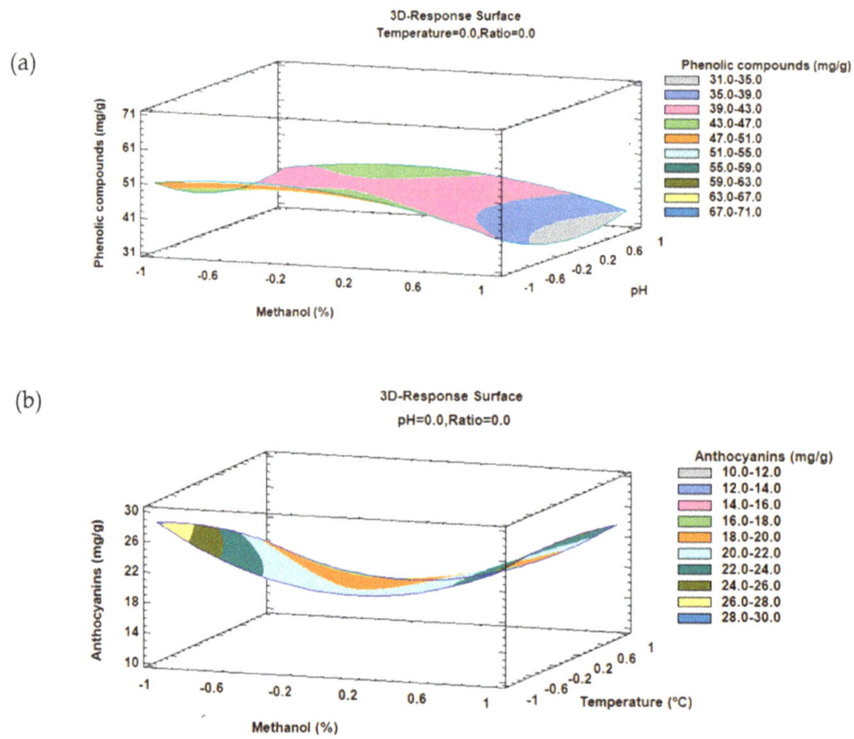

Figure 2. 3D surface plots of the Box–Behnken design using polynomial equations: (**a**) solvent composition and pH on the total phenolic compound extraction, and (**b**) temperature and solvent composition on the total anthocyanin extraction.

2.2. Optimal Conditions

From the Box–Behnken design, it is possible to extract information about the optimum values which show the maximum response for each factor. The optimum MAE conditions to extract the maximum amount of phenolic compounds are as follows: a solvent with 58.20% MeOH in water at pH 2, an extraction temperature of 100 °C, and a 0.5 g/20 mL sample–solvent ratio. The optimum

MAE conditions to extract the maximum amount of anthocyanins are as follows: a solvent with 50.4% MeOH in water at pH 3.33, an extraction temperature of 50 °C, and a 0.5 g/20 mL sample–solvent ratio. These results show that the optimal extraction of both phenolic compounds and anthocyanins occurs with values of methanol and pH closer to the lower end of the studied range (50% MeOH in water). With regard to temperature, numerous authors are in agreement that an increase in temperature favors extraction, but also that beyond a certain value, the compounds of interest can be denatured [30]. With respect to anthocyanins, high temperatures may reduce its recovery, since these compounds are thermally sensible and thus can be easily degraded [31]. With respect to phenolic compounds, although anthocyanins are also phenolic compounds, they are present at determinate levels in the overall mixture, so that the other compounds could be different phenolic compounds and less thermally sensible [32]. This possibility would explain why the optimal extraction temperature was high for the total phenolic compounds (100 °C), whereas for the anthocyanins, it was at the lower end of the range studied (50 °C). The phenolic compounds, which were less thermally sensible, increased the solubility in the solvent and the diffusion and mass transfer of the extracted molecules with high temperatures, favoring the extraction [33].

The results obtained at the optimum conditions using MAE were compared with those achieved by other extraction methods already developed from the same raw material (myrtle) [14,34–36]. Most of these studies employ traditional extraction techniques that imply long extraction times (in many cases, up to 24 h) without obtaining large recoveries. In comparison with traditional methods, such as maceration, MAE offers better extraction yields of the compounds of interest in a shorter time frame and with lower expense, regarding both solvents and costs. The higher extraction yield of the total phenolic compounds and total anthocyanins could be due to water dipole rotation and ionic conduction effects, which is the main mechanism of microwave heating.

2.3. Kinetics of the Extraction Process

Once the optimal values had been obtained, they were used to study the kinetics of the extraction process. Several extractions were carried out at different times, with fixed values for the factors already studied (percentage of methanol, temperature, sample–solvent ratio, and pH). The experiments were performed in triplicate, and the times employed were: 2, 5, 10, 15, 20, and 25 min. The results for the recovery of total anthocyanins and total phenolic compounds are represented in Figure 3.

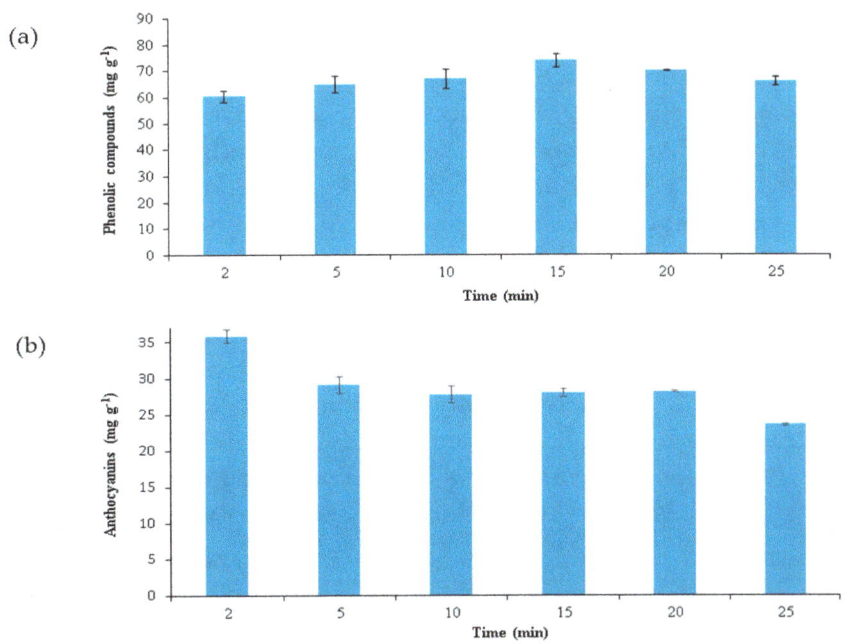

Figure 3. (**a**) Extraction kinetics of total phenolic compounds; (**b**) extraction kinetics of total anthocyanins.

It can be seen that for phenolic compounds (Figure 3a), the maximum recovery was achieved at 15 min, being lower when longer times were used. For anthocyanins (Figure 3b), it can be seen that the extracts that were subjected to microwave irradiation for 2 min gave better results—that is, the maximum quantity of anthocyanins was obtained. In addition, a longer extraction time of 5 min gave rise to worse results. This may be due to the degradation of anthocyanins when the extracts are subjected to prolonged microwave irradiation at that temperature [31]. Therefore, for anthocyanins, a shorter time of 2 min was selected as the optimum extraction time, and for phenolic compounds, a longer time of 15 min was chosen. In addition to these optimal extraction times, it is necessary to take into account the extra time required for the extracts to be tempered.

2.4. Precision of MAE Methods

The precision of the developed methods was evaluated in terms of repeatability and intermediate precision. Both parameters concern the precision of the MAE of myrtle samples under the same extraction conditions, but repeatability implies extractions carried out on the same day, whereas intermediate precision is related to different days. These terms were evaluated by following the methodology employed in several previous studies [32,37,38]. A total of 30 experiments were developed over three consecutive days by performing ten experiments each day. For repeatability, 10 extractions were performed on the first day of the study. For intermediate precision, 10 more extractions were carried out on each of the next two consecutive days. For phenolic compounds, the repeatability (RSD) was 3.98% and the intermediate precision was 4.54%. For anthocyanins, the repeatability (RSD) was 3.41% and the intermediate precision was 4.10%. Both methods were considered to have good repeatability and intermediate precision, since a maximum error of 5% is generally considered in this type of work [39].

2.5. Application to Real Sample

2.5.1. Study of Myrtle Berries from Different Locations

Both MAE methods for total phenolic compounds and for total and individual anthocyanins were applied to the entire myrtle ecotypes collected for this study. A total of 14 different ecotypes of myrtle were evaluated. 8 ecotypes were collected from local evergreen shrubs from the Puerto Real region (My-1, My-2, My-3, My-4, My-5, My-6, My-7, and My-8), and 6 ecotypes were collected from the San José del Valle region (My-9, My-10, My-11, My-12, My-13, and My-14). From each ecotype, the pulp was separated from the seed and processed in duplicate, using the optimum conditions for the developed MAE method for phenolic compounds and the optimum conditions for the extraction of anthocyanins. For the quantification of total phenolic compounds, the extracts were analyzed by the Folin–Ciocalteau (FC) spectrophotometric method. For the quantification of total and individual anthocyanins, the extracts were analyzed by UHPLC. The results of the analyses (total phenolic compounds, the eleven individual anthocyanins, and total anthocyanins) from the 14 myrtle ecotypes are shown in Table 3. First, it is noteworthy to highlight the high content of anthocyanins and total phenolic compounds found in this fruit, equaling or surpassing the contents of well-known superfruits [37,40,41]. The results indicate that myrtle ecotypes collected in the region of Puerto Real have a higher concentration of total phenolic compounds and total anthocyanins than the ecotypes collected in the region of San José del Valle. This information suggests, a priori, that the amount of bioactive compounds in myrtle is related to the location of the ecotype.

Table 3. Results of total phenolic compounds (mg g^{-1}) and total and individual anthocyanins (mg g^{-1}) for each myrtle ecotype ($n = 3$). Del-3,5-diGl: delphinidin 3,5-*O*-diglucoside; Del-3-Glu: delphinidin 3-*O*-glucoside; Cy-3-Ga: cyanidin 3-*O*-galactoside; Cy-3-Gl: cyanidin 3-*O*-glucoside; Cy-3-Ar: cyanidin 3-*O*-arabinoside; Pet-3-Gl: petunidin 3-*O*-glucoside; Del-3-Ara: delphinidin 3-*O*-arabinoside; Peo-3-Gl: peonidin 3-*O*-glucoside; Mal-3-Gl: malvidin 3-*O*-glucoside; Pet-3-Ar: petunidin 3-*O*-arabinoside; Mal-3-Ar: malvidin 3-*O*-arabinoside.

Compounds	Myrtle Ecotypes of Puerto Real								Myrtle Ecotypes of San José del Valle					
	MY-1	MY-2	MY-3	MY-4	MY-5	MY-6	MY-7	MY-8	MY-9	MY-10	MY-11	MY-12	MY-13	MY-14
Del-3,5-diGl	0.434 ± 0.012	0.440 ± 0.015	0.367 ± 0.013	0.352 ± 0.035	0.514 ± 0.0132	0.397 ± 0.018	0.210 ± 0.001	0.498± 0.023	0.156 ± 0.004	0.073 ± 0.003	0.1456 ± 0.004	0.181 ± 0.005	0.157 ± 0.004	0.180 ± 0.007
Del-3-Glu	9.405 ± 0.256	13.232 ± 0.369	9.555 ± 0.051	9.704 ± 0.159	15.110 ± 0.160	10.935 ± 0.001	10.164 ± 0.171	9.664 ± 0.497	8.049 ± 0.283	1.798 ± 0.072	7.897 ± 0.301	6.102 ± 0.234	5.432 ± 0.234	6.192 ± 0.236
Cy-3-Ga	0.159 ± 0.0006	0.288 ± 0.014	0.211 ± 0.010	0.309 ± 0.010	0.372 ± 0.010	0.326 ± 0.003	0.184 ± 0.004	0.426 ± 0.023	0.504 ± 0.023	0.047 ± 0.001	0.490 ± 0.023	0.133 ± 0.003	0.341 ± 0.013	0.154 ± 0.005
Cy-3-Gl	1.702 ± 0.063	2.326 ± 0.075	1.013 ± 0.019	1.067 ± 0.022	1.854 ± 0.048	1.142 ± 0.015	2.276 ± 0.016	1.626 ± 0.079	1.011 ± 0.039	0.321 ± 0.013	1.002 ± 0.035	1.191 ± 0.043	0.988 ± 0.038	1.235 ± 0.031
Cy-3-Ar	0.090 ± 0.002	0.098 ± 0.002	0.075 ± 0.002	0.145 ± 0.009	0.136 ± 0.0014	0.611 ± 0.656	2.124 ± 0.015	0.134 ± 0.005	0.943 ± 0.037	0.299 ± 0.012	0.898 ± 0.032	0.084 ± 0.002	0.765 ± 0.029	0.085 ± 0.038
Pet-3-Gl	4.738 ± 0.065	8.680 ± 0.239	6.512 ± 0.027	7.480 ± 0.159	9.958 ± 0.426	4.614 ± 0.196	2.036 ± 0.023	6.058 ± 0.330	0.094 ± 0.010	0.036 ± 0.001	0.091 ± 0.003	2.346 ± 0.087	1.247 ± 0.051	2.988 ± 0.002
Del-3-Ara	1.979 ± 0.073	1.838 ± 0.072	1.527 ± 0.069	1.601 ± 0.032	2.320 ± 0.062	4.974 ± 0.234	4.022 ± 0.052	1.627 ± 0.079	6.256 ± 0.222	1.634 ± 0.072	5.990 ± 0.189	0.900 ± 0.023	3.257 ± 0.138	0.912 ± 0.119
Peo-3-Gl	0.614 ± 0.021	0.856 ± 0.027	0.573 ± 0.023	0.608 ± 0.022	0.705 ± 0.019	0.7206 ± 0.056	0.869 ± 0.052	0.406 ± 0.022	0.627 ± 0.022	0.330 ± 0.011	0.599 ± 0.019	0.488 ± 0.023	0.178 ± 0.07	0.375 ± 0.029
Mal-3-Gl	6.893 ± 0.245	13.124 ± 0.417	14.346 ± 0.024	15.495 ± 0.331	0.750 ± 0.020	0.767 ± 0.060	0.925 ± 0.0004	6.227 ± 0.339	0.667 ± 0.024	0.352 ± 0.013	0.601 ± 0.023	9.832 ± 0.342	5.563 ± 0.190	6.877 ± 0.013
Pet-3-Ar	0.295 ± 0.0003	0.430 ± 0.010	0.411 ± 0.0174	0.504 ± 0.020	0.590 ± 0.013	0.656 ± 0.021	0.3891 ± 0.0004	0.378 ± 0.018	0.765 ± 0.023	0.203 ± 0.001	0.679 ± 0.021	0.184 ± 0.007	0.634 ± 0.022	0.199 ± 0.268
Mal-3-Ar	0.2451 ± 0.002	0.231 ± 0.007	0.356 ± 0.002	0.372 ± 0.002	0.328 ± 0.006	0.5402 ± 0.001	0.289 ± 0.002	0.154 ± 0.010	0.520 ± 0.024	0.263 ± 0.01	0.492 ± 0.012	0.299 ± 0.013	0.5757 ± 0.020	0.213 ± 0.007
Total anthocyanins	26.555 ± 0.395	41.544 ± 1.390	34.947 ± 0.183	37.638 ± 0.730	32.637 ± 0.854	25.682 ± 0.846	23.487 ± 0.234	35.846 ± 0.896	19.595 ± 0.577	5.355 ± 0.242	18.884 ± 0.762	21.740 ± 0.865	19.138 ± 0.645	19.409 ± 0.567
Total phenolic compounds	88.598 ± 2.983	124.684 ± 0.934	82.603 ± 0.343	63.457 ± 2.541	86.251 ± 2.934	81.225 ± 2.199	88.340 ± 3.899	90.682 ± 4.706	55.934 ± 2.743	69.550 ± 3.123	56.790 ± 2.065	60.654 ± 2.127	61.193 ± 2.356	59.898 ± 2.967

2.5.2. Multivariate Statistical Analysis

In order to assess whether the concentration of bioactive compounds in myrtle is related to the location of the ecotype, a non-supervised chemometric technique, namely, hierarchical cluster analysis (HCA), was carried out for the 14 myrtle ecotypes. HCA allows the trends in the myrtle ecotypes to be grouped according to the origin by using the concentrations of the studied bioactive compounds as independent variables for the formation of groups. Therefore, the variables employed in the differentiation were: the amount of total phenolic compounds (mg g^{-1}), the amount of each individual anthocyanins (mg g^{-1}), and the amount of total anthocyanins (mg g^{-1}). Ward´s method was used for the preparation of the clusters, and square Euclidean distance was employed to measure distances between clusters. The results of the HCA are graphically represented in the dendrogram in Figure 4, in which all of the ecotypes of myrtle are listed, along with the distance at which any of the two clusters are joined.

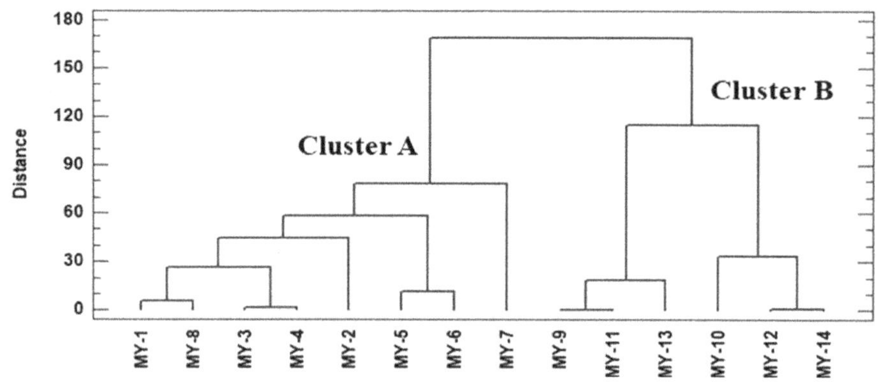

Figure 4. Dendrogram obtained by a hierarchical cluster analysis, based on the chemical parameters studied of the 14 samples by duplicated myrtle pulp extracts.

Based on these results, it can be observed that there are two main groups; Cluster A, which includes all of the myrtle ecotypes collected in Puerto Real, and Cluster B, which only includes the myrtle ecotypes collected in San José del Valle. Therefore, it can be concluded that the chemical information obtained—that is, the amounts of total phenolic compounds and anthocyanins, is related to the location of the ecotypes due to the tendency to be grouped according to their geographical origin. Specifically, samples from San José del Valle have a lower number of total phenolic compounds and total anthocyanins than samples collected in Puerto Real. This finding is consistent with bibliographic information, which highlights that myrtle is a shrub that prefers fertile and humid soils, and therefore, warm zones closer to sea level [42,43]. The wet climate of Puerto Real, due to its proximity to the sea, makes the myrtle shrubs of this region mature and grow more favorably than those located in inland areas, such as San José del Valle, where the climate is more variable and dry and the soil is less fertile.

3. Material and Methods

3.1. Myrtle Sample

Myrtle ecotypes were collected by the authors in December 2016 from local myrtle shrubs in their optimum ripeness stage from two areas (Puerto Real and San José del Valle) of the province of Cadiz, Andalusia, Spain. Specifically, 14 different myrtle ecotypes were collected: 8 ecotypes from the Puerto Real region, and 6 ecotypes from the San Jose del Valle region. Both regions are in the province of Cadiz but have different climatic characteristics, though they are fundamentally based in proximity to the sea. The area of Puerto Real is located on the coast, whereas the region of San José del Valle is located 50 km inland. Proximity to the coast results in very humid areas, with soils that have readily available water, and these provide more fertile conditions for the growth and maturation of many species. By contrast, San José del Valle, as an inland zone, experiences greater temperature changes and less water availability for plants, particularly in the summer. All this leads to the generation of myrtle ecotypes with different characteristics. In addition, a morphological characterization was made of both leaves and berries, applying the guidelines described by M., Mulas & M.R. Cani [43], to confirm that the samples collected came from different ecotypes.

The seeds of the myrtle berries were separated from the pulp. The pulp was lyophilized in a Virtis Benchtop K freeze-drier (SP Scientific, New York, NY, USA) and triturated in a spice grinder. The triturated and homogeneous sample was stored in a freezer at −20 °C prior to analysis.

3.2. Chemicals and Solvents

Methanol (HPLC grade) was purchased from Fischer Chemical (Loughborough, United Kingdom). Water was obtained from a Milli-Q water purification system from Millipore (Bedford, MA, USA). Hydrochloric acid and sodium hydroxide (both analytical grade) employed for the adjustment of pH were obtained from Panreac (Barcelona, Spain). The reagents necessary for the determination of total phenolic compounds were anhydrous sodium carbonate (Panreac, Barcelona, Spain), and Folin–Ciocalteu reagent (Merck Millipore, Darmstadt, Germany). The phenolic standard (gallic acid) and the anthocyanin standard (cyanidin chloride) were purchased from Sigma-Aldrich Chemical Co. (St. Louis, MO, USA).

3.3. Microwave-Assisted Extraction Procedure

The extraction of total phenolic compounds and total anthocyanins (as the sum of the individual components) from myrtle was performed by microwave-assisted extraction. The extracts were obtained using a MARS 6 One TouchTM Technology system (1800 W) (CEM Corporation, Matthews, NC, USA). Approximately 0.5 g of triturated myrtle was weighed into a MARSXpress vessel (CEM Corporation), and the appropriate volume of solvent was added. The vessel was closed securely and placed in the microwave with another eight tubes, which had the same solvent and volume. Each extraction was carried out under controlled MAE conditions. The parameters used were: solvent composition (50, 75, and 100% methanol in water), pH (2, 4.5, and 7), temperature (50, 75, and 100 °C) and sample–solvent ratio (0.5 g/10 mL, 0.5 g/15 mL, and 0.5 g/20 mL). The initial extraction time was 5 min, and this was followed by a set time to temper the sample. Once the samples had been warmed, the extracts were centrifuged (7500 rpm, 5 min) and the supernatant was added to a 25 mL volumetric flask. The precipitates from the extraction were subsequently redissolved in 5 mL of the same extraction solvent. The extracts were centrifuged again (7500 rpm, 5 min) and the supernatant was placed in the volumetric flask (25 mL). The volume was completed with the same solvent. The extracts were stored at −20 °C prior to analysis.

3.4. Identification of Anthocyanins

A chromatographic method using ultra-high performance liquid chromatography (UHPLC) coupled to quadrupole-time-of-flight mass spectrometry (Q-ToF-MS) (Xevo G2 QToF, Waters Corp., Milford, MA, USA) was developed for the identification of anthocyanins in MAE extracts. The injection volume was set to 3 μL. The chromatographic separation was performed on a reverse-phase C18 analytical column (1.7 μm, 2.1 mm × 100 mm, Acquity UPLC BEH C18, Waters). A gradient method, using acidified water (2% formic acid, solvent A) and methanol (solvent B) at a flow rate of 0.4 mL min^{-1} was used. The gradient was as follows (time, % solvent B): 0.00 min, 15%; 3.30 min, 20%; 3.86 min, 30%; 5.05 min, 40%; 5.35 min, 55%; 5.64 min, 60%; 5.95 min, 95%; 7.50 min, 95%. The total run time was 12 min, including 4 min for re-equilibration. The analyses were carried out using an electrospray source operating in positive ionization mode under the following conditions: desolvation gas flow = 700 L h^{-1}, desolvation temperature = 500 °C, cone gas flow =10 L h^{-1}, source temperature = 150 °C, capillary voltage = 700 V, cone voltage = 30 V and collision energy = 20 eV. The full-scan mode was used (m/z 100–1200). The following eleven anthocyanins were identified in the samples: delphinidin 3,5-*O*-diglucoside (m/z 627.1561), delphinidin 3-*O*-glucoside (m/z 465.1033), cyanidin 3-*O*-galactoside

(m/z 449.1084), cyanidin 3-*O*-glucoside (m/z 449.1084), cyanidin 3-*O*-arabinoside (m/z 419.0978), petunidin 3-*O*-glucoside (m/z 479.1189), delphinidin 3-*O*-arabinoside (m/z 435.0927), peonidin 3-*O*-glucoside (m/z 463.1240), malvidin 3-*O*-glucoside (m/z 493.1346), petunidin 3-*O*-arabinoside (m/z 449.1084), and malvidin 3-*O*-arabinoside (m/z 463.1240). Prior to chromatographic analysis, the extracts were filtered through a 0.20 μm nylon syringe filter (Membrane Solutions, Dallas, TX, USA). Information regarding the mass spectra, theoretical and measured masses, as well as the structure of the compounds, are included as Supplementary Material (Table S1 and Figure S1).

3.5. Determination of Anthocyanins

Once the anthocyanins had been identified, they were separated and quantified by ultra-high performance liquid chromatography (UHPLC) (Elite LaChrom Ultra System, VWR Hitachi, Tokyo, Japan) available in our research group. The UHPLC chromatogram representing the eleven anthocyanins is shown in Figure 5.

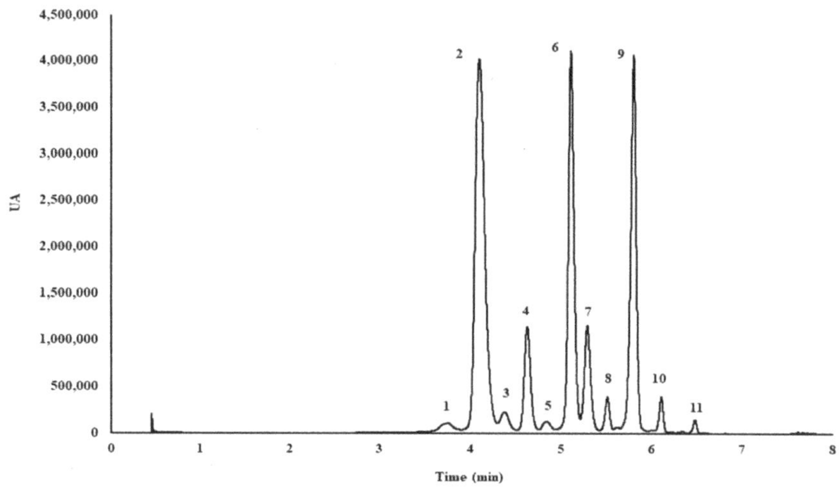

Figure 5. Ultra-high performance liquid chromatography (UHPLC) chromatogram of the eleven anthocyanins identified in the microwave-assisted extraction (MAE) extracts from myrtle berries. Peak assignment: (**1**) delphinidin 3,5-*O*-diglucoside; (**2**) delphinidin 3-*O*-glucoside; (**3**) cyanidin 3-*O*-galactoside; (**4**) cyanidin 3-*O*-glucoside; (**5**) cyanidin 3-*O*-arabinoside; (**6**) petunidin 3-*O*-glucoside; (**7**) delphinidin 3-*O*-arabinoside; (**8**) peonidin 3-*O*-glucoside; (**9**) malvidin 3-*O*-glucoside; (**10**) petunidin 3-*O*-arabinoside; (**11**) malvidin 3-*O*-arabinoside.

The UHPLC system was equipped with an autosampler (L-2200U), two pumps (L-2160U), a UV-vis detector (L-2420U) set at 520 nm for the analysis, and a column oven (L2300), set at 50 °C for the chromatographic analysis. Anthocyanins were analyzed on a "Fused Core" C18 column (2.6 μm, 2.1 × 100 nm, Phenomenex, Torrance, CA, USA). The separation and quantification were carried out using acidified water (5% formic acid, solvent A) and methanol (solvent B), working at a flow rate of 0.7 mL min^{-1}. After testing several methods and flows, amongst other parameters, the gradient employed was as follows (time, % solvent B): 0.00 min, 0%; 1.50 min, 5%; 3.30 min, 15%; 4.80 min, 25%; 5.40 min, 40%. The injection volume was set to 15 μL. This gradient provided optimum results in less than 7 min. Prior to chromatographic analysis, the extracts were filtered through a 0.20 μm nylon syringe filter (Membrane Solutions, Dallas, TX, USA). In order to quantify the eleven anthocyanins present in myrtle extracts, a calibration curve of cyanidin chloride ($y = 300568.8819x - 28462.4337$) was used as the anthocyanidin standard. The regression equation and the correlation coefficient ($R^2 = 0.9999$) were also calculated using Microsoft Office Excel 2013. The limit of detection (0.196 mg L^{-1}) was calculated as three times the standard deviation of the blank signal divided by the slope of the calibration curve. By analogy, the limit of quantification (0.653 mg L^{-1}) was calculated as ten times the standard deviation of the blank signal divided by the slope of the calibration curve.

The linear range studied was 0.06–35 mg L^{-1}. Each one of the eleven anthocyanins was quantified using the calibration curve for cyanidin chloride, and the molecular weight of the anthocyanins analyzed were taken into account. All analyses were carried out in duplicate, and the results were expressed as milligrams of anthocyanins per g of dry weight.

3.6. Determination of Total Phenolic Compounds

The Folin–Ciocalteu (FC) spectrophotometric method was used to determine the total phenolic compounds [44,45]. The FC method is based on the fact that phenolic compounds react at basic pH with the Folin–Ciocalteau reagent (a mixture of sodium tungstate and sodium molybdate). The products of the reduction have a blue color, and they have a broad absorption with a maximum of 765 nm. The FC assay was performed by transferring 0.25 mL of MAE extract, 1.25 mL of water, and 1.25 mL of the Folin–Ciocalteu reagent to a volumetric flask (25 mL). Then, 5 mL of a 20% aqueous sodium carbonate solution was also added, and the solution was made up to the mark with water. After 30 min, the absorbance of the solutions was measured at 765 nm. The range of absorbance obtained for the studied samples was 0.4–1.3. Prior to spectrophotometric analysis, the extracts were filtered through a 0.45 µm nylon filter (Membrane Solutions, Dallas, TX, USA). The absorbance was measured on a Heλios-γ Unicam UV-Vis Spectrophotometer (Thermo Scientific, Waltham, MA, USA). In order to quantify the phenolic compounds present in myrtle extracts, a calibration curve was developed under the same conditions, using gallic acid as the reference standard. The results are expressed as milligrams of gallic acid equivalent per g of fresh weight. The regression equation (y = 0.0010x + 0.0065) and correlation coefficient (R^2 = 0.9999) were calculated using Microsoft Office Excel 2013. The linear range studied was 100–2600 mg L^{-1}. All analyses were carried out in duplicate.

3.7. Optimization Procedure and Data Analyses

The spherical response surface Box–Behnken method was employed for experimental design in the optimization procedure. In this approach, the treatment combinations are at the midpoints of the edges of the process space and at the centre, which ensures that all experiments are in the region of interest [46]. To obtain the significant factors and the optimal MAE conditions, a Box–Behnken design with four factors and three levels for each factor was used: a low level (−1), a medium level (0), and a high level (1).

The four factors (independent variables) used in this work were solvent composition (methanol in water) (X_1), solvent pH (X_2), extraction temperature (X_3), and sample–solvent ratio (X_4), while the response variables (dependent variables) used were the total amount of phenolic compounds (Y_{TP}, mg g^{-1}) on the one hand, and the total amount of anthocyanins (Y_TA, mg g^{-1}), as the sum of individual ones, on the other. The experimental design consisted of 27 trials performed in duplicate, with three repetitions at the center point to calculate the pure error sum of squares. The whole experimental design matrix and the results obtained are shown in Table 4. My-9 from the San Jose del Valle location was the myrtle sample used for the optimization procedure.

A quadratic model was used for model construction, and this gave a second-order polynomial equation that correlated the relationship between independent variables and responses (Equation (3)):

$$Y = \beta_0 + \sum_{i=1}^{k} \beta_i X_i + \beta_{ii} X_i^2 + \sum_{i} \sum_{i=1}^{k} \beta_{ij} X_i X_j + r \tag{3}$$

where Y represents the aforementioned responses (Y_{TP} and Y_{TA}) for the extractions of total phenolic compounds and anthocyanins, respectively; β_0 is the model constant; X_i are the independent variables; β_i are the linear coefficients; β_{ij} are the interactive coefficients; β_{ii} are the quadratic coefficients; and r is the pure error sum of squares.

The statistical significances of the model, lack of fit, and regression terms were evaluated from the analysis of variance (ANOVA). The fitting quality of the polynomial model was evaluated by the determination coefficient (R^2). All experimental data were compiled by Design Expert software 11 (Trial Version, Stat- Ease Inc., Minneapolis, MN, USA). This software was employed for experimental design, data analysis, and model building.

Hierarchical clustering analysis (HCA) was carried out using the Statgraphic Centurion XVII software (Statgraphics Technologies, Inc., The Plains, VA, USA). Pareto Charts were used to identify factors and combinations of factors that are statistically significant at the selected confidence level (95%) for total phenolic compounds and anthocyanins. Regarding the HCA analysis, the Ward method and the Euclidean square distance were used.

Table 4. Box–Behnken design matrix of four variables, and measured and predicted responses.

Run	Factors				Responses			
	Solvent	pH	Temp.	Ratio	Y_{TP} (mg g^{-1})		Y_{TA} (mg g^{-1})	
					Measured	Predicted	Measured	Predicted
1	−1	−1	0	0	50.66	51.02	26.04	21.86
2	1	−1	0	0	34.64	40.29	24.74	26.30
3	−1	1	0	0	43.86	41.83	26.73	25.39
4	1	1	0	0	31.74	35.00	17.54	21.95
5	0	0	−1	−1	44.30	41.54	20.05	20.83
6	0	0	1	−1	42.60	46.26	20.50	18.83
7	0	0	−1	1	46.71	46.67	20.21	22.11
8	0	0	1	1	39.23	45.61	14.41	13.86
9	0	0	0	0	38.17	42.68	16.56	18.65
10	−1	0	0	−1	38.51	42.19	25.34	26.79
11	1	0	0	−1	31.50	31.55	26.42	26.49
12	−1	0	0	1	40.14	42.58	25.04	24.15
13	1	0	0	1	36.83	35.64	27.71	25.45
14	0	−1	−1	0	44.07	48.20	22.79	23.04
15	0	1	−1	0	46.49	48.10	18.40	16.20
16	0	−1	1	0	56.28	57.17	10.11	11.50
17	0	1	1	0	44.42	42.79	18.57	17.51
18	0	0	0	0	44.14	42.68	19.45	18.65
19	0	−1	0	−1	53.28	48.50	20.59	20.48
20	0	1	0	−1	46.14	46.27	19.82	19.32
21	0	−1	0	1	62.00	55.75	16.79	17.89
22	0	1	0	1	44.84	43.50	17.52	18.22
23	−1	0	−1	0	39.22	38.57	26.60	28.42
24	1	0	−1	0	39.00	36.70	26.86	24.31
25	−1	0	1	0	51.14	47.32	15.60	18.69
26	1	0	1	0	37.10	31.62	25.03	23.80
27	0	0	0	0	45.74	42.68	19.94	18.65

4. Conclusions

To the best of our knowledge, this is the first study in which MAE has been optimized for the extraction of bioactive compounds, like phenolic compounds and anthocyanins, from myrtle berries. For total phenolic compounds, the solvent methanol-water (58.20% MeOH in water) and pH (pH 2) were the most influential variables. For the recovery of anthocyanins, methanol composition (50.4% MeOH in water) and temperature (50 °C) were found to be the most efficient variables. The optimum extraction times were 2 and 15 min for phenolic compounds and anthocyanins, respectively. Both of the developed methods showed high repeatability and intermediate precision (RSD < 5%). The methods were successfully applied to different myrtle ecotypes collected in two different geographical areas. HCA analysis showed a correlation between the bioactive compounds studied and the location of the ecotype. Due to these previous findings, myrtle berries from more geographical locations could be interesting for further studies.

Based on these results, MAE (under optimum conditions) can be considered as a suitable technique for the extraction of bioactive compounds in myrtle berries. In addition, this technique presents several advantages in comparison to other extraction techniques, since it is fast, economic, and eco-friendlier, as it does not require the use of a high amount of solvent.

Supplementary Materials:
anthocyanins identified in myrtle berries: (**a**) Delphinidin 3,5-*O*-diglucoside; (**b**) Delphinidin 3-*O*-glucoside; (**c**) Cyanidin 3-*O*-galactoside; (**d**) Cyanidin 3-*O*-glucoside; (**e**) Cyanidin 3-*O*-arabinoside; (**f**) Petunidin 3-*O*-glucoside; (**g**) Delphinidin 3-*O*-arabinoside; (**h**) Peonidin 3-*O*-glucoside; (**i**) Malvidin 3-*O*-glucoside; (**j**) Petunidin 3-*O*-arabinoside; (**k**) Malvidin 3-*O*-arabinoside, Table S1: Mass spectra information of the eleven anthocyanins present in myrtle berries.

Author Contributions: Conceptualization, E.E.-B. and G.F.B.; methodology, A.V.G.d.P. and M.V.-E.; software, M.F.-G.; formal analysis, A.V.G.d.P., M.V.-E and A.A.-A.; investigation, A.V.G.d.P. and M.V.-E.; resources, A.J.-C., C.G.B. and M.P.; data curation, E.E.-B., M.F.-G. and G.F.B.; writing—original draft preparation, A.V.G.d.P. and M.V.-E.; writing—review and editing, G.F.B. and E.E.-B.; supervision, E.E.-B. and G.F.B.; project administration, G.F.B. and E.E.-B.

Acknowledgments: The authors express their acknowledgments to V. la Andaluza.

References

1. Bouzabata, A.; Casanova, J.; Bighelli, A.; Cavaleiro, C.; Salgueiro, L.; Tomi, F. The genus *Myrtus* L. in Algeria: Composition and biological aspects of essential oils from *M. communis* and *M. nivellei*: A review. *Chem. Biodivers.* **2016**, *13*, 672–680. [CrossRef] [PubMed]
2. Melito, S.; La Bella, S.; Martinelli, F.; Cammalleri, I.; Tuttolomondo, T.; Leto, C.; Fadda, A.; Molinu, M.G.; Mulas, M. Morphological, chemical, and genetic diversity of wild myrtle (*Myrtus communis* L.) populations in Sicily. *Turk. J. Agric. For.* **2016**, *40*, 249–261. [CrossRef]
3. Al-Yafeai, A.; Bellstedt, P.; Böhm, V. Bioactive compounds and antioxidant capacity of rosa rugosa depending on degree of ripeness. *Antioxidants* **2018**, *7*, 134. [CrossRef] [PubMed]
4. Yang, C.; Fu, L.; Du, B.; Chen, B.; Wang, F.; Wang, M. Isolation and characterization of new phenolic compounds with estrogen biosynthesis-inhibiting and antioxidation activities from *Broussonetia papyriferia* leaves. *PLoS ONE* **2014**, *9*. [CrossRef]
5. Wannes, A.; Mhamdi, B.; Sriti, J.; Ben Jemia, M.; Ouchikh, O.; Hamdaoui, G.; Kchouk, M.E.; Marzouk, B. Antioxidant activities of the essential oils and methanol extracts from myrtle (*Myrtus communis* var. italica L.) leaf, stem and flower. *Food Chem. Toxicol.* **2010**, *48*, 1362–1370. [CrossRef] [PubMed]
6. Aleksic, V.; Knezevic, P. Antimicrobial and antioxidative activity of extracts and essential oils of *Myrtus communis* L. *Microbiol. Res.* **2014**, *169*, 240–254. [CrossRef] [PubMed]
7. Wahid, N.; Chkichekh, A.; Bakry, M. Morphological traits and essential oil yield variation of three *Myrtus communis* L. populations: Implication for domestication. *J. Agric. Food Chem.* **2016**, *4*, 199–207.
8. Bouzabata, A.; Castola, V.; Bighelli, A.; Abed, L.; Casanova, J.; Tomi, F. Chemical variability of algerian *Myrtus communis* L. *Chem. Biodivers.* **2013**, *10*, 129–137. [CrossRef] [PubMed]
9. Fadda, A.; Mulas, M. Chemical changes during myrtle (*Myrtus communis* L.) fruit development and ripening. *Sci. Hortic.* **2010**, *125*, 477–485. [CrossRef]
10. Serreli, G.; Jerković, I.; Gil, K.A.; Marijanović, Z.; Pacini, V.; Tuberoso, C.I.G. Phenolic compounds, volatiles and antioxidant capacity of white myrtle berry liqueurs. *Plant Foods Hum. Nutr.* **2017**, *72*, 205–210. [CrossRef] [PubMed]
11. Messaoud, C.; Boussaid, M. *Myrtus communis* berry color morphs: A comparative analysis of essential oils, fatty acids, phenolic compounds, and antioxidant activities. *Chem. Biodivers.* **2011**, *8*, 300–310. [CrossRef] [PubMed]
12. De Oliveira, C.B.; Comunello, L.N.; Lunardelli, A.; Amaral, R.H.; Pires, M.G.S.; Da Silva, G.L.; Manfredini, V.; Vargas, C.R.; Gnoatto, S.C.B.; de Oliveira, J.R.; et al. Phenolic enriched extract of baccharis trimera presents anti-inflammatory and antioxidant activities. *Molecules* **2017**, *17*, 1113. [CrossRef] [PubMed]

13. Sarais, G.; D'Urso, G.; Lai, C.; Pirisi, FM.; Pizza, C.; Montoro, P. Targeted and untargeted mass spectrometric approaches in discrimination between *Myrtus communis* cultivars from Sardinia region. *J. Mass. Spectrom.* **2016**, *51*, 704–715. [CrossRef] [PubMed]
14. Tuberoso, C.I.G.; Rosa, A.; Bifulco, E.; Melis, M.P.; Atzeri, A.; Pirisi, F.M.; Dessì, M.A. Chemical composition and antioxidant activities of *Myrtus communis* L. berries extracts. *Food Chem.* **2010**, *123*, 1242–1251. [CrossRef]
15. Montoro, P.; Tuberoso, C.I.G.; Perrone, A.; Piacente, S.; Cabras, P.; Pizza, C. Characterisation by liquid chromatography-electrospray tandem mass spectrometry of anthocyanins in extracts of *Myrtus communis* L. berries used for the preparation of myrtle liqueur. *J. Chromatogr. A* **2006**, *1112*, 232–240. [CrossRef] [PubMed]
16. Tuberoso, C.I.G.; Melis, M.; Angioni, A.; Pala, M.; Cabras, P. Myrtle hydroalcoholic extracts obtained from different selections of *Myrtus communis* L. *Food Chem.* **2007**, *101*, 806–811. [CrossRef]
17. Spigno, G.; De Faveri, D.M. Microwave-assisted extraction of tea phenols: A phenomenological study. *J. Food Eng.* **2009**, *93*, 210–217. [CrossRef]
18. Zhao, C.-N.; Zhang, J.; Li, Y.; Meng, X.; Li, H. Microwave-assisted extraction of phenolic compounds from *Melastoma sanguineum* fruit: Optimization and identification. *Molecules* **2018**, *23*, 2498. [CrossRef] [PubMed]
19. Bouras, M.; Chadni, M.; Barba, F.J.; Grimi, N.; Bals, O.; Vorobiev, E. Optimization of microwave-assisted extraction of polyphenols from Quercus bark. *Ind. Crops. Prod.* **2015**, *77*, 590–601. [CrossRef]
20. Haddadi-Guemghar, H.; Janel, N.; Dairou, J.; Remini, H.; Madani, K. Optimisation of microwave-assisted extraction of prune (*Prunus domestica*) antioxidants by response surface methodology. *Int. J. Food Sci. Technol.* **2014**, *49*, 2158–2166. [CrossRef]
21. Liazid, A.; Guerrero, R.F.; Cantos, E.; Palma, M.; Barroso, C.G. Microwave assisted extraction of anthocyanins from grape skins. *Food Chem.* **2011**, *124*, 1238–1243. [CrossRef]
22. Li, H.; Deng, Z.; Wu, T.; Liu, R.; Loewen, S.; Tsao, R. Microwave-assisted extraction of phenolics with maximal antioxidant activities in tomatoes. *Food Chem.* **2012**, *130*, 928–936. [CrossRef]
23. Wen, Y.; Chen, H.; Zhou, X.; Deng, Q.; Zhao, Y.; Zhao, C.; Gong, X. Optimization of the microwave-assisted extraction and antioxidant activities of anthocyanins from blackberry using a response surface methodology. *RSC Adv.* **2015**, *5*, 19686–19695. [CrossRef]
24. Dahmoune, F.; Nayak, B.; Moussi, K.; Remini, H.; Madani, K. Optimization of microwave-assisted extraction of polyphenols from *Myrtus communis* L. leaves. *Food Chem.* **2015**, *166*, 585–595. [CrossRef] [PubMed]
25. Routray, W.; Orsat, V. Microwave-assisted extraction of flavonoids: A review. *Food Bioprocess Technol.* **2012**, *5*, 409–424. [CrossRef]
26. Dong, C.H.; Xie, X.Q.; Wang, X.L.; Zhan, Y.; Yao, Y.J. Application of Box–Behnken design in optimisation for polysaccharides extraction from cultured mycelium of Cordyceps sinensis. *Food Bioprod. Process* **2009**, *87*, 139–144. [CrossRef]
27. Hou, W.; Zhang, W.; Chen, G.; Luo, Y. Optimization of extraction conditions for maximal phenolic, flavonoid and antioxidant activity from melaleuca bracteata leaves using the response surface methodology. *PLoS ONE* **2016**, *11*. [CrossRef] [PubMed]
28. Razali, M.A.A.; Sanusi, N.; Ismail, H.; Othman, N.; Ariffin, A. Application of response surface methodology (RSM) for optimization of cassava starch grafted polyDADMAC synthesis for cationic properties. *Starch/Staerke* **2012**, *64*, 935–943. [CrossRef]
29. Castañeda-ovando, A.; Pacheco-hernández, M.D.L.; Páez-hernández, M.E.; Rodríguez, J.A.; Galán-vidal, C.A. Chemical studies of anthocyanins: A review. *Food Chem.* **2009**, *113*, 859–871. [CrossRef]
30. Spigno, G.; Tramelli, L.; De Faveri, D.M. Effects of extraction time, temperature and solvent on concentration and antioxidant activity of grape marc phenolics. *J. Food Eng.* **2007**, *81*, 200–208. [CrossRef]
31. Pereira, D.T.V.; Tarone, A.G.; Cazarin, C.B.B.; Barbero, G.F.; Martínez, J. Pressurized liquid extraction of bioactive compounds from grape marc. *J. Food Eng.* **2019**, *240*, 105–113. [CrossRef]
32. Espada-Bellido, E.; Ferreiro-González, M.; Barbero, G.F.; Carrera, C.; Palma, M.; Barroso, C.G. Alternative extraction method of bioactive compounds from Mulberry (*Morus nigra* L.) pulp using pressurized-liquid extraction. *Food Anal. Methods* **2018**, *11*, 2384–2395. [CrossRef]
33. D'Alessandro, L.G.; Kriaa, K.; Nikov, L.; Dimitrov, K. Ultrasound assited extraction of polyphenols from black chokeberry. *Sep. Purif. Technol.* **2012**, *93*, 42–47. [CrossRef]
34. Pereira, P.; Cebola, M.-J.; Oliveira, M.C.; Bernardo Gil, M.G. Antioxidant capacity and identification of bioactive compounds of *Myrtus communis* L. extract obtained by ultrasound-assisted extraction. *J. Food Sci. Technol.* **2017**, *54*, 4362–4369. [CrossRef] [PubMed]

35. Aidi Wannes, W.; Marzouk, B. Differences between myrtle fruit parts (*Myrtus communis* var. italica) in phenolics and antioxidant contents. *J. Food Biochem.* **2013**, *37*, 585–594. [CrossRef]
36. Scorrano, S.; Lazzoi, M.R.; Mergola, L.; Di Bello, M.P.; Del Sole, R.; Vasapollo, G. Anthocyanins profile by Q-TOF LC/MS in *Myrtus communis* berries from salento area. *Food Anal. Methods* **2017**, *10*, 2404–2411. [CrossRef]
37. Espada-Bellido, E.; Ferreiro-González, M.; Carrera, C.; Palma, M.; Barroso, C.G.; Barbero, G.F. Optimization of the ultrasound-assisted extraction of anthocyanins and total phenolic compounds in mulberry (*Morus nigra*) pulp. *Food Chem.* **2017**, *219*, 23–32. [CrossRef] [PubMed]
38. Stipcovich, T.; Barbero, G.F.; Ferreiro-González, M.; Palma, M.; Barroso, C.G. Fast analysis of capsaicinoids in Naga Jolokia extracts (*Capsicum chinense*) by high-performance liquid chromatography using fused core columns. *Food Chem.* **2018**, *239*, 217–224. [CrossRef] [PubMed]
39. AOAC Peer Verified Methods Advisory Committee. *AOAC Peer Verified Methods Program*; AOAC International: Gaithersburg, MD, USA, 1998; pp. 1–35.
40. Vázquez-Espinosa, M.; Espada-Bellido, E.; González de Peredo, A.V.; Ferreiro-González, M.; Carrera, C.; Palma, M.; Barroso, C.G.; Barbero, G.F. Optimization of microwave-assisted extraction for the recovery of bioactive compounds from the Chilean superfruit (*Aristotelia chilensis* (mol.) stuntz). *Agronomy* **2018**, *8*, 240. [CrossRef]
41. Garzón, G.A.; Narváez-Cuenca, C.E.; Vincken, J.P.; Gruppen, H. Polyphenolic composition and antioxidant activity of açai (*Euterpe oleracea* Mart.) from Colombia. *Food Chem.* **2017**, *217*, 364–372. [CrossRef] [PubMed]
42. González-Varo, J.P.; Albaladejo, R.G.; Aparicio, A. Mating patterns and spatial distribution of conspecific neighbours in the Mediterranean shrub *Myrtus communis* (Myrtaceae). *Plant Ecol.* **2009**, *203*, 207–215. [CrossRef]
43. Mulas, M.; Cani, M.R. Germplasm Evaluation of spontaneous Myrtle (*Myrtus communis* L.) for cultivar selection and crop development. *J. Herbs Spices Med. Plants* **1999**, *6*, 31–49. [CrossRef]
44. Singleton, V.L.; Orthofer, R.; Lamuela-Raventós, R.M. Analysis of total phenols and other oxidation substrates and antioxidants by means of folin-ciocalteu reagent. *Methods Enzymol.* **1999**, *299*, 152–178. [CrossRef]
45. Santos, U.P.; Campos, J.F.; Heron, F.V.; Paredes-Gamero, E.J.; Carollo, C.A.; Estevinho, L.M.; de Picol Souza, K.; Dos Santos, E.L. Antioxidant, antimicrobial and cytotoxic properties as well as the phenolic content of the extract from *Hancornia speciosa* Gomes. *PLoS ONE* **2016**, *11*. [CrossRef] [PubMed]
46. Ferreira, S.L.C.; Bruns, R.E.; Ferreira, H.S.; Matos, G.D.; David, J.M.; Brandão, G.C.; da Silva, E.G.P.; Portugal, L.A.; dos Reis, P.S.; Souza, A.S.; dos Santos, W.N.L. Box–Behnken design: An alternative for the optimization of analytical methods. *Anal. Chim. Acta* **2007**, *597*, 179–186. [CrossRef] [PubMed]

4

Spectrum-Effect Relationships between High-Performance Liquid Chromatography (HPLC) Fingerprints and the Antioxidant and Anti-Inflammatory Activities of Collagen Peptides

Junwen Wang [1,†], Dan Luo [2,†], Ming Liang [3], Ting Zhang [3], Xiquan Yin [3], Ying Zhang [3], Xiangliang Yang [1,2] and Wei Liu [1,2,*]

1 College of Life Science and Technology, Huazhong University of Science and Technology, Wuhan 430074, China; wangjw@hust.edu.cn (J.W.); yangxl@hust.edu.cn (X.Y.)
2 National Engineering Research Center for Nanomedicine, Wuhan 430075, China; Laurel565118@163.com
3 Infinitus (China) Co., Ltd., Guangzhou 510665, China; Fiona.Liang@infinitus-int.com (M.L.); Tinca.Zhang@infinitus-int.com (T.Z.); Xiquan.Yin@infinitus-int.com (X.Y.); sophia.zhang@infinitus-int.com (Y.Z.)
* Correspondence: wliu@hust.edu.cn
† These authors contributed equally to this work.

Academic Editors: Alessandra Gentili and Chiara Fanali

Abstract: A total of 13 batches of collagen peptide samples were extracted, isolated, and purified from chicken sternal cartilage under various process parameters. The fingerprint profiles of 13 batches of collagen peptides were established by high-performance liquid chromatography (HPLC). In addition, the amino acid profiles and molecular weight distributions of collagen peptides were investigated. The in vitro antioxidant activities of the peptide samples were measured using the 2,2′-Azinobis (3-ethylbenzothiazoline-6-sulphonic acid) diammonium salt (ABTS) assay, the 2,2-diphenyl-1-picrylhydrazyl (DPPH) assay, the ferric-reducing antioxidant power (FRAP) assay and an assay of the oxidative damage induced by hydrogen peroxide (H_2O_2) in the degenerative cartilage cells from the knee joint of rat C518 (C518 cell line). The anti-inflammatory activities of the peptide samples were assessed by measuring the inflammatory responses induced by lipopolysaccharides (LPS) in C518 cells. Subsequently, the spectrum-effect relationships between HPLC fingerprints and the antioxidant and anti-inflammatory activities of collagen peptides were investigated using grey relational analysis (GRA). Fifteen common peaks were obtained from the HPLC fingerprints of collagen peptides. Each collagen peptide sample had a characteristic set of amino acid types and contents. All of the hydrolysates of the collagen peptides were primarily composed of fractions II (500–1000 Da) and III (1000–3000 Da). Collagen peptides exhibited good scavenging activity on ABTS radical, DPPH radical, and ferric-reducing antioxidant power. Collagen peptides were also effective against H_2O_2-induced cellular oxidative damage in C518 cells. The antioxidant activity of collagen peptides was due to the low molecular weight and the presence of antioxidant and hydrophobic amino acid residues within its sequence. Collagen peptides significantly inhibited the secretion of inflammatory cytokines IL-1β, TNF-α, and PGE2 in C518 cells. The anti-inflammatory activity of collagen peptides may include increased synthesis of the key components of extracellular matrix (ECM) and inhibited apoptosis of chondrocytes. The GRA results showed that peaks 2, 3, and 8 were the main components contributing to the antioxidant activity of the collagen peptides, whereas peaks 11 and 14 were the main components contributing to the anti-inflammatory activity of the collagen peptides. The components of peaks 8 and 14 were identified as GPRGPPGPVGP and VAIQAVLSLYASGR by UPLC-MS/MS. Those identified collagen peptides offer a potential therapeutic strategy for the treatment of osteoarthritis (OA) due to their antioxidative stress and due to them disturbing the catabolism and anabolism processes in arthrodial cartilage.

Keywords: collagen peptide; HPLC fingerprint; antioxidant; anti-inflammatory; spectrum-effect relationship

1. Introduction

Osteoarthritis (OA), the most common degenerative joint disease worldwide found in elderly individuals, is a leading cause of physical disability. Type II collagen (CII) is a principal component of the extracellular matrix (ECM) and constitutes 90–95% of the total protein content in the articular cartilage [1]. Repeated oral administrations of CII induce oral tolerance and inhibit the development of OA. The immunologic response to orally administered antigens occurs in the gut-associated lymphoid tissue (GALT), which is composed of epithelium, lamina propria, Peyer's patch (PP), and mesenteric lymph nodes (MLN). Orally taken native CII antigens interact with PP in the GALT, resulting in turning off the T-cell attack to CII in the cartilage. This desensitization process in PP, also known as oral tolerance, avoids the recognition of endogenous CII in the cartilage as an antigen by the immune system [2]. The effects of oral administration of CII obtained from chicken, bovine, and sheep sources have been evaluated for the treatment of arthritis [3]. Chicken sternal cartilage containing a high amount of collagen is recognized as a potential source of CII [4]. Naturally derived CII from chick sternal cartilage is thought to be a potential oral alternative for OA treatment.

Although the etiology and underlying mechanisms of OA are complicated, much evidence has suggested that the progression of OA in patients is significantly associated with oxidative stress and the inflammatory factor network [5,6].

Recently, oxidative stress has been one of the research hotspots in the field of OA. There is increasing evidence that oxidative stress due to chronic production of endogenous reactive oxygen species (ROS) plays an important role in the physiology and pathophysiology of OA. Oxidative stress regulates intracellular signaling processes and induces chondrocyte senescence and apoptosis, which is characterized by degradation of ECM and a decrease in the synthesis of proteoglycan and CII [7]. Moreover, increased levels of ROS can damage DNA, including mitochondrial DNA, thereby affecting cell viability and contributing to the disruption of ECM homeostasis. Increased levels of ROS may also contribute to the senescent secretory phenotype and the reduced sensitivity of chondrocytes to insulin-like growth factor I (IGF-I) [8].

Antioxidant properties should be evaluated by a variety of methods because most natural antioxidants are multifunctional. Several assays have been frequently used to estimate antioxidant capacities in natural antioxidants, including the 2,2′-Azinobis (3-ethylbenzothiazoline-6-sulphonic acid) diammonium salt (ABTS) assay, the 2,2-diphenyl-1-picrylhydrazyl (DPPH) assay, and the ferric-reducing antioxidant power (FRAP) assay [9]. H_2O_2 is a common ROS produced in cellular metabolism. Intracellular steady-state concentrations of H_2O_2 above 1 μM cause oxidative stress that induces growth arrest and cell death. H_2O_2 induces chondrocyte apoptosis through the regulation of phosphatidylinositol 3-kinase (PI3K)/protein kinase B (Akt) and c-Jun N-terminal kinase (JNK) signaling pathways [5]. In experimental models used to investigate oxidative stress responses of cells, or cytoprotection by antioxidant agents, cultured cells are often exposed to H_2O_2 added as a bolus into the culture medium [10]. Therefore, the ABTS assay, DPPH assay, FRAP assay and an assay of cellular antioxidant activity (the protective effect of collagen peptides against H_2O_2-induced cellular oxidative damage in C518 cells) were performed to evaluate the antioxidant activity of collagen peptides.

Inflammation is another promoting factor in the OA process, including chondrocyte and synovium inflammation. The group of inflammatory cytokines is the most important group of compounds participating in the pathogenesis of OA. Among the many representatives of this group, the greatest importance is attributed to IL-1β, TNF-α, and PGE2, which have been shown to modulate ECM turnover, to accelerate the degradation of cartilage and to induce chondrocyte apoptosis in the development of OA [6,11].

IL-1β demonstrates potent bioactivities in inhibiting ECM synthesis and promoting cartilage breakdown [12]. The matrix metalloproteinases (MMPs) play important roles in cartilage degradation in OA, and IL-1β significantly upregulates the expression of MMPs, aggravating degradation processes in OA [13]. Moreover, it has been suggested that IL-1β induces the expression of the TNF-α gene in chondrocytes and upregulates the surface expression of the TNF receptor (TNFR) [14]. The effect of TNF-α in most cases coincides with the action of IL-1β, and in the case of many phenomena occurring in the course of OA there is a marked synergism between the two cytokines. This effect is the result of activation of the same group of intracellular signaling pathways, which in turn triggers similar effects that increase the inflammation and catabolism in joint tissues [6]. PGE2 is considered to be the major contributor to inflammatory pain in arthritic conditions. IL-1β has been shown to stimulate and produce high levels of PGE2 that may induce pain and degeneration with OA [12]. Pharmacologic inhibitors to these mediators may potentially be used as biological treatments in the future. Accordingly, the protective effect of collagen peptides against lipopolysaccharide (LPS)-induced proinflammatory mediators IL-1β, TNF-α, and PGE2 in C518 cells was investigated.

In virtue of the extreme complexity of collagen peptides, the components contributing to their effects are still unclear. Traditional methods for analysis of the active components of peptides are complex and time-consuming, and the spectrum-effect relationships of collagen peptides have not been reported. The aim of this study was to develop an efficient method for screening the active components of CII by establishing spectrum-effect relationships.

2. Results and Discussion

2.1. High-Performance Liquid Chromatography (HPLC) Fingerprints of the Collagen Peptides

The HPLC fingerprints of the collagen peptides and the reference chromatogram generated by the software are shown in Figure 1. The HPLC fingerprints of collagen peptide samples S1–S13 are shown in Figures S1–S13. Fifteen common peaks were obtained from all chromatograms, which were marked 1–15 in the reference chromatogram. Peak 8 (t = 9.087 min) was selected as the reference peak because it was a medium peak in the middle of the chromatogram. The relative retention time (RRT), relative peak area (RPA), and coefficient of variance (C.V.%) of the peak area of 15 common characteristic peaks are shown in Table 1. The majority of the C.V.% values were greater than 18.54%, which showed that the content of each common constituent in the samples varied significantly. The content of the unknown component, which was represented by peaks 5 and 7, showed a particularly large degree of variation.

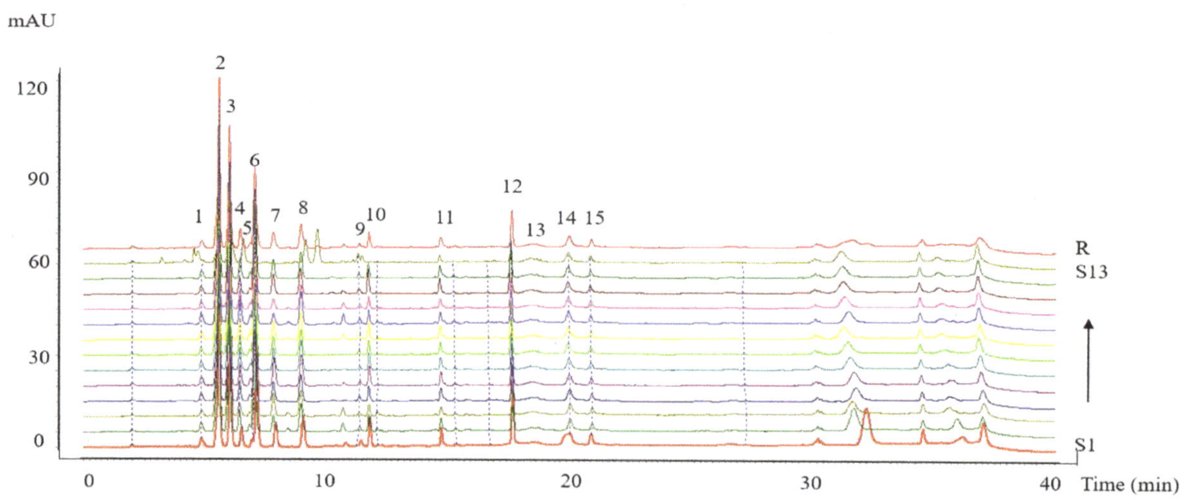

Figure 1. The high-performance liquid chromatography (HPLC) fingerprints of 13 batches of collagen peptide samples (S1–S13) and reference standard fingerprint (R).

Table 1. The relative retention time (RRT) and relative peak area (RPA) of the common peaks of the collagen peptides. C.V.%: coefficient of variance.

Peak No.	RRT	S1	S2	S3	S4	S5	S6	S7	S8	S9	S10	S11	S12	S13	C.V.%
1	0.538	0.333	0.376	0.241	0.329	0.307	0.364	0.372	0.355	0.400	0.371	0.348	0.337	0.499	18.54
2	0.620	4.766	4.808	5.099	4.516	4.653	4.700	4.821	4.745	5.031	4.780	4.769	4.543	1.761	22.17
3	0.669	3.585	3.401	4.693	3.406	3.429	3.467	3.502	3.478	3.449	3.420	3.535	3.395	2.997	13.60
4	0.720	0.533	0.617	0.453	0.570	0.559	0.594	0.608	0.590	0.750	0.625	0.527	0.575	0.880	20.92
5	0.768	0.182	0.201	0.098	0.183	0.156	0.178	0.193	0.191	0.226	0.209	0.180	0.176	0.000	37.45
6	0.789	2.972	2.862	2.683	2.828	2.825	2.857	2.893	2.898	3.043	2.964	2.950	2.834	0.081	31.60
7	0.875	0.751	0.547	0.791	0.664	0.845	0.675	0.696	0.598	0.315	0.414	0.703	0.655	0.000	41.04
8(s)	1.000	1.000	1.000	1.000	1.000	1.000	1.000	1.000	1.000	1.000	1.000	1.000	1.000	1.000	9.95
9	1.266	0.198	0.179	0.149	0.177	0.177	0.105	0.141	0.207	0.192	0.185	0.182	0.163	0.256	22.45
10	1.308	0.655	0.440	0.343	0.408	0.473	0.403	0.402	0.426	0.305	0.360	0.652	0.394	0.170	34.07
11	1.629	0.451	0.372	0.229	0.438	0.304	0.364	0.348	0.350	0.246	0.295	0.445	0.350	0.183	26.69
12	1.957	1.459	0.923	0.618	0.928	0.866	0.862	0.932	0.893	0.651	0.805	1.396	0.934	0.261	36.93
13	2.044	0.045	0.253	0.359	0.304	0.218	0.304	0.357	0.323	0.233	0.371	0.383	0.358	0.386	27.75
14	2.216	1.086	0.693	0.665	0.709	0.539	0.602	0.719	0.703	0.511	0.741	0.751	0.694	0.608	23.95
15	2.316	0.415	0.269	0.258	0.270	0.221	0.233	0.280	0.275	0.219	0.282	0.277	0.264	0.288	22.00

The similarity values ranged from 0.612 to 0.998 (Table 2). Compared to the reference chromatogram, the similarities between the reference fingerprint and the entire chromatographic profiles of 13 batches of collagen peptide samples were evaluated. The correlation coefficients of batches 1–13 were 0.777, 0.936, 0.972, 0.963, 0.977, 0.976, 0.969, 0.977, 0.972, 0.967, 0.961, 0.953, and 0.711, respectively. The similarity values of all batches exceeded 0.930, with the exception of S1 and S13, which indicated that there were relatively small differences among them.

Table 2. Similarities among collagen peptides from 13 origins.

	S1	S2	S3	S4	S5	S6	S7	S8	S9	S10	S11	S12	S13	R
S1	1.000													
S2	0.704	1.000												
S3	0.747	0.950	1.000											
S4	0.733	0.964	0.984	1.000										
S5	0.695	0.926	0.959	0.959	1.000									
S6	0.696	0.889	0.932	0.921	0.969	1.000								
S7	0.691	0.877	0.920	0.909	0.963	0.998	1.000							
S8	0.696	0.888	0.928	0.918	0.967	0.972	0.967	1.000						
S9	0.687	0.884	0.924	0.912	0.964	0.987	0.985	0.972	1.000					
S10	0.680	0.870	0.910	0.899	0.957	0.963	0.961	0.994	0.974	1.000				
S11	0.696	0.885	0.916	0.910	0.918	0.927	0.916	0.950	0.919	0.940	1.000			
S12	0.689	0.868	0.911	0.903	0.909	0.917	0.907	0.939	0.913	0.932	0.993	1.000		
S13	0.626	0.612	0.693	0.637	0.631	0.642	0.635	0.639	0.637	0.633	0.702	0.714	1.000	
R	0.777	0.936	0.972	0.963	0.977	0.976	0.969	0.977	0.972	0.967	0.961	0.953	0.711	1.000

2.2. *HPLC Fingerprints of Amino Acids*

The HPLC fingerprints of the amino acids and the reference chromatogram generated by the software are shown in Figure 2. Twenty-one common peaks were obtained from all chromatograms, which were marked 1–21 in the reference chromatogram. Peak 7 (Arg, t = 13.881 min) was selected as the reference peak because it was a medium peak in the middle of the chromatogram. The RRT, RPA, and C.V.% of the peak area of 21 common characteristic peaks are shown in Table 3.

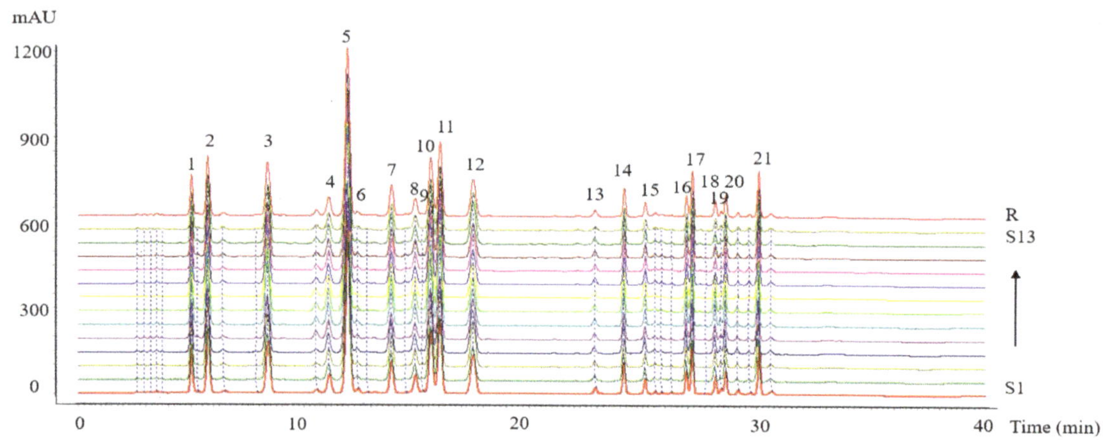

Figure 2. The HPLC fingerprints of the amino acids of 13 batches of collagen peptide samples (S1–S13) and amino acid reference standard fingerprint (R).

Table 3. The RRT and RPA of the common peaks of the amino acids.

Peak No.	RRT	S1	S2	S3	S4	S5	S6	S7	S8	S9	S10	S11	S12	S13	C.V.%
1	0.361	0.944	0.903	0.885	0.953	0.974	0.843	0.865	0.854	0.804	0.896	0.857	0.915	0.931	8.30
2	0.413	1.701	1.646	1.712	1.732	1.800	1.645	1.648	1.622	1.555	1.627	1.612	1.717	1.775	8.63
3	0.603	1.804	1.801	1.905	1.893	1.918	2.020	2.015	1.966	2.013	2.008	2.089	2.101	1.939	12.48
4	0.801	0.639	0.618	0.607	0.632	0.796	0.709	0.721	0.709	0.663	0.680	0.680	0.689	0.704	12.10
5	0.861	6.032	5.889	6.115	6.099	6.320	6.367	6.329	6.125	6.366	6.308	6.569	6.575	6.287	10.86
6	0.894	0.134	0.133	0.121	0.135	0.157	0.134	0.139	0.137	0.131	0.133	0.145	0.135	0.143	9.29
7(s)	1.000	1.000	1.000	1.000	1.000	1.000	1.000	1.000	1.000	1.000	1.000	1.000	1.000	1.000	10.39
8	1.075	0.631	0.611	0.599	0.626	0.705	0.650	0.659	0.641	0.627	0.632	0.666	0.656	0.661	9.96
9	1.098	0.065	0.061	0.068	0.067	0.063	0.058	0.052	0.050	0.052	0.057	0.064	0.070	0.081	10.37
10	1.125	2.148	2.076	2.126	2.137	2.204	2.096	2.114	2.047	2.082	2.083	2.120	2.136	2.143	9.40
11	1.156	2.398	2.372	2.440	2.441	2.539	2.529	2.524	2.461	2.566	2.551	2.683	2.661	2.549	10.60
12	1.263	1.638	1.608	1.948	1.833	2.146	1.391	1.322	1.339	1.443	1.556	1.545	1.926	1.974	10.40
13	1.649	0.212	0.197	0.173	0.191	0.205	0.182	0.187	0.188	0.181	0.187	0.216	0.198	0.152	12.71
14	1.741	0.732	0.665	0.627	0.670	0.736	0.636	0.641	0.609	0.590	0.601	0.662	0.639	0.624	9.65
15	1.809	0.328	0.328	0.334	0.361	0.379	0.335	0.348	0.343	0.354	0.335	0.353	0.355	0.328	10.43
16	1.938	0.479	0.481	0.398	0.455	0.498	0.448	0.453	0.429	0.425	0.021	0.484	0.462	0.438	30.24
17	1.957	0.943	0.920	0.833	0.922	0.982	0.914	0.936	0.905	0.897	0.426	0.969	0.944	0.943	18.25
18	2.027	0.366	0.345	0.349	0.357	0.394	0.448	0.445	0.442	0.252	0.008	0.289	0.255	0.318	40.55
19	2.045	0.111	0.110	0.125	0.128	0.135	0.147	0.126	0.123	0.079	0.284	0.101	0.104	0.144	39.35
20	2.061	0.476	0.464	0.443	0.474	0.528	0.480	0.485	0.472	0.470	0.473	0.515	0.513	0.447	10.51
21	2.166	1.026	0.985	0.949	0.985	1.090	0.942	0.966	0.951	0.955	0.951	1.003	0.971	1.059	7.48

Subsequently, eighteen common characteristic peaks in the chromatogram were identified by comparing the retention times and online UV spectra with those of the standards as follows: Peak 1 (Asp), peak 2 (Glu), peak 4 (Ser), peak 5 (Gly), peak 6 (His), peak 7 (Arg), peak 8 (Thr), peak 9 (HyP), peak 10 (Ala), peak 11 (Pro), peak 13 (Tyr), peak 14 (Val), peak 15 (Met), peak 16 (Ile), peak 17 (Leu), peak 19 (HyL), peak 20 (Phe), and peak 21 (Lys). Each collagen peptide sample had a characteristic set of amino acid types and contents. As a whole, the collagen peptide samples were rich in Gly, Ala, and Pro, which accounted for 23.98%, 16.20%, and 9.64% of the amino acid residues, respectively, whereas HyP (0.24%), HyL (0.51%), His (0.52%), and Tyr (0.73%) were relatively rare among the 18 quantified amino acids (Table 3).

It was found that the collagen peptide samples had a similar amino acid composition to the reference. The characteristic that collagen peptide samples had a high content of Gly and Pro was in accordance with previous findings, although the contents of Gly and Pro obtained in the current research were slightly lower than the values reported by Cao [4]. From these results it can be concluded that Gly, which constitutes approximately one-fourth of all residues in collagen, was present as a quarter residue in the sequence, and that high amounts of Pro could be accommodated while maintaining planar peptide bonds. These assumptions led to the construction of the correct model of sternal cartilage collagen as a (Gly-X-Y) n pattern. Indeed, it was the most commonly found triplet in collagen chains space, indicating that the collagen purified from chicken sternal cartilage was typical CII. Bioactive peptides usually contain 3–20 amino acid residues, and their activities are related to their amino acid compositions [15]. Generally, the antioxidant activities of the amino acid side chain, such as the imidazole group in His (peak 6), the phenolic hydroxyl group in Tyr (peak 13), and thioether in Met (peak 15), determine the antioxidant activities of peptides [16]. Additionally, antioxidative peptides containing aromatic amino acids at the C-terminus have high radical scavenging activities. Aromatic amino acids such as His (peak 6) and Tyr (peak 13) increase the antioxidant activities of peptides and protein hydrolysates because they easily donate protons to electron-deficient radicals and maintain their stabilities via resonance structures. The hydrophobic amino acids at the N-terminus of a peptide sequence, such as Gly (peak 5), Val (peak 14), and Leu (peak 17), are believed to contribute to antioxidative activity, as it is thought that hydrophobic amino acids can increase the abundance of peptides at the water–lipid interface, which may facilitate greater interaction between the peptide and fatty acids [17,18]. Collagen peptide samples S3 and S13 exhibited a lower abundance of Gly, His, Leu, Met, Tyr, and Val in comparison to the other samples. These findings were in accordance with the results of the ABTS and DPPH assays, which indicated that low antioxidant amino acid content in the collagen peptide samples might be related to low antioxidant activity.

The similarity values ranged from 0.995 to 1.000 (Table 4). Compared to the reference chromatogram, the similarities between the reference fingerprint and the entire chromatographic profiles of 13 batches of collagen peptide samples were evaluated: The correlation coefficients of batches 1–13 were 0.999, 1.000, 0.999, 0.999, 0.998, 0.999, 0.999, 0.999, 0.999, 1.000, 0.999, 0.999, and 0.999, respectively. The similarity values of all batches exceeded 0.998, which indicated that there were relatively small differences among them.

Table 4. Similarities among amino acids from 13 origins.

	S1	S2	S3	S4	S5	S6	S7	S8	S9	S10	S11	S12	S13	R
S1	1.000													
S2	1.000	1.000												
S3	0.999	0.999	1.000											
S4	0.999	1.000	1.000	1.000										
S5	0.998	0.998	0.999	0.999	1.000									
S6	0.998	0.998	0.997	0.997	0.995	1.000								
S7	0.998	0.998	0.996	0.997	0.995	1.000	1.000							
S8	0.998	0.998	0.996	0.997	0.995	1.000	1.000	1.000						
S9	0.998	0.998	0.997	0.997	0.995	0.999	0.999	0.999	1.000					
S10	0.999	0.999	0.998	0.999	0.997	0.999	0.999	0.999	1.000	1.000				
S11	0.998	0.998	0.997	0.998	0.996	0.999	0.999	0.999	1.000	1.000	1.000			
S12	0.998	0.999	0.999	0.999	0.998	0.998	0.997	0.998	0.999	0.999	0.999	1.000		
S13	0.999	0.999	1.000	1.000	0.999	0.997	0.996	0.997	0.997	0.998	0.998	0.999	1.000	
R	0.999	1.000	0.999	0.999	0.998	0.999	0.999	0.999	0.999	1.000	0.999	0.999	0.999	1.000

2.3. *Molecular Weight Distribution of Cartilage Hydrolysates*

The collagen peptides were fractionated according to molecular weight into four categories: I (below 500 Da), II (500–1000 Da), III (1000–3000 Da), and IV (heavier than 3000 Da). The relative content, obtained from their peak areas relative to the total peak area, of each peptide fraction is presented in Table 5. It was found that all of the hydrolysates were primarily composed of fractions II (500–1000 Da) and III (1000–3000 Da), which accounted for approximately 64.80% of the sample after enzymatic treatment.

Molecular weight reflects the hydrolysis of proteins and is highly correlated with the bioavailability and bioactivity of peptides [19,20]. Peptides of low molecular weight are more potent than those of high molecular weight because they cross the intestinal barrier more easily and exert biological effects. Moreover, short-chain peptides are more effective than long-chain peptides with regard to their effects on physiological processes, as they are less susceptible to gastrointestinal hydrolysis.

Table 5. Molecular weight distribution of collagen peptides.

Samples	<500 (Da)	500–1000 (Da)	1000–3000 (Da)	>3000 (Da)
S1	12.48	45.32	24.60	17.60
S2	8.89	39.18	30.77	21.16
S3	9.68	30.68	26.23	33.41
S4	12.89	32.34	25.76	29.01
S5	13.09	32.99	20.99	32.93
S6	12.37	34.21	22.21	31.21
S7	7.56	34.10	32.03	26.31
S8	7.03	33.55	30.48	28.94
S9	8.90	43.10	31.82	16.18
S10	9.51	44.95	24.41	21.13
S11	12.51	43.36	28.25	15.88
S12	10.24	42.43	31.54	15.79
S13	6.07	27.28	29.76	36.89

The peptide fraction with lower molecular weight was probably associated with higher antioxidant activity. Samples S1, S10, and S11 had greater percentages of fractions I (below 500 Da) and II (500–1000 Da) in comparison to the other samples. In contrast, S13 contained relatively low percentages of fractions I (below 500 Da) and II (500–1000 Da) (Table 5). These findings were consistent with observations from the in vitro antioxidant assay mentioned below and support the idea that the functional properties of antioxidative peptides are highly influenced by their molecular weight.

2.4. Antioxidant Activity

2.4.1. Comparative Antioxidant Activity of 13 Batches of Collagen Peptides

The ABTS radical scavenging activity of the collagen peptides varied from 34.91% to 48.14%, which represented a variation of approximately 1.4-fold. The results showed that the ABTS radical inhibitory capacity of the samples (at a concentration of 2 mg/mL) decreased in the following order: S1 > S11 > S7 > S8 > S5 > S12 > S2 > S4 > S10 > S6 > S9 > S3 > S13. S1 exhibited the most ABTS radical scavenging activity, followed by S11 and S7 (Table 6).

The DPPH radical scavenging activity of the collagen peptides varied from 18.46% to 55.06%, which represented a variation of approximately three-fold. The results showed that the DPPH radical inhibitory capacity of the samples (at a concentration of 25 mg/mL) decreased in the following order: S12 > S11 > S10 > S5 > S7 > S9 > S6 > S8 > S4 > S2 > S1 > S3 > S13. S12 exhibited the most DPPH radical scavenging activity, followed by S11 and S10 (Table 6).

The ferric radical scavenging activity of collagen peptides varied from 31.56% to 49.10%, which represented a variation of approximately 1.5-fold. The results showed that the ferric radical scavenging activity of the samples (at a concentration of 50 mg/mL) decreased in the following order: S10 > S9 > S3 > S11 > S6 > S1 > S2 > S7 > S12 > S8 > S4 > S5 > S13. S10 exhibited the most ferric radical scavenging activity, followed by S9 and S3 (Table 6).

The results of these experiments suggested that all of the collagen peptide samples possessed good in vitro antioxidant efficacy. Furthermore, previous results suggested that the chicken sternal cartilage hydrolysates had great oxygen radical absorption capacity (ORAC), with a half maximal inhibitory concentration (IC50) value of 0.48 ± 0.02 μM TE/mg peptides. The exposure of abundant antioxidant residues was responsible for their ORAC antioxidant activity [21]. It is interesting to note that S13 exhibited the lowest antioxidant capacity in all three antioxidant assays, although the antioxidant activity of the collagen peptides varied according to the assay used.

The enzyme/substrate (E/S) ratio, pH, and temperature of digestion had a more significant effect on the hydrolysis of collagen peptides compared to digestion time, as no significant difference of the DPPH radical scavenging activity of S12, S11, and S10 was observed with the extension of the hydrolysis time.

It was clear that enzyme dosage was an important parameter for collagen peptide hydrolysis, since there were relatively significant differences in the antioxidant activity of S10 and S13, which was particularly evident in the FRAP assay.

S11 exhibited the greatest antioxidant capacity in the ABTS assay, DPPH assay, and FRAP assay (the top 4). Thus, the optimal enzymatic hydrolysis conditions for collagen peptide, papain treatment at an E/S ratio of 0.50% at 50 °C and at pH 6.5 for 5 h, could be used as a basis for production scale-up. The results of the optimal enzymatic hydrolysis conditions were basically consistent with a previous study [22].

Table 6. The antioxidant activity of collagen peptides in the 2,2′-Azinobis (3-ethylbenzothiazoline-6-sulphonic acid) diammonium salt (ABTS) assay, the 2,2-diphenyl-1-picrylhydrazyl (DPPH) assay, and the ferric-reducing antioxidant power (FRAP) assay.

Samples	ABTS	DPPH	FRAP
S1	48.14 ± 0.15	44.38 ± 1.28	37.02 ± 0.57
S2	43.21 ± 0.19	45.94 ± 2.50	35.19 ± 0.72
S3	36.78 ± 0.25	40.16 ± 1.17	42.65 ± 1.00
S4	42.65 ± 1.23	46.05 ± 2.18	33.06 ± 0.51
S5	43.30 ± 1.38	53.50 ± 4.36	31.56 ± 0.17
S6	41.93 ± 0.43	49.83 ± 2.10	37.65 ± 0.22
S7	46.07 ± 0.39	51.50 ± 1.12	34.77 ± 0.34
S8	44.72 ± 0.17	49.28 ± 2.93	33.83 ± 0.25
S9	38.79 ± 0.35	50.39 ± 2.06	45.58 ± 0.54
S10	42.50 ± 0.37	53.62 ± 1.21	49.10 ± 0.56
S11	47.84 ± 0.21	54.23 ± 2.39	38.08 ± 0.47
S12	43.28 ± 0.18	55.06 ± 2.49	34.00 ± 0.35
S13	34.91 ± 0.22	18.46 ± 1.64	18.15 ± 0.90

2.4.2. Protective Effect of Collagen Peptides against H_2O_2-Induced Oxidative Damage

The results of the Cell Counting Kit-8 (CCK-8) assay showed that the collagen peptides were not cytotoxic ($P < 0.01$). Exposure to exogenous H_2O_2 decreased the viability of C518 cells, which indicated that exposure to H_2O_2 was cytotoxic for C518 cells. In contrast, a significant increase in C518 cell viability was observed in all samples, especially for S1, S2, and S9, when C518 cells were pretreated with collagen peptides (Figure 3). Taken together, these findings clearly show that collagen peptides conferred significant protection against H_2O_2-induced cellular oxidative damage and can thus be used as a natural antioxidant.

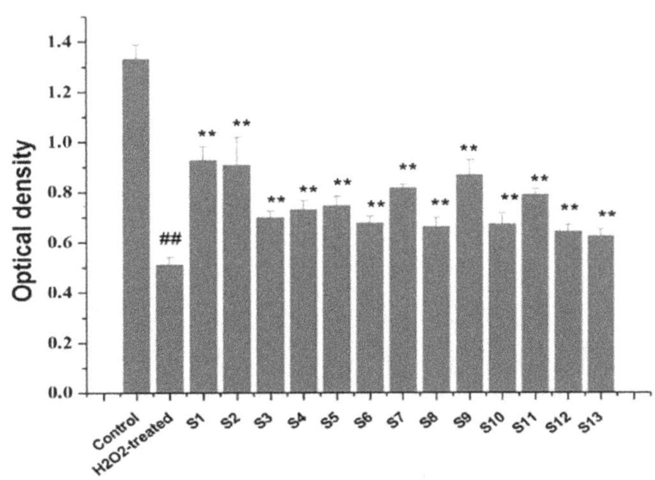

Figure 3. Collagen peptides prevented H_2O_2-induced oxidative damage. Data are expressed as means ± SD ($n = 3$); ** $p < 0.01$ versus H_2O_2-treated group; ## $p < 0.01$ versus the control group.

2.5. Inhibitory Effects of Collagen Peptides on LPS-Induced Release of Proinflammatory Mediators IL-1β, TNF-α, and PGE2

The effects of the collagen peptide samples on IL-1β secretion varied significantly. S2, S10, and S11 exhibited extremely strong inhibitory effects on IL-1β expression, whereas those of S1 and S12 were slightly weaker. S8, S9, and S13 showed strong inhibitory effects on IL-1β expression, whereas S3, S4, S5, S6, and S7 did not show anti-inflammatory activity (Table 7).

The levels of released TNF-α were increased in the LPS-stimulated culture media. Pretreatment with collagen peptides at a concentration of 2.0 mg/mL for 24 h substantially blocked TNF-α generation in C518 cells. S1, S2, and S3 showed the strongest inhibitory effects on TNF-α generation (Table 7).

All collagen peptide samples significantly attenuated LPS-induced PGE2 production, which was not consistent with the data from the experiments assessing IL-1β and TNF-α secretion. S1, S2, and S8 showed the strongest inhibitory effects on LPS-induced PGE2 production (Table 7).

Taken together, our findings show that collagen peptides significantly inhibited the secretion of IL-1β, TNF-α, and PGE2. Notably, the inhibitory effect of collagen peptides on PGE2 was particularly apparent. Furthermore, previous studies also confirmed that CII treatment dose-dependently inhibited the overproduction of inflammatory cytokines levels (IL-1β and TNF-α) in synoviocytes in the collagen-induced rat arthritis (CIA) model. In conclusion, CII extracted from chick sternal cartilage possessed antiarthritis activity, which may have been a result of its regulation of the humoral and cellular immune systems [23]. Thus, downregulation of proinflammatory mediators by collagen peptides may be a promising therapy for OA. Nevertheless, the molecular mechanisms underlying the anti-inflammatory effects of collagen peptides are not completely understood, so further investigation is required.

Table 7. Inhibitory effects of collagen peptides on lipopolysaccharide (LPS)-induced IL-1β, TNF-α, and PGE2 release in C518 cells.

Group	IL-1β (ng/L)	TNF-α (ng/L)	PGE2 (ng/L)
Control	69.66 ± 4.32	169.65 ± 6.85	90.35 ± 5.15
LPS-treated	86.62 ± 2.44 #	211.33 ± 11.43 #	115.10 ± 5.20 #
S1	61.69 ± 1.37 **	152.62 ± 7.55 *	40.79 ± 9.56 **
S2	44.42 ± 0.81 **	128.97 ± 3.09 *	47.10 ± 2.81 **
S3	78.73 ± 1.28	132.91 ± 10.76 *	59.56 ± 0.02 **
S4	81.94 ± 0.28	172.71 ± 14.85 *	52.57 ± 5.22 **
S5	79.22 ± 1.84	168.35 ± 8.69 *	74.69 ± 2.90 **
S6	85.00 ± 2.99	158.69 ± 5.57 *	49.68 ± 2.81 **
S7	75.99 ± 2.51	199.97 ± 0.78 *	55.84 ± 4.92 **
S8	69.37 ± 4.56 *	193.53 ± 8.59 *	40.79 ± 9.97 **
S9	72.15 ± 5.55 *	185.42 ± 7.68 *	52.57 ± 7.47 **
S10	42.46 ± 4.70 **	176.09 ± 1.32 *	55.84 ± 4.92 **
S11	54.56 ± 2.10 **	193.05 ± 5.25 *	49.68 ± 6.29 **
S12	59.48 ± 3.28 **	184.95 ± 6.76 *	52.57 ± 9.80 **
S13	74.5 ± 3.15 *	182.56 ± 7.17 *	68.82 ± 2.89 **

Data are expressed as means ± SD (n = 3); ** $p < 0.01$ and * $p < 0.05$ versus the LPS-treated group; # $p < 0.05$ versus the control group.

2.6. GRA Results

The ranking results of the GRA were as follows (the top 10): ABTS scavenging activity, 14 > 8 > 15 > 3 > 6 > 2 > 1 > 13 > 5 > 4; DPPH scavenging activity, 2 > 8 > 6 > 13 > 3 > 12 > 10 > 14 > 5 > 11; ferric reducing antioxidant power, 2 > 3 > 8 > 1 > 4 > 14 > 6 > 15 > 13 > 11; protection against H_2O_2-induced oxidative damage, 10 > 11 > 15 > 14 > 9 > 2 > 12 > 1 > 4 > 5; inhibited production of LPS-induced IL-1β, 11 > 12 > 10 > 14 > 15 > 9 > 3 > 8 > 5 > 1; inhibited production of LPS-induced TNF-α, 3 > 14 > 8 > 11 > 15 > 2 > 7 > 12 > 10 > 9; inhibited production of LPS-induced PGE2, 6 > 14 > 8 > 2 > 3 > 15 > 5 > 11 > 1 > 12 (Table 8). Overall, peaks 2, 3, and 8 were the main contributors to the antioxidant activity of collagen peptides, whereas peaks 11 and 14 were the main contributors to the anti-inflammatory activity of collagen peptides.

Table 8. The correlation coefficient between the common characteristic peaks and the efficacy of collagen peptides.

Peak No.	r						
	ABTS	DPPH	FRAP	Cell Viability	IL-1β	TNF-α	PGE2
1	0.7732	0.7338	0.7722	0.7315	0.6441	0.6492	0.7622
2	0.7788	0.8110	0.8102	0.7334	0.6435	0.6733	0.8079
3	0.7795	0.7744	0.8081	0.7065	0.6531	0.7025	0.7983
4	0.7497	0.7451	0.7584	0.7294	0.6303	0.6537	0.7368
5	0.7503	0.7467	0.6933	0.7168	0.6450	0.6313	0.7748
6	0.7795	0.7853	0.7456	0.7166	0.6349	0.6501	0.8282
7	0.6673	0.7068	0.6722	0.6819	0.6279	0.6617	0.7029
8	0.8040	0.7982	0.7783	0.7085	0.6507	0.6852	0.8159
9	0.7285	0.7181	0.6870	0.7345	0.6569	0.6555	0.7486
10	0.7008	0.7639	0.6960	0.7534	0.6802	0.6574	0.7271
11	0.7225	0.7461	0.6985	0.7517	0.6943	0.6827	0.7707
12	0.7161	0.7728	0.6950	0.7331	0.6811	0.6581	0.7486
13	0.7544	0.7842	0.7083	0.6555	0.6238	0.6039	0.7446
14	0.8080	0.7530	0.7561	0.7428	0.6729	0.6853	0.8254
15	0.7947	0.7428	0.7397	0.7492	0.6587	0.6789	0.7932

Generating peptides via proteolysis or peptidolysis results in complex combinations of peptides with different masses and physicochemical properties. Classical strategies commonly used to achieve isolation and subsequent characterization employ chromatographic techniques coupled to mass spectrometry, in particular liquid chromatography (LC) coupled to tandem MS (MS/MS) [24,25]. The accuracy and speed of peptide identification are some of the key features that set MS/MS apart from the other methodologies used to analyze protein mixtures [26]. Moreover, the probability-based search engine Mascot has found widespread use as a tool in correlating tandem mass spectra with peptides in a sequence database [27,28].

The amino acid sequences of the collagen peptides were tentatively identified using ultra-performance liquid chromatography-tandem mass spectrometry (UPLC-MS/MS), and all the resulting fragmentation spectra (MS/MS) were searched against the *Gallus gallus* database (NCBI) with the Mascot database searching software (version 2.1, Matrix Science, London, UK). In addition, we also compared the ultraviolet spectra and retention times of the peaks with those of standard peptides obtained by the solid phase peptide synthesis (SPPS) method, and they matched each other. The primary structures of the peptide fractions were determined as GPRGPPGPVGP (peak 8, 987.12 Da) and VAIQAVLSLYASGR (peak 14, 1447.67 Da). Peaks 8 and 14 probably each represented more than one peptide, but only the structure of the main component of each peak was identified in this study. This is the first report of the isolation and identification of these peptides from chicken sternal cartilage. However, further study is needed to identify the structures of other collagen peptide components, particularly the effective components.

3. Materials and Methods

3.1. Materials

HPLC-grade acetonitrile was purchased from Tedia (Fairfield, OH, USA). Amino acid standards, ABTS, DPPH, 2,4,6-tripyridyl-s-triazine (TPTZ), LPS, CCK-8, rat IL-1β, PGE2, and the TNF-α ELISA kit were purchased from Sigma-Aldrich (St. Louis, MO, USA). Dulbecco's Modified Eagle's Medium (DMEM), fetal bovine serum (FBS), phosphate-buffered saline (PBS), and 1% penicillin/streptomycin were bought from Gibco (Carlsbad, CA, USA). The synthesized peptides (purity $\geq$ 95%) were obtained from Bankpeptide Biological Technology Co., Ltd. (Hefei, China). Other reagents were of analytical grade.

3.2. *Preparation of Collagen Peptides*

Collagen peptide was prepared according to the methods reported by Xie with some modifications [22]. Fresh chicken sternal cartilage was removed from the periosteum and the calcified portion, cut into slices, minced into paste, and stored at −20 °C. Then, the paste was treated with 5 M guanidine hydrochloride in Tris-HCl (0.05 M, pH 7.5) at a sample/solution ratio of 1:10 (*w*/*v*), with gentle stirring for 24 h at 4 °C to remove proteoglycans [29]. The precipitate was washed using Tris-HCl (0.05 M, pH 7.5) and acetic acid (0.5 M). In the enzymatic hydrolysis process of collagen peptides, the four parameters including the enzyme (papain) to substrate ratio (E/S: 0.25%, 0.50%, and 0.75% *w*/*w*), pH (6.0, 6.5, and 7.0), temperature (50, 55, and 60 °C) and time (4, 5, and 6 h), were investigated in single-factor experiments (Table 9). At the end of the enzymatic hydrolysis, the solution was heated in a boiling water bath for 10 min to inactivate the enzymes. The extracts were collected by centrifugation (Avanti J-26XP, Beckman Coulter, Brea, CA, USA) at 10,000× *g* for 30 min at 4 °C. The supernatant was collected and salted out overnight at 4 °C by adding 2.5 M NaCl in Tris-HCl (0.05 M, pH 7.5) at a ratio of 1:20 (*w*/*v*). The resulting precipitate was collected by centrifugation (Avanti J-26XP, Beckman Coulter, USA) at 10,000× *g* for 20 min at 4 °C. Subsequently, the precipitate was dissolved with acetic acid (0.1 M) and allowed to reach dialysis equilibrium with NaCl (0.2 M). The dialysate was freeze-dried and referred to as "collagen peptides".

Table 9. The processing conditions of collagen peptides. E/S: enzyme to substrate ratio.

Samples	E/S (*w*/*w*)	pH	Temperature (°C)	Time (h)
S1	0.75%	6.0	50	6
S2	0.50%	6.0	50	6
S3	0.25%	6.0	50	6
S4	0.50%	6.0	55	6
S5	0.50%	6.5	55	6
S6	0.50%	7.0	55	6
S7	0.50%	7.0	50	4
S8	0.50%	7.0	55	4
S9	0.50%	7.0	60	4
S10	0.50%	6.5	50	6
S11	0.50%	6.5	50	5
S12	0.50%	6.5	50	4
S13	0.25%	6.5	50	6

3.3. *HPLC Fingerprints*

3.3.1. HPLC Conditions

Chromatographic analyses of collagen peptides were performed using a Shimadzu LC-20A HPLC system (Kyoto, Japan), including a quaternary solvent delivery system, online degasser, autosampler manager, column compartment and diode-array detector (DAD), as well as LCsolution software. The HPLC operating conditions were as follows: The column was an Inertsil ODS-SP C18 column (4.6 × 250 mm, 5 μm). A binary gradient elution system, comprising buffer (A) 0.05% TFA/acetonitrile and buffer (B) 0.1% TFA/water, was applied as follows: Initial, 98% B; 30 min, 40% B; 30.01 min, 98% B; 40 min, 98% B. The detection wavelength was set at 214 nm, the flow rate was 0.8 mL/min, the column temperature was maintained at 40 °C, the sample temperature was ambient, and the injection volume was 10 μL.

3.3.2. Preparation of the Sample Solution

Collagen peptides were dissolved in deionized water at a concentration of 10 mg/mL and then filtered through a 0.22 μm micropore film to yield the sample solution.

3.3.3. Analysis of HPLC Fingerprints

Validation of Methodology

The chromatographic fingerprinting methodology was validated to assess its precision, repeatability, and stability. Precision was evaluated by the analysis of six injections of the same testing sample consecutively. Repeatability was examined by determination of six different samples prepared from the same collagen peptide sample. Stability was examined by analysis of the sample solution at 0, 2, 4, 6, 8, 12, and 24 h. The methodology validation showed that the relative standard deviation (RSD) for precision was in the range of 1.28–1.75%, whereas that of reproducibility was less than 1.18% and that of storage stability was 1.46–1.83%. All results indicated that the HPLC fingerprint analysis method was valid and satisfactory.

Similarity Analysis

The HPLC fingerprints were matched automatically using the Similarity Evaluation System for Chromatographic Fingerprint of Traditional Chinese Medicine software developed by the Chinese Pharmacopoeia Committee (Version 2012) (Beijing, China). The reference fingerprint was generated using the median method [30], after which the similarity values between the reference fingerprint and the entire chromatographic profiles of 13 batches of collagen peptides were calculated.

3.4. Amino Acid Composition

Amino acid composition of collagen peptides was determined according to the method of Sun with a slight modification [31]. A high-performance liquid chromatography equipped with a PICO.TAG column (Waters, Milford, MA, USA) was used. The amino acid composition of collagen peptides was determined after hydrolysis at 150 °C for 1 h with 6 M hydrochloric acid prior to derivatization with phenyl isothiocyanate. A binary gradient elution system, comprising (A) 0.1 M ammonium acetate (the pH was adjusted to 6.5 with acetic acid)/acetonitrile (93:7 ratio) and (B) 80% acetonitrile/water, was applied as follows: Initial, 0% B; 15 min, 15% B; 18 min, 24% B; 25 min, 40% B; 30 min, 40% B; 30.01 min, 100% B; 40 min, 100% B. The amino acid standards included L-alanine (Ala), L-arginine (Arg), L-aspartic acid (Asp), L-glutamic acid (Glu), L-glycine (Gly), L-histidine (His), hydroxylysine (HyL), hydroxyproline (HyP), L-isoleucine (Ile), L-leucine (Leu), L-lysine (Lys), L-methionine (Met), L-phenylalanine (Phe), L-proline (Pro), L-serine (Ser), L-threonine (Thr), L-tyrosine (Tyr), and L-valine (Val).

3.5. Molecular Weight Distribution

Molecular weight distribution of collagen peptides was determined by gel permeation chromatography on a Superdex Peptide HR 10/300 GL (10 × 300 mm, Amersham Biosciences Co., Piscataway, NJ, USA) with UV detection at 214 and 280 nm. The mobile phase was 5 mM phosphate buffer containing 10 mM NaCl (pH 7.4), at a flow rate of 0.5 mL/min, which corresponded to an operating pressure of 1.8 MPa. A molecular weight calibration curve was obtained from the following standards: Glycine trimer (189 Da), oxidized glutathione (612 Da), vitamin B12 (1355 Da), aprotinin (6500 Da), cytochrome C (12500 Da), and bovine serum albumin (66430 Da) (Sigma Co., St. Louis, MO, USA). UNICORN 5.0 software (Amersham Biosciences Co., Piscataway, NJ, USA) was used to analyze the chromatographic data.

3.6. Antioxidant Activity Determination

3.6.1. ABTS Assay

The ABTS assay was conducted following the method of Arnao with some modifications [32]. The stock solutions included 7.0 mM $ABTS^{+}$ solution and 2.45 mM $K_2S_8O_2$ solution. The working solution was prepared by mixing the two stock solutions in equal quantities and allowing them to react for 12 h at room temperature in the dark. Before analysis, this stock solution was diluted with PBS to obtain an absorbance of 0.70 ± 0.02 at 734 nm. Next, 100 μL samples were allowed to react with 2 mL $ABTS\cdot^{+}$ solution in the dark. After 6 min, the absorbance was recorded at 734 nm. Results were determined using the following Equation (1):

$$\text{ABTS radical scavenging capacity (\%)} = [(A_{blank} - A_{sample})/A_{blank}] \times 100\%, \quad (1)$$

where A_{sample} and A_{blank} are the absorbance of the test sample and blank sample, respectively.

3.6.2. DPPH Assay

The DPPH assay was performed according to the method of Brand–Williams with some modifications [33]. First, 1 mL of 0.2 mM DPPH solution was added to an equal volume of the sample solution. The mixture was vortexed for a few seconds and incubated for 30 min at room temperature, after which the absorbance of the mixture was measured at 517 nm. Results were calculated according to the following Equation (2):

$$\text{DPPH radical scavenging activity (\%)} = [(A_{blank} - A_{sample})/A_{blank}] \times 100\%, \quad (2)$$

where A_{sample} and A_{blank} are the absorbance of the test sample and blank sample, respectively.

3.6.3. FRAP Assay

The FRAP assay was performed according to the methods reported by Benzie with some modifications [34]. The FRAP solution was prepared by mixing 10 volumes of acetate buffer (300 mM, pH 3.6) with 1 volume of TPTZ (10 mM dissolved in 40 mM HCl) and 1 volume of $FeCl_3 \cdot 6H_2O$ (20 mM in water). Next, 200 μL of the sample were allowed to react with 2 mL FRAP solution in the dark for 30 min at 37 °C, and the absorbance of the sample was recorded at 593 nm. Aqueous standard solutions of $FeSO_4 \cdot 7H_2O$ (100–1000 μM) were used to generate a calibration curve. The final results were expressed as millimole $FeSO_4 \cdot 7H_2O$ equivalents per gram of collagen peptides (mM $FeSO_4 \cdot 7H_2O$/g, collagen peptides).

3.6.4. In Vitro H_2O_2-Induced Oxidative Damage Model

Cell Lines and Culture

C518 cells were purchased from Saiqi Biological Engineering Co., Ltd. (Shanghai, China). C518 cells were cultured and maintained in DMEM supplemented with 10% FBS, 100 U/mL penicillin, and 100 μg/mL streptomycin at 37 °C under a humidified atmosphere containing 5% CO_2.

Protective Effects of Collagen Peptides against H_2O_2-Induced Oxidative Stress

C518 cells were seeded in 96-well plates (2×10^4 cells/mL, 100 μL) and incubated for 24 h. They were pretreated with various collagen peptide samples (1.0 mg/mL) and cultured for 30 min. The control was pretreated with DMEM under the same conditions. Thereafter C518 cells were exposed to H_2O_2 (1 mM) in the presence or absence of collagen peptides for 24 h. Finally, cell viability was determined using the CCK-8 assay.

3.7. *Anti-Inflammatory Activity Determination*

C518 cells were seeded in 96-well plates (2×10^4 cells/mL, 100 μL) and incubated for 24 h. They were pretreated with various collagen peptide samples (2.0 mg/mL) and cultured for 1 h. The control was pretreated with DMEM under the same conditions. Thereafter C518 cells were exposed to LPS (10 μg/mL) in the presence or absence of collagen peptides for 24 h. Finally, the abundance of IL-1β, PGE2, and TNF-α in cell-free supernatants was determined using ELISA kits according to the manufacturer's instructions.

3.8. *Spectrum-Effect Relationships*

The spectrum-effect relationships between HPLC fingerprints and antioxidant and anti-inflammatory effects were established by GRA. Grey system theory is an interdisciplinary scientific area that was first introduced in the early 1980s [35], and it has been applied in decision making in an extremely wide range of multiple-attribute group decision-making problems [36].

3.8.1. Data Preprocessing

In GRA, initial data preprocessing is performed to transform the original data sequences, with different measurement units, into comparable sequences. In this study, the mean value normalization preprocessing method was applied in the data series treatment to obtain a dimensionless matrix, because this method can easily accommodate a wide range of units among the factors used [37].

3.8.2. Definition of Reference Sequences and Comparison Sequences

The antioxidant and anti-inflammatory activities of the collagen peptides were utilized as reference sequences, and the characteristic peak areas obtained by HPLC fingerprints were utilized as comparison sequences. The original reference sequences and comparison sequences are represented by $\{X_o(k)\}$ and $\{X_i(k)\}$, $i = 1, 2, \ldots, m$; $k = 1, 2, \ldots, n$ respectively, where m is the number of experiments and n is the total number of observations of data.

3.8.3. Grey Relational Grade Calculation

The grey relational coefficient was calculated from the deviation sequence using the following relation:

$$\gamma\{X_o(k), X_i(k)\} = \frac{\Delta\text{min} + \xi\Delta\text{max}}{\Delta_{oi}\text{min}(k) + \xi\Delta\text{max}} 0 < \gamma\{X_o(k), X_i(k)\} \ll 1, \quad (3)$$

where $\Delta_{oi}(k)$ is the deviation sequence of the reference sequence $\{X_o(k)\}$ and comparability sequence $\{X_i(k)\}$, and ξ is the resolution coefficient, usually $\xi \in (0,1)$. The resolution coefficient is typically chosen to be 0.5. A grey relational grade is the weighted average of the grey relational coefficient and is defined as follows:

$$\gamma_{oi} = \frac{1}{\text{n}} \gamma_{oi}\{X_o(k), X_i(k)\}, \quad (4)$$

3.8.4. Grey Relational Coefficient Calculation and Ranking

The grey relational grade between the reference sequence and comparison sequences was calculated. The grey relational grade was proportional to the similarity of the developing trends (i.e., the greater the grade, the more similarity) [38].

3.9. *Purification and Identification of Collagen Peptides*

3.9.1. Purification of Collagen Peptides by RP-HPLC

The collagen peptide was dissolved in distilled water and loaded onto a semipreparative C18 RP-HPLC column (10.0 mm × 250 mm, 5 μm, Agilent Technologies, Santa Clara, CA, USA). The HPLC 1200 system (Agilent Technologies, USA) was equipped with a quaternary pump solvent delivery system and a diode-array detector (DAD). The sample injection volume and concentration were 250 μL and 10 mg/mL, respectively. The column was eluted by a linear gradient of acetonitrile (2–60%) containing 0.1% TFA at a flow rate of 1.0 mL/min. The UV absorbance of the eluent was monitored at 214 nm. This step was repeated several times, and the different elution fractions were pooled, concentrated, and lyophilized for sequence identification.

3.9.2. Identification of Collagen Peptides by UPLC-MS/MS Analysis

A Waters Acquity UPLC system (Waters, Milford, MA, USA) coupled with a Thermo Q Exactive mass spectrometer (Thermo Fisher Scientific, Bremen, Germany) was used for peptide separation and identification. The UPLC-MS/MS operating conditions were as follows: The column was an Eksigent C18 trap column (75 μm × 250 mm, 3 μm). A binary gradient elution system, comprising (A) 0.1% TFA/acetonitrile and (B) 0.1% TFA/water, was applied as follows: Initial, 90% B; 60 min, 40% B. The detection wavelengths were 214 and 280 nm, the flow rate was 3.0 μL/min, the freeze-dried sample was dissolved in 0.1% aqueous formic acid, and the injection volume was 1.0 μL. The mass spectrometer was fitted with an electrospray ionization (ESI) source used in the positive ion mode.

3.10. Statistical Analysis

All assays were carried out using at least three independent sets of experiments, and all results were expressed as average values with their corresponding standard deviations. One-way analysis of variance by Tukey's test was performed using SPSS 22.0 (IBM Co., Armonk, NY, USA).

4. Conclusions

In this work, HPLC fingerprints and a series of in vitro antioxidant and anti-inflammatory assays were combined to investigate the spectrum-effect relationship of collagen peptides. The results showed that peaks 2, 3, and 8 were the main components contributing to the antioxidant activity of the collagen peptides, whereas peaks 11 and 14 were the main components contributing to the anti-inflammatory activity of the collagen peptides. Subsequently, the amino acid sequences of peaks 8 and 14 were identified as GPRGPPGPVGP and VAIQAVLSLYASGR by UPLC-MS/MS. These identified peptides may have potential as drugs or functional foods for the treatment or prevention of OA. This report establishes a new platform for identifying the functional components of collagen peptides by the spectrum-effect relationship, which may lead to the development of new directions in the utilization of bioactive peptides in the future. However, further studies are required to investigate the cellular and molecular mechanism by which peptide pretreatment exerts its effects in the animal OA models.

Author Contributions: J.W. and D.L. designed the research; J.W., D.L., M.L., T.Z., X.Y. (Xiquan Yin) and Y.Z. performed the study and analyzed the data; J.W., X.Y. (Xiangliang Yang) and W.L. drafted and revised the manuscript; all authors approved the final version.

Acknowledgments: The authors thank the analytical and testing center of the Huazhong University of Science and Technology for their help in the testing of HPLC-MS/MS analysis.

References

1. Jeevithan, E.; Zhang, J.Y.; Wang, N.P.; He, L.; Bao, B.; Wu, W.H. Physico-chemical, antioxidant and intestinal absorption properties of whale shark type-II collagen based on its solubility with acid and pepsin. *Process. Biochem.* **2015**, *50*, 463–472. [CrossRef]

2. Bakilan, F.; Armagan, O.; Ozgen, M.; Tascioglu, F.; Bolluk, O.; Alatas, O. Effects of native type II collagen treatment on knee osteoarthritis: A Randomized Controlled Trial. *Eurasian J. Med.* **2016**, *48*, 95. [CrossRef] [PubMed]
3. Jeevithan, E.; Wu, W.H.; Wang, N.P.; He, L.; Bao, B. Isolation, purification and characterization of pepsin soluble collagen isolated from silvertip shark (*Carcharhinus albimarginatus*) skeletal and head bone. *Process. Biochem.* **2014**, *49*, 1767–1777. [CrossRef]
4. Cao, H.; Xu, S.Y. Purification and characterization of type II collagen from chick sternal cartilage. *Food Chem.* **2008**, *108*, 439–445. [CrossRef] [PubMed]
5. Lepetsos, P.; Papavassiliou, A.G. ROS/oxidative stress signaling in osteoarthritis. *BBA-Mol. Basis Dis.* **2016**, *1862*, 576–591. [CrossRef]
6. Wojdasiewicz, P.; Poniatowski, Ł.A.; Szukiewicz, D. The role of inflammatory and anti-Inflammatory cytokines in the pathogenesis of osteoarthritis. *Mediat. Inflamm.* **2014**, *2014*, 1–19. [CrossRef] [PubMed]
7. Zhuang, C.; Wang, Y.J.; Zhang, Y.K.; Xu, N.W. Oxidative stress in osteoarthritis and antioxidant effect of polysaccharide from *angelica sinensis*. *Int. J. Biol. Macromol.* **2018**, *115*, 281–286. [CrossRef] [PubMed]
8. Heijink, A.; Vanhees, M.; van den Ende, K.; van den Bekerom, M.P.; van Riet, R.P.; van Dijk, C.N.; Eygendaal, D. Biomechanical considerations in the pathogenesis of osteoarthritis of the elbow. *Knee Surg. Sport Tr. A* **2012**, *20*, 423–435. [CrossRef] [PubMed]
9. Thaipong, K.; Boonprakob, U.; Crosby, K.; Cisneros-Zevallos, L.; Hawkins Byrne, D. Comparison of ABTS, DPPH, FRAP, and ORAC assays for estimating antioxidant activity from guava fruit extracts. *J. Food Compos. Anal.* **2006**, *19*, 669–675. [CrossRef]
10. Gulden, M.; Jess, A.; Kammann, J.; Maser, E.; Seibert, H. Cytotoxic potency of H_2O_2 in cell cultures: Impact of cell concentration and exposure time. *Free Radic. Biol. Med.* **2010**, *49*, 1298–1305. [CrossRef]
11. Shen, C.L.; Smith, B.J.; Lo, D.F.; Chyu, M.C.; Dunn, D.M.; Chen, C.H.; Kwun, I.S. Dietary polyphenols and mechanisms of osteoarthritis. *J. Nutr. Biochem.* **2012**, *23*, 1367–1377. [CrossRef] [PubMed]
12. Lee, A.S.; Ellman, M.B.; Yan, D.Y.; Kroin, J.S.; Cole, B.J.; van Wijnen, A.J.; Im, H.-J. A current review of molecular mechanisms regarding osteoarthritis and pain. *Gene* **2013**, *527*, 440–447. [CrossRef]
13. Gu, Y.T.; Chen, J.; Meng, Z.L.; Ge, W.Y.; Bian, Y.Y.; Cheng, S.W.; Xing, C.K.; Yao, J.L.; Fu, J.; Peng, L. Research progress on osteoarthritis treatment mechanisms. *Biomed. Pharmacother.* **2017**, *93*, 1246–1252. [CrossRef] [PubMed]
14. Qin, J.; Shang, L.; Ping, A.S.; Li, J.; Li, X.J.; Yu, H.; Magdalou, J.; Chen, L.B.; Wang, H. TNF/TNFR signal transduction pathway-mediated anti-apoptosis and anti-inflammatory effects of sodium ferulate on IL-1β-induced rat osteoarthritis chondrocytes in vitro. *Arthritis Res. Ther.* **2012**, *14*, R242. [CrossRef] [PubMed]
15. Roufik, S.; Gauthier, S.F.; Turgeon, S.L. In vitro digestibility of bioactive peptides derived from bovine β-lactoglobulin. *Int. Dairy J.* **2006**, *16*, 294–302. [CrossRef]
16. Matsui, R.; Honda, R.; Kanome, M.; Hagiwara, A.; Matsuda, Y.; Togitani, T.; Ikemoto, N.; Terashima, M. Designing antioxidant peptides based on the antioxidant properties of the amino acid side-chains. *Food Chem.* **2018**, *245*, 750–755. [CrossRef]
17. Chi, C.F.; Hu, F.Y.; Wang, B.; Li, Z.R.; Luo, H.Y. Influence of amino acid compositions and peptide profiles on antioxidant capacities of two protein hydrolysates from skipjack tuna (*Katsuwonus pelamis*) dark muscle. *Mar. Drugs* **2015**, *13*, 2580–2601. [CrossRef]
18. Sarmadi, B.H.; Ismail, A. Antioxidative peptides from food proteins: A review. *Peptides* **2010**, *31*, 1949–1956. [CrossRef]
19. Jin, J.; Ma, H.L.; Wang, B.; Aelg, Y.; Wang, K.; He, R.; Zhou, C. Effects and mechanism of dual-frequency power ultrasound on the molecular weight distribution of corn gluten meal hydrolysates. *Ultrason. Sonochem.* **2016**, *30*, 44–51. [CrossRef]
20. Yang, B.; Yang, H.S.; Li, J.; Li, Z.X.; Jiang, Y.M. Amino acid composition, molecular weight distribution and antioxidant activity of protein hydrolysates of soy sauce lees. *Food Chem.* **2010**, *124*, 551–555. [CrossRef]
21. Lin, X.L.; Yang, L.; Wang, M.; Zhang, T.; Liang, M.; Yuan, E.D.; Ren, J.Y. Preparation, purification and identification of cadmium-induced osteoporosis-protective peptides from chicken sternal cartilage. *J. Funct. Foods* **2018**, *51*, 130–141. [CrossRef]
22. Xie, Y.; Pang, X.; Zhao, R.B. Isolation and identification of type-II collagen peptide from chicken sternal cartilage. *Food Sci.* **2011**, *32*, 26–29.

23. Cao, H.; Xu, S.Y.; Ge, H.S.; Xu, F. Molecular characterisation of type II collagen from chick sternal cartilage and its anti-rheumatoid arthritis activity. *Food Agr. Immunol.* **2014**, *25*, 119–136. [CrossRef]
24. Maux, S.L.; Nongonierma, A.B.; Fitzgerald, R.J. Improved short peptide identification using HILIC-MS/MS: Retention time prediction model based on the impact of amino acid position in the peptide sequence. *Food Chem.* **2015**, *173*, 847–854. [CrossRef] [PubMed]
25. Zhu, C.Z.; Zhang, W.G.; Zhou, G.H.; Xu, X.L.; Kang, Z.L.; Yin, Y. Isolation and identification of antioxidant peptides from jinhua ham. *J. Agric. Food Chem.* **2013**, *61*, 1265–1271. [CrossRef] [PubMed]
26. Xu, C.J.; Ma, B. Software for computational peptide identification from MS-MS data. *Drug Discov. Today* **2006**, *11*, 595–600. [CrossRef]
27. Lam, H.; Deutsch, E.W.; Eddes, J.S.; Eng, J.K.; King, N.; Stein, S.E.; Aebersold, R. Development and validation of a spectral library searching method for peptide identification from MS/MS. *Proteomics* **2007**, *7*, 655–667. [CrossRef]
28. Weatherly, D.B.; Atwood, J.A.; Minning, T.A.; Cavola, C.; Tarleton, R.L.; Orlando, R. A heuristic method for assigning a false-discovery rate for protein identifications from Mascot database search results. *Mol. Cell. Proteom.* **2005**, *4*, 762–772. [CrossRef]
29. Wang, H. Zaocys type II collagen regulated the balance of Treg/Th17 cells in mice with collagen-induced arthritis. *J. South Med. Univ.* **2014**. [CrossRef]
30. Meng, J.M.; Jiang, H.M.; Lu, J.Q.; Zhou, K.; Xiao, Y.S. Preliminary Study on HPLC Fingerprint of *Thlaspi arvense* L. *Med. Plant* **2017**, *8*, 15–19. [CrossRef]
31. Sun, W.Z.; Zhao, H.F.; Zhao, Q.Z.; Zhao, M.M.; Yang, B.; Wu, N.; Qian, Y.L. Structural characteristics of peptides extracted from Cantonese sausage during drying and their antioxidant activities. *Innov. Food Sci. Emerg. Technol.* **2009**, *10*, 558–563. [CrossRef]
32. Arnao, M.B.; Cano, A.; Acosta, M. The hydrophilic and lipophilic contribution to total antioxidant activity. *Food Chem.* **2001**, *73*, 239–244. [CrossRef]
33. Brand-Williams, W.; Cuvelier, M.E.; Berset, C. Use of a free radical method to evaluate antioxidant activity. *LWT-Food Sci. Technol.* **1995**, *28*, 25–30. [CrossRef]
34. Benzie, I.F.; Strain, J.J. The ferric reducing ability of plasma (FRAP) as a measure of "antioxidant power": The FRAP assay. *Anal. Biochem.* **1996**, *239*, 70–76. [CrossRef] [PubMed]
35. Deng, J.L. Control problems of grey systems. *Syst. Control Lett.* **1982**, *1*, 288–294. [CrossRef]
36. Wu, H.H. A comparative study of using grey relational analysis in multiple attribute decision making problems. *Qual. Eng.* **2002**, *15*, 209–217. [CrossRef]
37. Kadier, A.; Abdeshahian, P.; Simayi, Y.; Ismail, M.; Hamid, A.A.; Kalil, M.S. Grey relational analysis for comparative assessment of different cathode materials in microbial electrolysis cells. *Energy* **2015**, *90*, 1556–1562. [CrossRef]
38. Kuo, Y.Y.; Yang, T.; Huang, G.W. The use of grey relational analysis in solving multiple attribute decision-making problems. *Comput. Ind. Eng.* **2008**, *55*, 80–93. [CrossRef]

5

Simultaneous Identification and Dynamic Analysis of Saccharides during Steam Processing of Rhizomes of *Polygonatum cyrtonema* by HPLC–QTOF–MS/MS

Jian Jin [1,2,3], Jia Lao [4], Rongrong Zhou [5], Wei He [4], You Qin [1,2], Can Zhong [1,3], Jing Xie [1,3], Hao Liu [1,3], Dan Wan [1,3], Shuihan Zhang [1,3,*] and Yuhui Qin [1,2,*]

1 Institute of Chinese Materia Medica, Hunan Academy of Chinese Medicine, Changsha 410013, China; jinjian2016@163.com (J.J.); 20162010@stu.hnucm.edu.cn (Y.Q.); canzhong651@163.com (C.Z.); axxj2057@163.com (J.X.); 350013@hnucm.edu.cn (H.L.); 350017@hnucm.edu.cn (D.W.)
2 School of Chinese Medicine, Hunan University of Chinese Medicine, Changsha 410208, China
3 2011 Collaboration and Innovation Center for Digital Chinese Medicine in Hunan, Changsha 410208, China
4 Resgreen Group International Inc., Changsha 410329, China; laojia1973@163.com (J.L.); hewei3218@126.com (W.H.)
5 College of Pharmacy, Changchun University of Chinese Medicine, Changchun 130117, China; rz172@georgetown.edu
* Correspondence: zhangshuihan0220@126.com (S.Z.); dlqyh@sohu.com (Y.Q.);

Abstract: The sweet rhizomes of *Polygonatum cyrtonema* are widely used as a tonic and functional food. A sensitive and rapid analytical method was developed for simultaneous identification and dynamic analysis of saccharides during steam processing in *P. cyrtonema* using HPLC–QTOF–MS/MS. Fructose, sorbitol, glucose, galactose, sucrose, and 1-kestose were identified, as well as a large number of oligosaccharides constituted of fructose units through β-(2→1) or β-(2→6). Polysaccharides and oligosaccharides were decomposed to monosaccharides during a steaming process, since the contents of glucose, galactose, and fructose were increased, while those of sucrose, 1-kestose, and polysaccharides were decreased. The high content of fructose was revealed to be the main determinant for increasing the level of sweetness after steaming. The samples of different repeated steaming times were shown to be well grouped and gradually shift along the PC1 (72.4%) axis by principal component analysis. The small-molecule saccharides, especially fructose, could be considered as markers for the steaming process of rhizomes of *P. cyrtonema*.

Keywords: *Polygonatum cyrtonema*; saccharides; oligosaccharides; fructose; HPLC–QTOF–MS/MS; steaming

1. Introduction

The rhizomes of *Polygonatum cyrtonema* are wildly used as a tonic and functional food in China. The use of this plant is documented in the well-known ancient Pharmacopoeia "*Shennong Bencao Jing*", and it is considered as a "Top grade" herb. Its main efficacy includes replenishing energy, strengthening immunity, and treating fatigue, weakness, loss of appetite, and so on. The genus *Polygonatum* has also been used in some other regions, such as India, Japan, Europe, and North America [1].

There are various ways to cook the rhizomes of *Polygonatum*. They are often cooked with meats or porridges, made into tea or medicated wine, or consumed as fruits or vegetables [2]. It is worth noting that raw rhizomes without processing treatment are rarely directly used. Raw rhizomes are often processed to enhance their tonic function by repeated steaming and drying. In particular, according to its traditional uses, the rhizomes of *Polygonatum* are processed by steaming nine times, which is recently combined with autoclave as an efficient approach. Traditional quality control is performed by qualitative index, described as changes such as rhizomes becoming black, soft, and especially, gaining a sweet taste. However, the changes of the chemical composition during the steaming process are not understood.

With respect to chemical composition, there could be steroidal saponins, triterpenoid saponins, homoisoflavanones, polysaccharides, and lectins in the *Polygonatum* plants [1]. Polysaccharides are the main chemical components, and possess various pharmacological actions, such as antioxidant protection and immune-regulation [3]. The quality of *P. cyrtonema* is evaluated depending on the content of polysaccharides, with a minimal content of 7% required by Chinese Pharmacopoeia. Polysaccharides from *Polygonatum* plants are composed of different ratios of monosaccharides, mainly including mannose, galactose, glucose, fructose, rhamnose, arabinose, and galacturonic acid [1]. A branched fructan was also isolated from *P. cyrtonema* [4]. A large number of monosaccharides, oligosaccharides, and polysaccharides were found in rhizomes of *Polygonatum* plants, making it worthwhile to try to evaluate the processed products of rhizomes of *P. cyrtonema* with the help of saccharides in order to clarify the reason why the processed rhizomes of *Polygonatum* gradually take on a sweet taste.

Commonly-used analytical modes of saccharides include nuclear magnetic resonance (NMR) [5], gas chromatography (GC) [6], and liquid chromatography (LC) [7,8]. LC is widely used thanks to its various detectors, while NMR cannot be used to detect trace amounts of saccharides, and GC requires tedious derivatization [9]. However, because the absence of chromophore and fluorophore groups, avoiding direct detection by ultraviolet and fluorescence or diode array detectors, pre-column derivatization is often employed in LC, which is time consuming and could introduce extra variations into the final data [10]. LC equipped with an evaporative light scattering detector (ELSD) could be used to detect the monosaccharides; for instance, glucose, fructose, and sucrose were examined in the raw materials of *agave* leaf [11]. Structural information can even be provided by well-developed analytical techniques; for example, mass spectrometry (MS) coupled with LC is currently a more powerful analytical tool for saccharide analysis [10]. Sucrose, fructose, and glucose in plant tissues can be unambiguously determined by LC–MS [10]. Three major tri–saccharides in wheat flour were also determined by LC–MS [12]. It is worth mentioning that most previous study focused on fructose, glucose, sucrose and other oligosaccharides, but seldom distinguished glucose and galactose, which are the most common isomeric hexoses present in physiology. LC coupled with atmospheric pressure chemical ionization (APCI)–MS was developed to identify most of the standard saccharides, but it is still hard to separate glucose and galactose [13]. It seems that no simple method is available for the differentiation of these two isomers, since current methods required either derivatization techniques based on the reaction of reducing sugars with 1-phenyl-3-methyl-5-pyrazolone, phenylhydrazones [14], or the help of a zinc diethylenetriamine (Zndien) metal-ligand system [15].

In *Polygonatum* plants, polyphenolic antioxidants have ever been characterized by high-performance liquid chromatography (HPLC)–quadrupole-time-of-flight (QTOF)–MS, a useful technology for structural identification, even when standard compounds are not available [16]. Nevertheless, few studies of simultaneous analysis of saccharides in rhizomes of *Polygonatum* were reported, let alone characterization of the changes of the saccharide composition during the nine times of repeated steaming process. In the present study, an efficient and sensitive method was developed by HPLC–QTOF–MS/MS to profile and identify saccharides, including glucose and galactose, in fresh and processed rhizomes of *P. cyrtonema* by nine-time steaming. The present study will provide not only a useful tool for the analysis of saccharides, but also information for the improvement of *Polygonatum* steam process.

2. Results and Discussion

2.1. Appearance Change

The rhizomes of *P. cyrtonema* were processed by repeated steaming for one to nine times, shown in Figure 1.

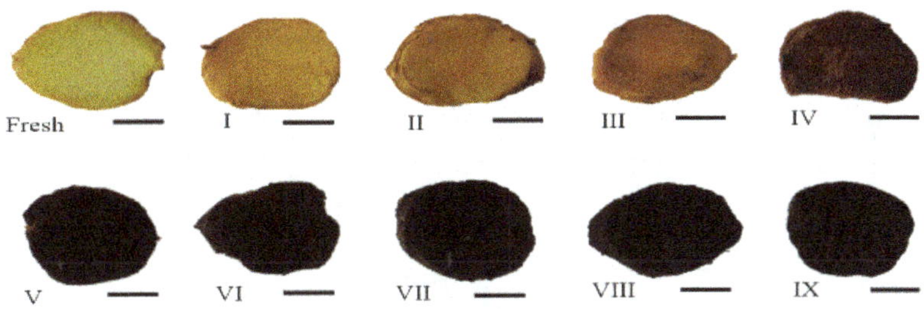

Figure 1. Samples of the rhizomes of *P. cyrtonema* are processed by different times of repeated steaming; "fresh" implies without steam process; roman numerals indicate the number of times of steaming; bar, 1 cm.

Steaming treatment is a usual food processing technology for the pretreatment of lignocellulosic materials [17]. In recent decades, an autoclaving steam process has emerged as an industrially-adopted method, not only to deliver a wide range of high quality products to meet the traditional Chinese requirements, but also to extend shelf life as a technique of food pasteurization to kill food pathogens and ensure food safety. Heavy metals in *Polygonatum* roots, such as Pb and Cd, could also be eliminated by exposure to high temperature, pressurized steaming conditions [18]. The well-known medicinal herb black ginseng is also usually treated with steaming; ginsenoside composition changes during the steaming process [19,20]. Traditionally, rhizomes of *Polygonatum* were used mainly as a crude drug, while in recent years, they are more popular as food when cooked with meats or porridges, made into tea or medicated wine, or consumed as fruits or vegetables [2]. The color and taste are two of the most crucial sensory properties of food. From fresh to processed samples with different numbers of repeated steaming, the samples changed to black, and became sweeter. But after steaming four times, the color appeared stable. The Maillard reaction, during which sugars react with amino acids under thermal conditions [21], would be largely responsible for the dark-colored appearance of processed rhizomes of *P. cyrtonema*. For the sweet taste, the present study was designed to systematically assess the saccharides, especially the small-molecular saccharides, between fresh and processed rhizomes of *P. cyrtonema*.

2.2. Optimization and Validation of the HPLC–QTOF–MS/MS Method

For simultaneous determination of saccharides and their derivatives, the optimal chromatographic conditions were investigated. Various mixtures of water, methanol, and acetonitrile at flow rate 1.2 mL/min, 1.0 mL/min, 0.8 mL/min, 0.5 mL/min and 0.3 mL/min were tested as the mobile phase. The detection temperature was tested at 15, 25, 35, and 45 °C. Eventually, the optimal chromatographic conditions were as follows: Prevail Carbohydrate column, with acetonitrile 75% and ultrapure water 25% as mobile phase at 1.0 mL/min at 25 °C. As shown in Figure 2A, the peaks in fresh and processed samples are separated by HPLC–QTOF–MS/MS. This condition was suitable for most of the small molecular standards (Figure 2B), except for glucose and galactose, which appeared at the same retention time 11.60 min. Surprisingly, when mobile phase flow rate decreased to 0.3 mL/min, only the standard glucose appeared at retention time 39.56 min, while no peak of the galactose appeared using sole standard solutions (Figure 2C). Then, the quantitative determination of glucose in samples was carried out with the mobile phase flow rate at 0.3 mL/min for the peak at retention time 39.56 min. The content of galactose was calculated by subtracting the glucose from the peak at retention time 11.60 min with mobile phase flow rate at 1.0 mL/min.

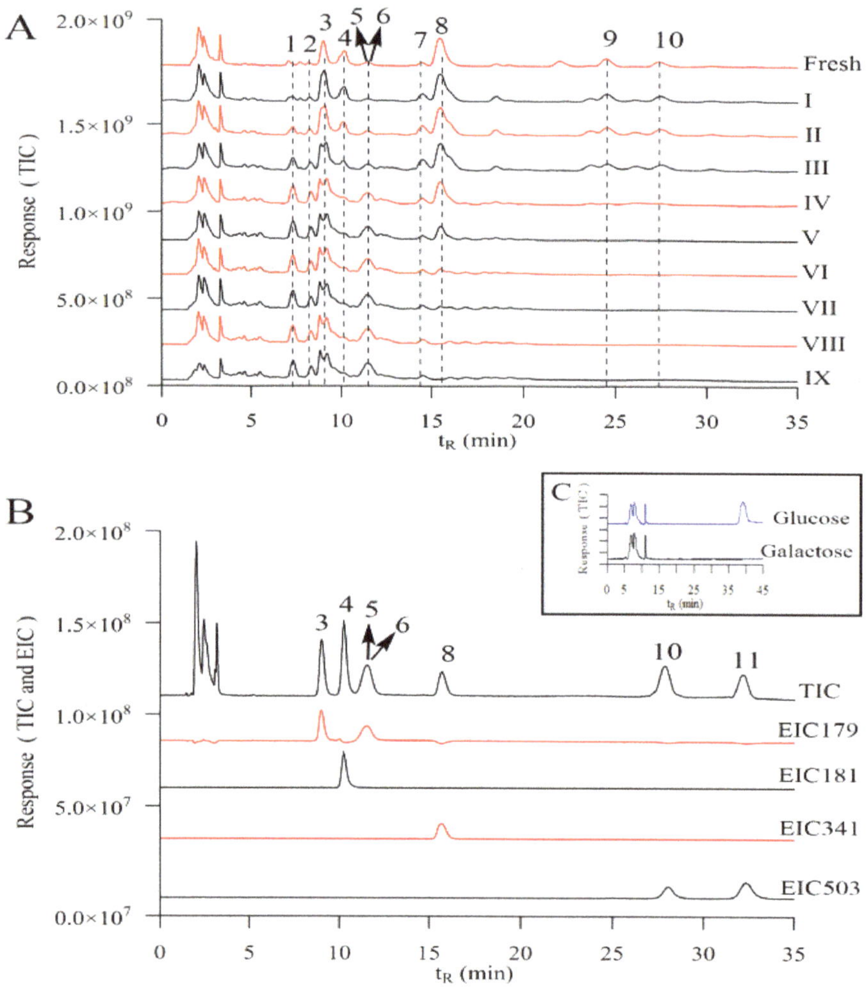

Figure 2. Chromatogram of the extract of the rhizomes of *P. cyrtonema* are processed by different times of repeated steaming (**A**), standards of small–molecular saccharides (**B**) by HPLC–QTOF–MS/MS at solvent flow rate 1.0 mL/min, glucose and galactose at solvent flow rate 0.3 mL/min (**C**); 1, fructose derivative; 2, glucose derivative; 3, fructose; 4, sorbitol; 5, glucose; 6, galactose; 7, hexose; 8, sucrose; 9, trisaccharide; 10, 1-kestose; 11, raffinose; TIC, total ion chromatogram; EIC, extracted ion chromatogram; "fresh" means without steaming; roman numerals indicate the number of times of steaming.

Noting that the peak of TIC and EIC will be saturated and even lead to bifurcation when the standard concentration is too high (Figure S1), the standard stock solutions containing fructose, glucose, galactose, sucrose and 1-kestose were prepared and diluted to a series of appropriate concentrations for the construction of calibration curves. These curves showed good linearity, and the correlation coefficients were found to be in the range with R^2 0.99 for all of the compounds in a certain concentration range (Table 1), consistent with saccharides determined by hydrophilic interaction liquid chromatography (HILIC)–TOF–MS technology [7]. The limits of detection (LOD) and limits of quantification (LOQ) were in the range of 0.41–3.63 μg/mL and 1.33–12.10 μg/mL, respectively. The intraday and interday variations of the analytes (RSDs) were within 0.93–1.91% and 1.82–3.51%, respectively. The RSDs for stability were lower than 3.01%. The average recoveries of the standard compounds ranged from 94.3% to 107.5%, and their RSDs were within 6.39% (Table 2). All these results demonstrated that the developed quantitative HPLC–QTOF–MS method was linear, precise, stable, sensitive, and accurate enough for the determination of small-molecular saccharides in rhizomes of *P. cyrtonema*. There are studies using HPLC system coupled with MS to investigate saccharides in plant tissues or products, but previous reports mainly investigated glucose, fructose and sucrose [10,22]. LC–APCI–MS was developed to distinguish saccharides with different types of columns, but it is still hard to separate glucose and galactose [13]. To the best of our knowledge, this is the first time that small-molecule saccharides, including glucose and galactose, were separated and determined by the highly sensitive and rapid HPLC–QTOF–MS/MS method using a common column without derivatization.

Table 1. Linear regression, LOD and LOQ of investigated compounds.

Compounds	Range (μg/mL)	Regression Equation [a]	R^2	LOD [b] (μg/mL)	LOQ [c] (μg/mL)
Fructose [d]	6.4–64	y = 71886x + 682691	0.996	0.48	1.60
Glucose [d]	15.2–152	y = 238502x + 2043001	0.991	0.90	2.99
Glucose [e]	15.2–152	y = 457396x + 2818970	0.992	0.87	2.89
Galactose [d]	15.7–157	y = 65662x + 479229	0.992	3.63	12.10
Sucrose [d]	7.1–71	y = 296456x + 1182954	0.994	0.41	1.33
1-Kestose [d]	73–366	y = 71886x + 922640	0.995	0.92	3.06

[a] y: peak area, x: concentration of the analyte (μg/mL); [b] LOD, limit of detection; [c] LOQ, limit of quantification; [d] solvent flow rate at 1.0 mL/min; [e] solvent flow rate at 0.3 mL/min.

Table 2. Precision, stability and recovery of investigated compounds.

Compounds	Precision (n = 6)		Stability (48 h) (RSD, %)	Recovery (n = 3)	
	Intra-Day (RSD, %)	Inter-Day (RSD, %)		Mean (%)	RSD (%)
Fructose [a]	1.37	1.82	2.29	96.7	2.37
Glucose [a]	1.24	2.56	3.01	94.3	2.46
Glucose [b]	1.62	2.29	1.48	98.3	2.13
Galactose [a]	1.91	3.51	2.12	96.7	3.13
Sucrose [a]	1.84	3.27	2.14	107.5	6.39
1-Kestose [a]	0.93	2.22	1.29	96.1	3.71

[a] solvent flow rate at 1.0 mL/min; [b] solvent flow rate at 0.3 mL/min.

2.3. *Identification of Small-Molecule Saccharides*

Since the chromatographic peaks could not be identified unambiguously by HPLC retention time alone, HPLC–QTOF–MS/MS was used to confirm identities by comparing the retention time and molecular ions for each peak. In this experiment, the mass spectral conditions were optimized in negative-ion mode, and the six compounds exhibiting distinct quasi-molecular ions $[M-H]^-$ and $[M+COOH]^-$ were optimized in this mode: fructose (peak 3), molecular weight (MW) 180 m/z = 179 $[M-H]^-$, sorbitol (peak 4) (MW 182) m/z = 181 $[M-H]^-$, glucose (peak 5) (MW 180) m/z = 179 $[M-H]^-$, galactose (peak 6) (MW 180) m/z = 179 $[M-H]^-$, sucrose (peak 8) (MW 342) m/z = 341 $[M-H]^-$ and m/z = 387 $[M+COOH]^-$, 1-kestose (peak 10) (MW 504) m/z = 503 $[M-H]^-$ and m/z = 549 $[M+COOH]^-$ and raffinose (peak 11) (MW 504) m/z = 503 $[M-H]^-$ and m/z = 549 $[M+COOH]^-$. For the monosaccharide, the fragment ions demonstrated the same m/z value at 179, 161, 143, 113 and 101. For the disaccharide sucrose, the fragment ions are at m/z 341, 179, 161, 143, 119 and 101. For the trisaccharides, i.e., 1-kestose and raffinose, the fragment ions are at m/z 503, 341, 323, 221, 179, 161, 143, 113 and 101 (Table 3).

Table 3. Identification of small–molecule saccharides by HPLC–QTOF–MS/MS technology

No	t_R (min)	$[M-H]^-$ (m/z) (Δppm)	Fragment Ions (m/z)	Molecular Formula	Compound
1	7.09	323.1003 (−4.98)	323.0990; 179.0564; 161.0461; 143.0364; 113.0258; 101.0251	$C_{12}H_{20}O_{10}$	Hexose derivative
2	8.16	323.1010 (−6.13)	323.1011; 179.0567; 161.0464; 143.0386; 113.0230; 101.0267	$C_{12}H_{20}O_{10}$	Hexose derivative
3	9.12	179.0581 (−0.58)	179.0533; 161.0458; 143.0354; 113.0242; 101.0245	$C_6H_{12}O_6$	Fructose
4	10.43	181.0739 (−0.36)	181.0715; 163.0629; 149.0436; 119.0338; 101.0247	$C_6H_{14}O_6$	Sorbitol
5 [a]	11.60	179.0566 (−2.94)	179.0583; 161.0433; 143.0350; 113.0248; 101.0248	$C_6H_{12}O_6$	Glucose
6 [a]	11.60	179.0585 (−0.08)	179.0574; 161.0480; 143.0336; 112.9853; 101.0246	$C_6H_{12}O_6$	Galactose
7	14.4	179.0587 (−0.65)	179.0583; 161.0451; 143.0378; 112.9846; 101.0241	$C_6H_{12}O_6$	Hexose
8	15.24	341.1125 (−1.14)	341.1090; 179.0555; 161.0465; 143.0338; 119.0350; 101.0243	$C_{12}H_{22}O_{11}$	Sucrose
9	23.71	503.1666 (−1.1)	503.1668; 341.1101; 323.0988; 221.0670; 179.0554; 161.0441; 143.0328; 113.0220; 101.0240	$C_{18}H_{32}O_{16}$	Trisaccharide
10	26.49	503.1670 (−4.19)	503.1668; 341.1054; 323.0986; 221.0655; 179.0553; 161.0467; 143.0368; 113.0255; 101.0244	$C_{18}H_{32}O_{16}$	1-Kestose
11	33.15	503.1653 (2.62)	503.1653; 341.1031; 323.0936; 221.0616; 179.0513; 161.0412; 143.0308; 113.0216; 101.0219	$C_{18}H_{32}O_{16}$	Raffinose

[a] Characterization with sole standard glucose or galactose.

In the HPLC–QTOF–MS/MS spectra, the molecular ions of each compound agreed well with the chemical structures. Basing on the above results, the standard compounds of fructose, sorbitol, glucose, galactose, sucrose, and 1-kestose were identified in the extract of the rhizomes of *P. cyrtonema*, and the identities of each peak were clearly confirmed. However, raffinose, a common oligosaccharide in food [23], was not found in fresh or processed samples.

It is worth noting that there were also obvious peaks 1, 2, 7, and 9 in the rhizomes of *P. cyrtonema*. Peak 1 and 2 were shown to be due to the same fragment ions to monosaccharide with m/z value at 179, 161, 143, 113, and 101, indicating a component of hexose. The fragment ion at m/z 323 could be due to $[341-H_2O]^-$, where 341 is the characterized ion of disaccharide. Then, peak 1 and 2 could be hexose derivatives formed by dehydration from disaccharides. The m/z values of peak 7 fragment ions were at 179, 161, 143, 113, and 101, the same with that of monosaccharide, indicating a hexose structure. Peak 9 demonstrated the same fragment ions with those of 1-kestose and raffinose, with *m/z* value at 503, 341, 323, 221, 179, 161, 143, 113, and 101, also indicating a trisaccharide. 1-kestose is formed of two molecules of fructose linked by β-(2→1) D-fructosyl-fructose bonds, and terminated with one glucose unit [24]. Noting that for the retention time, the peak of fructose is 2.48 min before glucose, peak 9 could be a trisaccharide composed of three molecules of fructose, because the retention time of peak 9 was 2.78 min before that of 1-kestose. There could be two possibilities for the linkage of this identified trisaccharide. One is that the molecules of fructose are linked by β-(2→1) and β-(2→6) as mixed type F3 fructan in inulin [25], namely MF3 as is shown in Figure 3A peak 9-1. The other could be simpler that a molecule of fructose replaces the glucose in 1-kestose, keeping the straightforward structure, namely SF3 (Figure 3A peak 9-2).

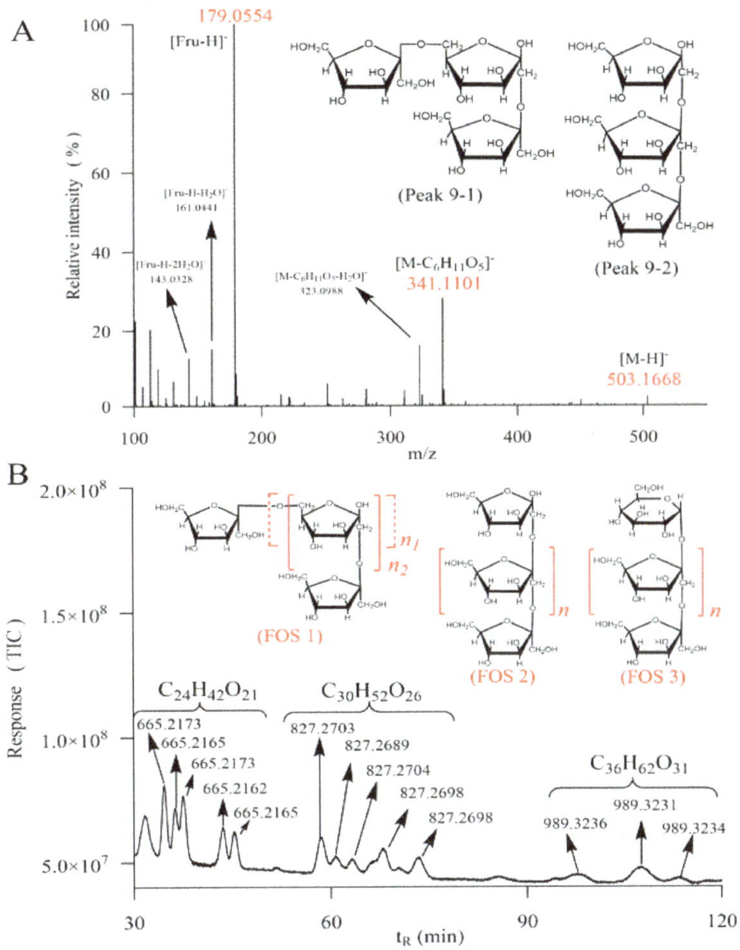

Figure 3. MS/MS spectrum, the two speculated structures of peak 9 (**A**) and LC spectrum with *m/z* values, speculated molecule formulas and structures in fresh rhizomes of *P. cyrtonema* by HPLC–QTOF–MS/MS with mobile phase flow rate at 1.2 mL/min (**B**). FOS, fructo-oligosaccharides.

In order to further characterize the fructo-oligosaccharides (FOS), the mobile phase increased to 1.2 mL/min for 120 min to overview the outcome. It could be speculated that the peaks appearing at *m/z* value 665, 827, and 989 could have molecular formulas $C_{24}H_{42}O_{21}$ (tetrasaccharide), $C_{30}H_{52}O_{26}$ (pentasaccharide) and $C_{36}H_{62}O_{31}$ (hexasaccharide) (Figure 3B). These FOS should be derived from the trisaccharides, i.e., 1-kestose, MF3 and SF3, by lengthening the chains with adding a variable number of fructose units through β-(2→1) or β-(2→6), as is shown in Figure 3B FOS 1, FOS 2 and FOS 3. Fibers of *P. odoratum* have been found to be composed of arabinose, xylose, sorbose, mannose and galactose [26] and lectin in *P. cyrtonema* could be binding to mannose [27]. Another monosaccharide analysis study showed that all the complete hydrolytic polysaccharides of *P. cyrtonema* are determined to be heteropolysaccharides containing rhamnose, arabinose, xylose, mannose, galactose and glucose [28].

However, the present investigation found that oligosaccharides in *P. cyrtonema* were mainly made up of fructose and glucose. The speculated structures of FOS in the present work are consistent with those of the previous study, in which the neutral polysaccharide in *P. cyrtonema* was found to be a branched fructan, composed of (2→6)-linked β-D-fructofuranosyl residues and (2→1)-linked β-D-fructofuranosyl residues in the backbone [4]. Since the intake of prebiotic fructans could sustain health and overall well-being [29], the observation of a large number of FOS in *P. cyrtonema* could be an evidence to explain its functional benefits.

2.4. Dynamic Change of Saccharides

The contents of fructose, glucose, galactose, sucrose, and 1-kestose in three samples treated with different times of individual processing steps were measured by the developed method, as well as the total small-molecule saccharides and polysaccharides. Even though the saccharides in different *Polygonatum* samples usually varied [30], the present results demonstrated that there was an obvious influence on the quantities of small-molecule saccharides between fresh and steam processed samples. The dynamic changes and proposed model are illustrated in Figure 4.

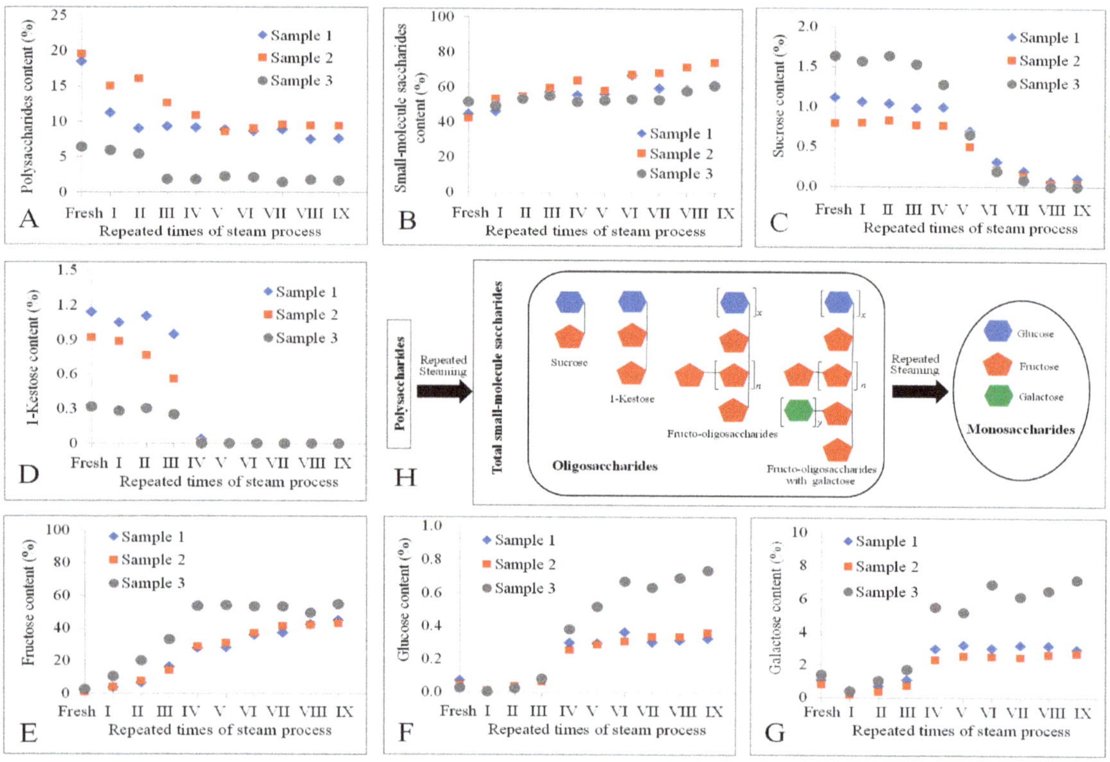

Figure 4. The content of polysaccharides (**A**), total small–molecule saccharides (**B**), sucrose (**C**), 1-kestose (**D**), fructose (**E**), glucose (**F**), galactose (**G**), in the rhizomes of *P. cyrtonema* processed by different times of repeated steaming and the proposed model of saccharides change during repeated steaming (**H**); "fresh" means without steam process; roman numerals indicate the number of times of steaming.

The content of polysaccharides decreased, while the total small-molecule saccharides increased, indicating the polysaccharides were decomposed to small-molecule saccharides with the repeated times of steam process (Figure 4A,B). The measured oligosaccharides sucrose and 1-kestose were broke down forming glucose and fructose, because sucrose and 1-kestose were decreased (Figure 4C,D), whereas glucose and fructose increased (Figure 4E,F). Unexpectedly, the content of galactose also increased (Figure 4G), indicating this monosaccharide could be a branch linked to FOS in *P. cyrtonema*, as proposed in the model Figure 4H. Interestingly, steaming four times seems to be a landmark for these small-molecule saccharides, since their content appeared to be relatively stable after that. All these dynamic changes of polysaccharides and small-molecule saccharides caused by the decomposition of the glycosidic linkages, which could be broken by steam processing [26], resulting in an increase in the contents of glucose, galactose, and fructose, and a decrease in oligosaccharide and polysaccharides. It is reported that the fructose in Radix Rehmanniae gradually decreased, and the glucose remained relatively stable after steaming [9]. These could be species dependent, since the glycosidic linkages might be different in different kinds of plants.

Surprisingly, the content of fructose gradually increased to an extremely high value of about 50% of dry sample weight, while that of most food is less than 10% [31]. Only in a few foods, for instance honey, fructose, and glucose together account for 85–95%, and fructose usually predominates, which is responsible for the sweetness [32]. The change of fructose could be the main reason why the taste of the rhizomes of *P. cyrtonema* became sweeter after the steaming processing, since fructose is a major determinant of sweetness in food [10]. The high content of fructose is also in accordance with our speculation that oligosaccharides were mainly composed of fructose. It is reported that fructans isolated from the rhizomes of *P. odoratum* could be composed of 29 units of fructose and 1 unit of glucose [33]. A neutral polysaccharide with an average degree of polymerization of 28 from *P. cyrtonema* Hua is also mainly made up of fructose [4]. Therefore, it could be concluded that the high concentration of fructose probably came from the decomposition of fructo-oligosaccharides and polysaccharides.

2.5. PCA Statistical Analysis

To overview the difference between fresh and processed rhizomes of *P. cyrtonema*, unsupervised PCA was performed. The contents of monosaccharides, oligosaccharides, identified saccharides derivatives, as well as polysaccharides, in all the samples of fresh and processed rhizomes were considered as the variables of PCA. The PCA biplot displayed the scores and loadings of the first two components (PC1 and PC2), revealing the projection of an observation on the subspace with score points (Figure 5). 72.4% and 15.8% of the variation in the pattern of concentrations of identified saccharides and derivatives were explained by PC1 and PC2, respectively. These two components together explained 88.2% of the variation.

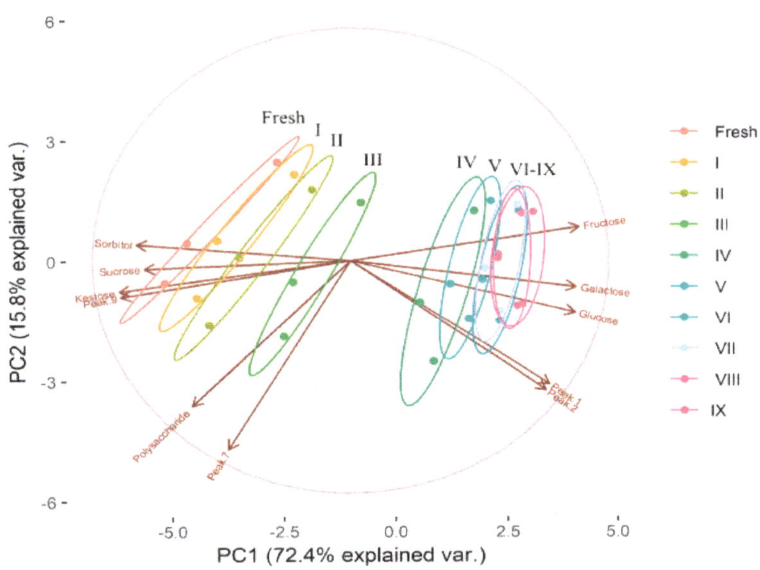

Figure 5. A biplot showing the samples of fresh and processed rhizomes of *P. cyrtonema* by different times of repeated steaming on a two-dimensional space derived from principal component analysis basing on identified saccharides and their derivatives; the arrows indicated the projections of the original features onto the principal components; "fresh" means without steam process; roman numerals indicate the number of times of steaming.

The variables of fructose, glucose, galactose, peak 1, and peak 2 shown in the same direction correlated along PC1, whereas sucrose, 1-kestose, sorbitol and peak 9 were oppositely correlated to PC1. Polysaccharides and peak 7 were primarily correlated with PC2. The samples gradually shifted along PC1 axis with the increasing times of repeated steaming. PCA results could substantially separate fresh and processed rhizomes by 1–4 times of repeated steaming, but failed to distinguish samples processed with 4–9 times steaming, as there were a large number of samples overlapping between these groups. This result was in consistent with the previous observation that the color and aforementioned compounds were relatively stable after steaming four times. To a certain degree, present PCA study was in accordance with previous report that the raw and steamed samples were separated into two groups by PCA in potato strips [34] and in Radix Rehmanniae samples [7,9]. To date, no compound has been used as a marker compound for processed *P. cyrtonema*. Then, these small–molecule saccharides might be considered as marker compounds. In particular, fructose, due to its high content, could be given special consideration as an important marker compound.

3. Materials and Methods

3.1. Chemicals and Reagents

Galactose (CAS 387116-33-2), 1-kestose (CAS 470-69-9) and D-raffinose (CAS 512-69-6) were obtained from National Institutes for Food and Drug Control (Beijing, China). D-fructose (CAS 57-48-7), D-glucose (CAS 50-99-7), D-sorbitol (CAS 50-70-4), and sucrose (CAS 57-50-1) were purchased from Hefei Bomei Biotechnology. Co., Ltd. (Hefei, China). Acetonitrile (HPLC–grade) was purchased from Merck (Darmstadt, Germany). Ultrapure water was purchased from Wahaha (Changsha, China). Other reagents were of analytical grade.

3.2. Preparation of Samples

Fresh rhizomes of *P. cyrtonema* were collected from Xinhua County, Hunan Province, China and authenticated by Professor Zhaoming Xie and Hao Liu from Hunan Academy of Chinese Medicine. The voucher specimens were kept in our department for future reference. Rhizomes of *P. cyrtonema* were cleaned, dried and then cut into thin slices (3 mm $\pm$ 1 mm) right after the harvest. The rhizomes in a glass bottle were processed by nine times repeated steaming with autoclave (121 °C, 0.12 MPa, 30 min); then, samples were kept in the autoclave until cooled to room temperature. Samples were collected after each cycle of steaming process.

3.3. Extraction of Saccharides

The steaming-processed rhizomes of *P. cyrtonema* were dried at 60 °C to a constant weight. Then, samples were ground by passing a 40-mesh screen and extracted by distilled water with ultrasonication (KM-500DB, 40 KHz, Kunshan Meimei Ultrasonic Instrument Co., Ltd., Jiangsu, China) for 30 min followed by maintenance at 90 °C for 60 min. After that, the extracts were centrifuged (3000$\times$ g for 15 min) to separate the supernatant from residues. Then, 10 mL of ethanol (95%) was added to 2 mL of the supernatant for precipitating the polysaccharides. The small-molecule saccharides, including monosaccharides and oligosaccharides, which were easier to dissolve in ethanol, were still dissolved in upper layer [35]. Then, the polysaccharides and small-molecule saccharides were separated by centrifugation (3000$\times$ g, 10 min). The precipitated polysaccharides were redissolved in distilled water for colorimetric quantitative measurement. The up layer was collected for colorimetric quantitative measurement of the total small-molecule saccharides and HPLC–QTOF–MS/MS analysis to identify monosaccharides and oligosaccharide compounds.

3.4. Spectrophotometric Quantitative Measurement

The concentration of total small-molecule saccharides and polysaccharides were determined with the phenol/sulfuric acid colorimetric method [36]. Briefly, 2 mL sample solution and 1 mL of 5%

aqueous solution of phenol were mixed in a test tube. Subsequently, 5 mL of concentrated sulfuric acid was added rapidly to the mixture. Test tubes were shaken in an ultrasonic bath for 10 min and then left at room temperature for 20 min for color development. The absorbance of the acquired solution was measured at 490 nm on a UV–Vis spectrophotometer (UV 2450, Shimadzu Corporation, Kyoto, Japan). A reference solution was prepared in an identical manner as that explained above, except that the 2 mL sample solution was replaced by deionized water. The quantification was done based on a calibration curve obtained with glucose (linear range 1.2–12 μg/mL, $R^2 = 0.998$).

3.5. *HPLC–QTOF–MS/MS Analysis*

The extracts of small-molecule saccharides were passed through 0.22 μm filter for HPLC–QTOF–MS/MS analysis. Chromatographic analysis was carried out on an Agilent 1200 liquid chromatography system coupled with QTOF–MS/MS, which was equipped with an electrospray interface (Agilent 6530, Agilent Technologies, Santa Clara, CA, USA), in accordance with our previous studies [37]. Various mixtures of water, methanol, and acetonitrile at flow rate 1.2 mL/min, 1.0 mL/min, 0.8 mL/min, 0.5 mL/min, and 0.3 mL/min were tested as the mobile phase. The detection temperature was tested at 15, 25, 35, and 45 °C. For most of the measurements, acetonitrile and ultrapure water were used as mobile phase A and B, respectively. Solvent flow rate was 1.0 mL/min and the column (Prevail Carbohydrate ES 5 μm, 250 mm × 4.6 mm, Alltech, New Westminster, Canada) was operated at 25 °C. The solvent gradient was used as follows: 75% A and 25% B. The condition of the QTOF was as follows: scan range 100–1500 *m/z*; drying gas (N_2) flow rate, 8.0 L/min; drying gas temperature, 320 °C; sheath gas temperature, 320 °C; capillary voltage, 3.5 kV; fragmentor, 110 V; collision energy at 10, 15 and 20 eV. The operation and acquisition of data were controlled by Agilent Mass Hunter LC/MS Acquisition console, while the analysis of data was done by Qualitative Analysis B.05.00. Monosaccharides and oligosaccharides were determined according to precursor ions, the fragment ions and retention time comparing to those of the standard compounds.

3.6. *Validation of the Methods for Quantitative Characterization of Monosaccharides and Oligosaccharides*

Monosaccharides and oligosaccharides were quantitatively determined as described previously, but with minor modifications [38]. Briefly, standard solutions of monosaccharides and oligosaccharides at a series of appropriate concentrations were prepared for the construction of calibration curves, which were constructed by plotting the extracted ion chromatograms (EIC) peak area versus the concentration. The LOD and LOQ were calculated at approximately three-fold and ten-fold of the signal-to-noise (S/N) ratios, respectively. The measurement of the interday and intraday variabilities was used to determine the precision of this method. The standard solution was analyzed for six replicates within the same day, and additionally, on three consecutive days for evaluating intraday and interday variabilities, respectively. For the stability assessment, the sample extract was analyzed at 0, 2, 4, 8, 12, 24, and 48 h at room temperature. Recovery was determined by adding an accurately-measured amount of the standard compounds to the sample. Relative standard deviation (RSD) was used to assess the results.

3.7. *PCA Statistical Analysis*

The experiments were done in triplicate with three samples, and statistical analyses were performed using the statistical software R (https://www.r--project.org). A 30 × 11 matrix was constructed for the multivariate data treatment. The determined peak area value of the EIC responses of monosaccharides, oligosaccharides, polysaccharides, and saccharides derivatives were defined as variables, and were therefore placed in the columns. The thirty extracts were defined as samples and placed in rows. The data were imported by R software and treated using ggbiplot package to perform Principal Component Analysis (PCA) [39].

4. Conclusions

A sensitive and rapid analytical method was developed for the simultaneous identification and dynamic analysis of saccharide compounds during steam processing of *P. cyrtonema* using HPLC–QTOF–MS/MS. Fructose, sorbitol, glucose, galactose, sucrose, and 1-kestose were identified in the extract of the rhizomes of *P. cyrtonema*. Additionally, a large number of oligosaccharides constituted of fructose units through β-(2→1) or β-(2→6). Polysaccharides and oligosaccharides were decomposed to monosaccharides during repeated steaming processes, since the contents of glucose, galactose, and fructose increased, while those of sucrose, 1-kestose and polysaccharides decreased. Fructose was revealed to be the main determinant for the increasing sweetness after steaming of *P. cyrtonema*. Principal component analysis using saccharides as variables could substantially separate fresh and processed rhizomes of *P. cyrtonema* by repeated steaming 1–4 times, while samples with 4–9 times repeated steaming were grouped together. The samples gradually shifted along the PC1 (72.4%) axis with increases in the number of repeated times of steaming. The small-molecule saccharides, especially fructose, could be considered as markers for the steam process of rhizomes of *P. cyrtonema*.

Author Contributions: J.J. performed the experiments, analyzed the data and wrote the manuscript. J.L., R.Z. and W.H. contributed to the design of experiments. Y.Q., C.Z. and J.X. contributed to the chemical experiments and data analysis. H.L. and D.W. helped on the collection of the samples and data analysis. Y.Q. and S.Z. conceived the project and directed the research.

Acknowledgments: We thank Hunan Engineering Research Center of Anti-tumor TCM Creation & Technology for technical support.

References

1. Zhao, P.; Zhao, C.; Li, X.; Gao, Q.; Huang, L.; Xiao, P.; Gao, W. The genus *Polygonatum*: A review of ethnopharmacology, phytochemistry and pharmacology. *J. Ethnopharmacol.* **2018**, *214*, 274–291. [CrossRef] [PubMed]
2. Wujisguleng, W.; Liu, Y.; Long, C. Ethnobotanical review of food uses of *Polygonatum* (Convallariaceae) in China. *Acta Soc. Bot. Pol.* **2012**, *81*, 239–244. [CrossRef]
3. Liu, N.; Dong, Z.; Zhu, X.; Xu, H.; Zhao, Z. Characterization and protective effect of *Polygonatum sibiricum* polysaccharide against cyclophosphamide-induced immunosuppression in Balb/c mice. *Int. J. Biol. Macromol.* **2018**, *107*, 796–802. [CrossRef] [PubMed]
4. Liu, F.; Liu, Y.; Meng, Y.; Yang, M.; He, K. Structure of polysaccharide from *Polygonatum cyrtonema* Hua and the antiherpetic activity of its hydrolyzed fragments. *Antivir. Res.* **2004**, *63*, 183–189. [CrossRef] [PubMed]
5. Blunder, M.; Orthaber, A.; Bauer, R.; Bucar, F.; Kunert, O. Efficient identification of flavones, flavanones and their glycosides in routine analysis via off-line combination of sensitive NMR and HPLC experiments. *Food Chem.* **2017**, *218*, 600–609. [CrossRef] [PubMed]
6. Ruiz-Matute, A.I.; Brokl, M.; Soria, A.C.; Sanz, M.L.; Martínez-Castro, I. Gas chromatographic–mass spectrometric characterisation of tri- and tetrasaccharides in honey. *Food Chem.* **2010**, *120*, 637–642. [CrossRef]
7. Liu, Z.; Lou, Z.; Ding, X.; Li, X.; Qi, Y.; Zhu, Z.; Chai, Y. Global characterization of neutral saccharides in crude and processed *Radix Rehmanniae* by hydrophilic interaction liquid chromatography tandem electrospray ionization time-of-flight mass spectrometry. *Food Chem.* **2013**, *141*, 2833–2840. [CrossRef] [PubMed]
8. Wu, X.; Jiang, W.; Lu, J.; Yu, Y.; Wu, B. Analysis of the monosaccharide composition of water-soluble polysaccharides from *Sargassum fusiforme* by high performance liquid chromatography/electrospray ionisation mass spectrometry. *Food Chem.* **2014**, *145*, 976–983. [CrossRef] [PubMed]
9. Xue, S.; Wang, L.; Chen, S.; Cheng, Y. Simultaneous analysis of saccharides between fresh and processed Radix Rehmanniae by HPLC and UHPLC-LTQ-Orbitrap-MS with multivariate statistical analysis. *Molecules* **2018** *23*, 541. [CrossRef] [PubMed]

10. Georgelis, N.; Fencil, K.; Richael, C.M. Validation of a rapid and sensitive HPLC/MS method for measuring sucrose, fructose and glucose in plant tissues. *Food Chem.* **2018**, *262*, 191–198. [CrossRef] [PubMed]
11. Corbin, K.R.; Byrt, C.S.; Bauer, S.; Debolt, S.; Chambers, D.; Holtum, J.A.; Karem, G.; Henderson, M.; Lahnstein, J.; Beahan, C.T. Prospecting for energy-rich renewable raw materials: *Agave* leaf case study. *PLoS ONE* **2015**, *10*, e0135382. [CrossRef] [PubMed]
12. Liu, Z.; Rochfort, S. Identification and quantitative analysis of oligosaccharides in wheat flour using LC–MS. *J. Cereal Sci.* **2015**, *63*, 128–133. [CrossRef]
13. Ricochon, G.; Paris, C.; Girardin, M.; Muniglia, L. Highly sensitive, quick and simple quantification method for mono and disaccharides in aqueous media using liquid chromatography–atmospheric pressure chemical ionization–mass spectrometry (LC–APCI–MS). *J. Chromatogr. B* **2011**, *879*, 1529–1536. [CrossRef] [PubMed]
14. Lattova, E.; Perreault, H. Labelling saccharides with phenylhydrazine for electrospray and matrix-assisted laser desorption–ionization mass spectrometry. *J. Chromatogr. B* **2003**, *793*, 167–179. [CrossRef]
15. Gaucher, S.P.; Leary, J.A. Stereochemical differentiation of mannose, glucose, galactose, and talose using zinc(II) diethylenetriamine and ESI-ion trap mass spectrometry. *Anal. Chem.* **1998**, *70*, 3009–3014. [CrossRef] [PubMed]
16. Hu, X.; Zhao, H.; Shi, S.; Li, H.; Zhou, X.; Jiao, F.; Jiang, X.; Peng, D.; Chen, X. Sensitive characterization of polyphenolic antioxidants in *Polygonatum odoratum* by selective solid phase extraction and high performance liquid chromatography-diode array detector-quadrupole time-of-flight tandem mass spectrometry. *J. Pharm. Biomed. Anal.* **2015**, *112*, 15–22. [CrossRef] [PubMed]
17. Wang, L.; Xu, H.; Yuan, F.; Fan, R.; Gao, Y. Preparation and physicochemical properties of soluble dietary fiber from orange peel assisted by steam explosion and dilute acid soaking. *Food Chem.* **2015**, *185*, 90–98. [CrossRef] [PubMed]
18. Kyung-Tae, K.; Jungeun, N.; Jungeun, L.; Jung-Ok, K.; Gee-Dong, L.; Joong-Ho, K. Elimination of heavy metals (Pb, Cd) by steaming and roasting conditions of *Polygonatum odoratum* Roots. *Korean J. Food Preserv.* **2005**, *12*, 209–215.
19. Kim, S.N.; Kang, S.-J. Effects of black Ginseng (9 times-steaming Ginseng) on hypoglycemic action and changes in the composition of ginsenosides on the steaming process. *Korean J. Food Sci. Technol.* **2009**, *41*, 77–81.
20. Wangyu, K.; Jongmoon, K.; Han, S.B.; Seungki, L.; Nakdoo, K.; Manki, P.; Chongkook, K.; Jeonghill, P. Steaming of ginseng at high temperature enhances biological activity. *J. Nat. Prod.* **2000**, *63*, 1702–1704.
21. Arena, S.; Renzone, G.; D'Ambrosio, C.; Salzano, A.M.; Scaloni, A. Dairy products and the Maillard reaction: A promising future for extensive food characterization by integrated proteomics studies. *Food Chem.* **2017**, *219*, 477–489. [CrossRef] [PubMed]
22. Zhang, T.-B.; Yue, R.-Q.; Xu, J.; Ho, H.-M.; Ma, D.-L.; Leung, C.-H.; Chau, S.-L.; Zhao, Z.-Z.; Chen, H.-B.; Han, Q.-B. Comprehensive quantitative analysis of Shuang-Huang-Lian oral liquid using UHPLC–Q-TOF-MS and HPLC-ELSD. *J. Pharm. Biomed. Anal.* **2015**, *102*, 1–8. [CrossRef] [PubMed]
23. Martínez-Villaluenga, C.; Frías, J.; Vidal-Valverde, C. Raffinose family oligosaccharides and sucrose contents in 13 Spanish lupin cultivars. *Food Chem.* **2005**, *91*, 645–649. [CrossRef]
24. Apolinário, A.C.; Damasceno, B.P.G.D.L.; Beltrão, N.E.D.M.; Pessoa, A.; Converti, A.; Silva, J.A.D. Inulin-type fructans: A review on different aspects of biochemical and pharmaceutical technology. *Carbohydr. Polym.* **2014**, *101*, 368–378. [CrossRef] [PubMed]
25. Laere, A.V.; Ende, W.V.D. Inulin metabolism in dicots: Chicory as a model system. *Plant Cell Environ.* **2002**, *25*, 803–813. [CrossRef]
26. Lan, G.; Chen, H.; Chen, S.; Tian, J. Chemical composition and physicochemical properties of dietary fiber from *Polygonatum odoratum* as affected by different processing methods. *Food Res. Int.* **2012**, *49*, 406–410. [CrossRef]
27. Liu, B.; Cheng, Y.; Bian, H.J.; Bao, J.K. Molecular mechanisms of *Polygonatum cyrtonema* lectin-induced apoptosis and autophagy in cancer cells. *Autophagy* **2009**, *5*, 253–255. [CrossRef] [PubMed]
28. Li, L.; Liao, B.Y.; Thakur, K.; Zhang, J.G.; Wei, Z.J. The rheological behavior of polysaccharides sequential extracted from *Polygonatum cyrtonema* Hua. *Int. J. Biol. Macromol.* **2018**, *109*, 76–771. [CrossRef] [PubMed]
29. Peshev, D.; Ende, W.V.D. Fructans: Prebiotics and immunomodulators. *J. Funct. Foods* **2014**, *8*, 348–357. [CrossRef]
30. Ervin, E.L.; Syperda, G. Seasonal effects on soluble sugars and cytological aspects of *Polygonatum*

canaliculatum rhizomes. *Bull. Torrey Bot. Club* **1971**, *98*, 162–167. [CrossRef]

31. Chumpitazi, B.P.; Lim, J.; McMeans, A.R.; Shulman, R.J.; Hamaker, B.R. Evaluation of FODMAP carbohydrates content in selected foods in the United States. *J. Pediatr.* **2018**, *199*, 252–255. [CrossRef] [PubMed]
32. Cavia, M.M.; Fernández-Muiño, M.A.; Gömez-Alonso, E.; Montes-Pérez, M.J.; Huidobro, J.F.; Sancho, M.T. Evolution of fructose and glucose in honey over one year: Influence of induced granulation. *Food Chem.* **2002**, *78*, 157–161. [CrossRef]
33. Tomoda, M.; Satoh, N.; Sugiyama, A. Isolation and characterization of fructans from *Polygonatum odoratum* var.japonicum rhizomes. *Chem. Pharm. Bull. (Tokyo)* **1973**, *21*, 1806–1810. [CrossRef]
34. Aisala, H.; Sinkkonen, J.; Kalpio, M.; Sandell, M.; This, H.; Hopia, A. In situ quantitative (1)H nuclear magnetic resonance spectroscopy discriminates between raw and steam cooked potato strips based on their metabolites. *Talanta* **2016**, *161*, 245–252. [CrossRef] [PubMed]
35. Qi, X.; Cheng, L.; Li, X.; Zhang, D.; Wu, G.; Zhang, H.; Wang, L.; Qian, H.; Wang, Y.N. Effect of cooking methods on solubility and nutrition quality of brown rice powder. *Food Chem.* **2018**. [CrossRef] [PubMed]
36. Ravber, M.; Knez, Z.; Škerget, M. Simultaneous extraction of oil- and water-soluble phase from sunflower seeds with subcritical water. *Food Chem.* **2015**, *166*, 316–323. [CrossRef] [PubMed]
37. Jin, J.; Kang, W.; Zhong, C.; Qin, Y.; Zhou, R.; Liu, H.; Xie, J.; Chen, L.; Qin, Y.; Zhang, S. The pharmacological properties of *Ophiocordyceps xuefengensis* revealed by transcriptome analysis. *J. Ethnopharmacol.* **2018**, *219*, 195–201. [CrossRef] [PubMed]
38. Mohammat, A.; Yili, A.; Aisa, H.A. Rapid quantification and quantitation of alkaloids in Xinjiang *Fritillaria* by ultra performance liquid chromatography-quadrupole time-of-flight mass spectrometry. *Molecules* **2017**, *22*, 719. [CrossRef] [PubMed]
39. Wu, T.; Taylor, C.; Nebl, T.; Ng, K.; Bennett, L.E. Effects of chemical composition and baking on in vitro digestibility of proteins in breads made from selected gluten-containing and gluten-free flours. *Food Chem.* **2017**, *233*, 514–524. [CrossRef] [PubMed]

6

Small Molecular Weight Aldose (D-Glucose) and Basic Amino Acids (L-Lysine, L-Arginine) Increase the Occurrence of PAHs in Grilled Pork Sausages

Wen Nie [1], Ke-zhou Cai [1,2,*], Yu-zhu Li [1], Shuo Zhang [1], Yu Wang [1], Jie Guo [1], Cong-gui Chen [1,2] and Bao-cai Xu [1,3]

1 School of Food Science and Engineering, Hefei University of Technology, Hefei 230009, Anhui Province, China; 17356511419@163.com (W.N.); 1932526946@163.com (Y.L.); 18856023156@163.com (S.Z.); wyll_ah92@163.com (Y.W.); hfutkaylee@163.com (J.G.); chencg1629@hfut.edu.cn (C.C.); baocaixu@163.com (B.X.)

2 Key Laboratory on Deep Processing of Agricultural Products for Anhui Province, Hefei 230009, Anhui Province, China

3 Engineering Research Centre of Bio-Process, Ministry of Education, Hefei 230009, Anhui Province, China

* Correspondence: kzcai@hfut.edu.cn

Academic Editor: Alessandra Gentili

Abstract: (1) Background: Amino acids and carbohydrates are widely used as additives in the food industry. These compounds have been proven to be an influencing factor in the production of chemical carcinogenic compounds polycyclic aromatic hydrocarbons (PAHs). However, the effect of the properties of the amino acids and carbohydrates on the production of PAHs is still little known. (2) Methods: We added different (i) R groups (the R group represents an aldehyde group in a glucose molecule or a ketone group in a fructose molecule); (ii) molecular weight carbohydrates; (iii) polarities, and (iv) acid-base amino acids to pork sausages. The effects of the molecular properties of carbohydrates and amino acids on the formation of PAHs in grilled pork sausages were investigated. (3) Results: The results showed that a grilled sausage with aldehyde-based D-glucose was capable of producing more PAHs than a sausage with keto-based D-fructose. A higher PAH content was determined in the grilled pork sausage when the smaller molecular weight, D-glucose, was added compared with the sausage where the larger molecular weight, 4-(α-D-glucosido)-D-glucose and cellulose were added. The addition of basic amino acids (L-lysine, L-arginine) was capable of producing more PAHs compared with the addition of acidic amino acids (L-glutamic acid, L-aspartate). When amino acid containing a benzene ring was added, a smaller volume of PAHs was produced compared with the addition of other amino acids. (4) Conclusions: Our study suggests that systematic consideration of molecule properties is necessary when using food additives (amino acids and carbohydrates) for food processing.

Keywords: amino acids; carbohydrates; acidity; polarity; molecular weight

---

1. Introduction

Polycyclic aromatic hydrocarbons (PAHs), formed through incomplete combustion of wood or gasoline, are regarded as potentially genotoxic and carcinogenic to humans [1,2]. The occurrence of PAHs in food, such as edible oils, meat, and dairy products, has been regarded as a consequence of high-temperature processing [3]. Previous epidemiological studies have speculated that food consumption might contribute 88–98% of PAHs exposure. This is especially the case in populations that are non-smoking and not subject to occupational exposure [4]. Among the sources of PAHs exposure, meat products account for the largest proportion. The formation and control of PAHs during the processing of meat products is topical in current research.

The definite mechanism for the formation of PAHs is not well understood. Some researchers proposed that they might be formed through free radical reactions, intramolecular addition or the polymerization of small molecules [5–7]. In meat products, PAHs are formed during processing at a high temperature, such as by smoking, drying, roasting, and grilling [8–10]. Grilled sausages are a common food and can be contaminated with high concentrations of PAHs [11]. Amino acids, a kind of food additive, are often used to improve the sensory properties (color, taste, and texture) of meat products to meet the needs of consumers [12,13]. Some scholars have studied the pyrolysis of aliphatic α-amino acids, glutamine, glutamic acid, and aspartic acid and the formation of PAHs. The results show that single chemical components of glutamine, glutamic acid, and aspartic acid can pyrolyze to form PAHs at high temperatures. The author speculates that all three acids appear to break down to smaller "building blocks" and then pyrosynthesize the PAHs [14]. Other scholars have explored whether glucose can enhance the pyrolysis of proline to form PAHs and found that glucose can provide a low-temperature pathway for the decomposition of proline, thus, enhancing the pyrolysis of proline to form PAHs [15]. The above studies indicate that some amino acids can be pyrolyzed to form PAHs in a single simulated system, and glucose can promote the pyrolysis of proline to form PAHs. However, meat products are a complex matrix. The formation of PAHs in meat products by glucose and amino acids is significantly different from the formation of PAHs by the pyrolysis of glucose and amino acids in a single chemical system. There is still no information on the effect of amino acid types on the PAHs formation in cooked meat.

In addition to nutritional properties, carbohydrates can also be used as a sweetener, gel, thickener, or stabilizer in the meat industry. The physical and chemical properties, the degree of polymerization, solubility, viscosity, film formation, and gelation greatly influence the retention and release of flavor [16,17]. In addition, carbohydrates retain aromatic components and reduce flavor loss during meat processing [17]. Other studies considered using polysaccharide to promote water holding capacity and texture [18]. Although many carbohydrates have been successfully used to increase the flavor, color, and texture of meat, there is still limited research on their safety including the effect of PAHs production in cooked meat.

Therefore, it is of interest to specifically study the effects of amino acids and carbohydrate types on PAHs formation in cooked meat. In this paper, we tested the effects of carbohydrates with a different R-base (an aldehyde group in a glucose molecule and a ketone group in a fructose molecule) and different molecular weights, and the property of amino acids (with different polarities and a different acid-base) on the formation of twelve kinds of PAHs in grilled sausages. All of these PAHs are noted as potential food contaminants by the United States Environmental Protection Agency [19].

2. Results and Discussion

2.1. Validation

The detection limits, recoveries, and precision for twelve kinds of PAHs data are presented in Table 1. Linearity was determined using external standard plot method. The HPLC of all 12 PAHs in the master sample is presented in Figure 1. The high R^2 values indicated good linearity over the concentration range. As can be seen, the limit of detection (LOD) and limit of quantification (LOQ) of all PAHs ranged from 0.03 to 0.18 μg kg^{-1} and 0.10 to 0.91 μg kg^{-1} and met the Commission of the European Communities (2011b). Relative standard deviations (RSD) and recovery in our study ranged from 1.23 to 9.71% and 79.22 to 105.78%, respectively. According to the regulation [20], recovery sets should be in the range of 50–120%; recovery in the experiments varied between 79.22% and 105.78%, which is in good accordance with the regulation. The precision was adequate, with RSD $< 10\%$. It has been well established that the presence of some impurities in meat samples, such as aliphatic hydrocarbons, fatty acids, phenols, and polycyclic organic compounds, may greatly reduce the extraction efficiency of PAHs [21].

Table 1. Parameters of validation of 12 polycyclic aromatic hydrocarbons (PAHs) in sausages.

PAHs	Linear Range (ng mL^{-1})	Correlation Coefficients (R^2)	LOD (μg kg^{-1})	LOQ (μg kg^{-1})	Recovery [a] (%)	RSD [b] (%)
NA	0.10–10.00	0.9999	0.05	0.17	105.78	1.23
AC	0.10–10.00	0.9996	0.18	0.60	93.46	1.46
FLU	0.20–10.00	0.9999	0.06	0.20	84.54	5.54
FLT	0.20–10.00	0.9993	0.16	0.53	83.83	7.89
BaA	0.25–5.00	0.9998	0.09	0.30	90.52	4.37
CHR	0.30–7.50	0.9999	0.10	0.91	92.56	3.42
BbFA	1.00–20.00	0.9995	0.06	0.20	96.74	1.90
BkFA	1.00–20.00	0.9999	0.08	0.27	95.33	2.05
BaP	0.20–10.00	0.9999	0.11	0.37	91.31	4.55
DBahA	0.10–10.00	0.9991	0.03	0.10	84.53	9.71
BPE	0.20–10.00	0.9998	0.04	0.20	88.34	4.56
IPY	0.30–10.00	0.9997	0.06	0.30	79.22	6.18

Explanations: LOD–limit of detection; LOQ–limit of quantification; RSD–relative standard deviation of repeatability; [a] mean recoveries of four different spiking levels in triplicate in the same day; [b] mean relative standard deviations (RSD) of two different spiking levels in triplicates on two different day; Naphthalene (NA), Acenaphthene (Ac), Fluorene (FLU), Fluoranthene (FLT), Benzo[a]anthracene (BaA), Chrysene (CHR), Benzo[b]fluoranthene (BbFA), Benzo[k]fluoranthene (BkFA), Benzo[a]pyrene (BaP), Dibenzo[a,h]anthracene (DBahA), Benzo[g,h,i]perylene (BPE), Indeno[1,2,3-c,d]pyrene (IPY).

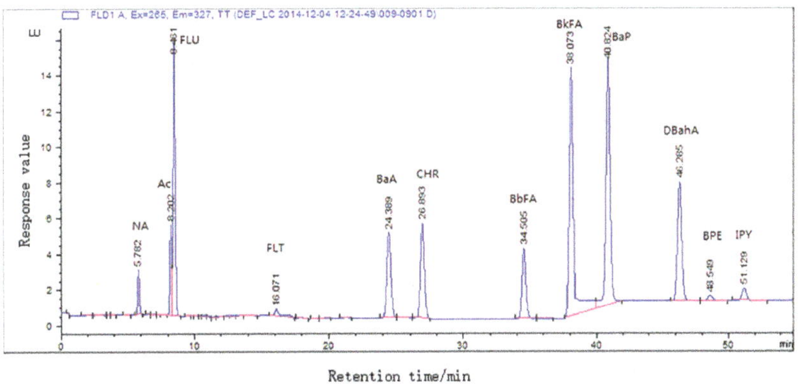

Figure 1. Standard chromatogram of 12 polycyclic aromatic hydrocarbons.

2.2. *Effect of Carbohydrate Characteristics on PAHs Content*

Carbohydrate, as an important nutrient of meat and a common additive in meat products, was reported to have a notable influence on PAHs production during meat processing [22,23]. In this study, the effects of aldose (D-glucose) and ketose (D-fructose) on PAHs production during meat processing were investigated by adding D-glucose and D-fructose to grilled pork sausages. The D-glucose, 4-(α-D-Glucosido)-D-glucose, and cellulose were used to explore the effect of carbohydrate molecular weight on the formation of PAHs during meat processing. The results showed there was no significant difference in the content of PAHs between the grilled pork sausage with D-fructose and the blank control sausage ($p > 0.05$). However, compared with these two groups, the occurrence of PAHs in the grilled pork sausages with D-glucose was significantly increased ($p < 0.05$) (details are shown in Table 2). The reason for this phenomenon might be due to D-glucose containing free aldehyde groups, which can be decomposed to smaller molecular aldehyde compounds at high temperatures, and these aldehydes can be further subjected to complex cracking, polymerization or condensation reaction to ultimately produce PAHs. However, the ketone group in D-fructose needs to be converted into an aldehyde group at a higher temperature to further generate PAHs by a cleavage reaction [24].

Table 2. Concentrations (n = 3) of 12 PAHs produced as a consequence of adding different carbohydrates to grilled sausages.

Species	NA	Ac	FLU	FLT	BaA	CHR	BbFA	BkFA	DBahA	BPE	BaP	IPY	$\sum PAH_{12}$
Control group	17.93 ± 1.35 c	12.99 ± 1.02 c	2.79 ± 0.14 c	ND	2.21 ± 0.16 c	3.62 ± 0.10b c	ND	7.52 ± 0.36 b	0.71 ± 0.05 c	1.25 ± 0.08 c	0.88 ± 0.12 d	ND	49.90 ± 3.46 c
D-Fructose	18.36 ± 1.41 c	11.15 ± 1.14 c	3.13 ± 0.26 c	1.06 ± 0.15 c	3.04 ± 0.27 bc	2.47 ± 0.16 d	ND	6.01 ± 0.33 c	1.04 ± 0.11 b	1.37 ± 0.09 b	1.26 ± 0.06 c	ND	48.89 ± 4.05 c
D-Glucose	34.25 ± 1.22 a	23.45 ± 1.68 a	7.66 ± 0.23 a	4.24 ± 0.18 a	7.41 ± 0.36 a	6.75 ± 0.36 a	0.47 ± 0.03 a	13.76 ± 1.28 a	3.16 ± 0.30 a	3.07 ± 0.26 a	5.59 ± 0.22 a	ND	109.81 ± 6.49 a
4-(α-D-Glucosido)-D-glucose	25.17 ± 2.23 b	17.42 ± 1.35 b	5.06 ± 0.44 b	2.31 ± 0.18 b	4.01 ± 0.23 b	4.48 ± 0.24 b	0.29 ± 0.01 b	8.47 ± 0.31 b	1.35 ± 0.10 b	1.54 ± 0.10 b	3.46 ± 0.25 b	ND	73.56 ± 5.67 b
Cellulose	17.16 ± 1.27 c	11.56 ± 1.06 c	3.03 ± 0.21 c	ND	2.44 ± 0.17 c	3.53 ± 0.28 c	ND	6.46 ± 0.31 c	0.82 ± 0.03 bc	1.39 ± 0.11 b	1.12 ± 0.18 cd	ND	47.51 ± 3.38 c

Explanations: the results are expressed as the mean and standard deviation; ND—Means no PAHs detected; a, b, c, d, different letters indicate significant differences in PAHs content between any two rows in Table 2 ($p < 0.05$); Naphthalene (NA), Acenaphthene (Ac), Fluorene (FLU), Fluoranthene (FLT), Benzo[a]anthracene (BaA), Chrysene (CHR), Benzo[b]fluoranthene (BbFA), Benzo[k]fluoranthene (BkFA), Benzo[a]pyrene (BaP), Dibenzo[a,h]anthracene (DBahA), Benzo[g,h,i]perylene (BPE), Indeno[1,2,3-c,d]pyrene (IPY).

When comparing the influence of the carbohydrate's molecular weight on the content of PAHs in grilled pork sausages, it was found that the concentration of PAHs in sausages supplemented with D-glucose or 4-(α-D-Glucosido)-D-glucose was significantly higher than that of the blank control group. However, the content of PAHs in the grilled sausage to which cellulose had been added was not significantly different from that of the control group. In addition, the PAHs content in the grilled sausage with D-glucose was significantly higher than that of in the grilled pork sausage with 4-(α-D-Glucosido)-D-glucose ($p < 0.05$) (details are shown in Table 2). A higher PAHs content was determined in the grilled sausage with smaller molecular weight, D-glucose, when compared with the larger molecular weight, 4-(α-D-Glucosido)-D-glucose and cellulose. However, there was no significant difference in the content of PAHs in grilled pork sausages with the addition of small molecular weight D-fructose and large molecular weight cellulose ($p > 0.05$). This result may be due to the effect of carbohydrates on PAHs production being affected by both the R group and molecular weight, and the effect of R group on PAHs formation was greater than the effect of molecular weight. Some scholars explored the formation of PAHs from the pyrolysis of carbohydrates, amino acids, and fatty acids, and it was found that high-temperature pyrolysis of D-glucose produced more PAHs than starch, which result is consistent with our research [25]. Other scholars studied the mechanism of dehydration, carbonization, decarbonylation, decarboxylation, dehydrogenation, and cross-linking of cellulose at high temperatures when studying the mechanism of pyrolysis of cellulose to form PAHs. It was proposed that PAHs could be formed through transformation and rearrangement [26]. The study also found that the yield of PAHs produced by pyrolysis of D-glucose and sucrose was higher than that of cellulose. This result is consistent with our findings. However, the current mechanism of the molecular weight of carbohydrates on the formation of PAHs during high-temperature processing is still unclear.

2.3. Effect of Amino Acid Properties on PAHs Content

To explore the effects of amino acid polarity and acidity/alkalinity on the occurrence of PAHs in meat processing under high temperature, several different polar amino acids, non-polar amino acids, acidic amino acids, and basic amino acids were chosen to be added to grilled pork sausages. When investigating the effects of three non-polar amino acids on the content of PAHs in grilled pork sausages, it was found that the content of PAHs in sausages supplemented with L-proline was significantly increased compared with the blank control group ($p < 0.05$). However, the content of PAHs in grilled pork sausages added L-tryptophan or L-phenylalanine was not significantly different from that of the blank control group ($p > 0.05$) (details are shown in Table 3). The effects of three polar amino acids on the content of PAHs in grilled pork sausages were investigated. The result showed that the addition of L-threonine or L-serine significantly increased the PAHs content in grilled pork sausages when compared with the blank control group ($p < 0.05$). However, there was no significant difference between the sausage supplemented with L-tyrosine and the blank control group ($p > 0.05$) (details are shown in Table 3). The authors did not find obvious rules when comparing the effects of polar and non-polar amino acids on the formation of PAHs in grilled pork sausages. This indicates that the formation of PAHs in grilled pork sausages is not affected by the polarity of the amino acids.

Table 3. Concentrations ($n = 3$) of 12 PAHs produced as a consequence of adding different amino acids to grilled sausages.

	Species	NA	Ac	FLU	FLT	BaA	CHR	BbFA	BkFA	DBahA	BPE	BaP	IPY	∑PAH12
	Control group	17.93 ± 1.35 de	12.99 ± 1.02 cd	2.79 ± 0.14 cd	ND	2.21 ± 0.16 c	3.62 ± 0.10 c	ND	7.52 ± 0.36 c	0.71 ± 0.05 c	1.25 ± 0.08 cd	0.88 ± 0.12 b	ND	49.90 ± 3.46 d
Non-polar amino acid	L-proline	30.42 ± 2.26 c	27.10 ± 1.54 b	6.17 ± 0.33 b	3.16 ± 0.18 a	5.06 ± 0.27 ab	7.83 ± 0.24 b	1.01 ± 0.04 a	17.43 ± 1.33 a	1.64 ± 0.11 b	2.73 ± 0.13 b	2.23 ± 0.20 a	ND	104.78 ± 8.72 b
	L-tryptophan	18.35 ± 1.42 de	11.43 ± 1.24 d	3.01 ± 0.26 c	ND	1.88 ± 0.13 d	3.59 ± 0.21 c	ND	7.17 ± 0.45 c	0.75 ± 0.07 c	1.16 ± 0.11 d	1.01 ± 0.16 b	ND	48.35 ± 3.18 d
	L-phenylalanine	16.77 ± 1.37 e	13.21 ± 1.10 cd	2.85 ± 0.24 cd	ND	2.35 ± 0.17 c	3.81 ± 0.26 c	ND	8.02 ± 0.43 c	0.67 ± 0.05 c	1.36 ± 0.14 cd	0.85 ± 0.11 b	ND	49.89 ± 4.05 d
Polar amino acid	L-tyrosine	18.19 ± 1.52 de	12.15 ± 1.13 d	2.66 ± 0.22 d	ND	1.99 ± 0.14 cd	3.77 ± 0.28 c	ND	7.86 ± 0.46 c	0.69 ± 0.04 c	1.22 ± 0.11 d	0.94 ± 0.06 b	ND	49.47 ± 4.44 d
	L-threonine	39.45 ± 3.26 b	28.58 ± 2.19 ab	6.82 ± 0.31 b	ND	4.86 ± 0.20 b	7.96 ± 0.30 b	0.86 ± 0.03 a	16.54 ± 1.42a	1.56 ± 0.12 b	2.75 ± 0.21 b	1.96 ± 0.08 a	ND	111.34 ± 9.43 b
	L-serine	37.49 ± 3.33 b	34.01 ± 2.64 a	8.12 ± 0.36 a	3.72 ± 0.20 a	5.78 ± 0.26 ab	9.47 ± 0.82 ab	1.02 ± 0.01 a	19.68 ± 1.54 a	1.85 ± 0.11 b	3.27 ± 0.20 a	2.35 ± 0.17 a	ND	126.76 ± 10.17 a
Basic amino acid	L-lysine	50.48 ± 2.57 a	32.08 ± 1.88 a	8.74 ± 023 a	ND	8.76 ± 0.18 a	12.15 ± 0.32 a	ND	12.78 ± 1.14 b	3.21 ± 0.08 a	3.13 ± 0.10 a	1.52 ± 0.05 ab	ND	132.85 ± 6.13 a
	L-arginine	52.67 ± 3.36 a	30.91 ± 2.49 a	9.64 ± 0.33 a	ND	7.26 ± 0.21 a	11.61 ± 1.40 a	ND	17.89 ± 1.42 a	3.69 ± 0.33 a	3.98 ± 0.36 a	2.17 ± 0.14 a	ND	139.82 ± 8.45 a
Acidic amino acid	L-glutamic acid	20.96 ± 3.61 d	17.67 ± 2.20 c	2.37 ± 0.27 d	ND	5.45 ± 0.22 ab	7.27 ± 0.31 b	ND	10.17 ± 1.26 b	1.62 ± 0.11 b	1.86 ± 0.14 c	2.09 ± 0.10 a	ND	69.46 ± 9.27 c
	L-aspartate acid	22.76 ± 2.18 d	15.37 ± 1.06 c	3.54 ± 0.16 c	ND	3.03 ± 0.17 bc	4.59 ± 0.20 c	ND	9.54 ± 0.96 bc	2.92 ± 0.16 a	1.58 ± 0.10 c	1.14 ± 0.04 b	ND	64.47 ± 3.47 c

Explanations: the results are expressed as the mean and the standard deviation; ND—Means no PAHs detected; a, b, c, d, e, different letters indicate significant differences in PAHs content between any two rows in Table 3 ($p < 0.05$); Naphthalene (NA), Acenaphthene (Ac), Fluorene (FLU), Fluoranthene (FLT), Benzo[a]anthracene (BaA), Chrysene (CHR), Benzo[b]fluoranthene (BbFA), Benzo[k]fluoranthene (BkFA), Benzo[a]pyrene (BaP), Dibenzo[a,h]anthracene (DBahA), Benzo[g,h,i]perylene (BPE), Indeno[1,2,3-c,d]pyrene (IPY).

When investigating the effects of two basic amino acids on the content of PAHs in grilled pork sausage, it was found that the content of PAHs in the sausage with L-lysine and L-arginine was significantly higher than that in the blank control group. However, there was no significant difference in the content of PAHs in the grilled pork sausage with L-arginine or L-lysine ($p > 0.05$). When investigating the effects of two acidic amino acids on the content of PAHs in grilled pork sausage, it was found that the content of PAHs in the sausage to which L-glutamic acid and L-aspartic acid were added was significantly higher than that in the blank control group ($p < 0.05$). However, there was no significant difference in the PAHs content between the sausages supplemented with L-glutamic acid or L-aspartic acid ($p > 0.05$). In addition, when comparing the effects of acidic amino acids (L-lysine and L-arginine) and basic amino acids (L-glutamic acid and L-aspartate) on the formation of PAHs in grilled pork sausages, the authors found that the content of PAHs in sausages with basic amino acids was significantly higher than that in grilled pork sausages with acidic amino acids ($p < 0.05$) (details are shown in Table 3). The reason may be due to the basic amino acids promoting the Maillard reaction during the grilling of the sausages. Previous studies have shown that the Maillard reaction was increased with the increase of Ph [23], and the Maillard reaction has a great relationship with the formation of PAHs [23,27]. The pyrolysis of proteins produces free amino acids and these amino acids can react with reducing sugars (such as glucose) to form the Amalido compound (1-amino-1-deoxy-2-ketosaccharide) [14]. These compounds then undergo pyrolysis to produce PAHs. But the detailed relationship between Maillard reaction and PAHs production is still unclear.

2.4. Effect of Benzene Ring in Amino Acid on PAHs Content

The authors found a very interesting phenomenon when investigating the effects of three non-polar amino acids and three polar amino acids on the content of PAHs in grilled pork sausages. When investigating the effects of three non-polar amino acids on the content of PAHs in grilled pork sausages, it was found that the content of PAHs in sausages supplemented with L-tryptophan and L-phenylalanine was significantly lower than proline (as shown in Figure 2A). When investigating the effects of three polar amino acids on the content of PAHs in grilled pork sausages, the results showed that the content of PAHs in sausages supplemented with L-tyrosine was significantly lower than the sausage with L-serine or L-threonine ($p < 0.05$), but there was no significant difference compared with the control group ($p > 0.05$) (as shown in Figure 2B). Aspartic acid and proline degrade at high temperatures to produce PAHs, such as phenanthrene and aromatics. Tryptophan could not be degraded to produce PAHs when treated at high temperature [15]. This is consistent with the results of our present experiment. It is well known that L-tryptophan, L-phenylalanine, and L-tyrosine are all aromatic amino acids, and each contains a benzene ring. Therefore, it is reasonable to speculate that the benzene ring in the amino acid molecule is a significant reason that these aromatic amino acids hardly influence the production of PAHs. This may be due to the benzene ring in the amino acid molecule being structurally stable and not easily cleaved into small molecular substances, for example, an aldehyde, an enyne, etc. at a high temperature.

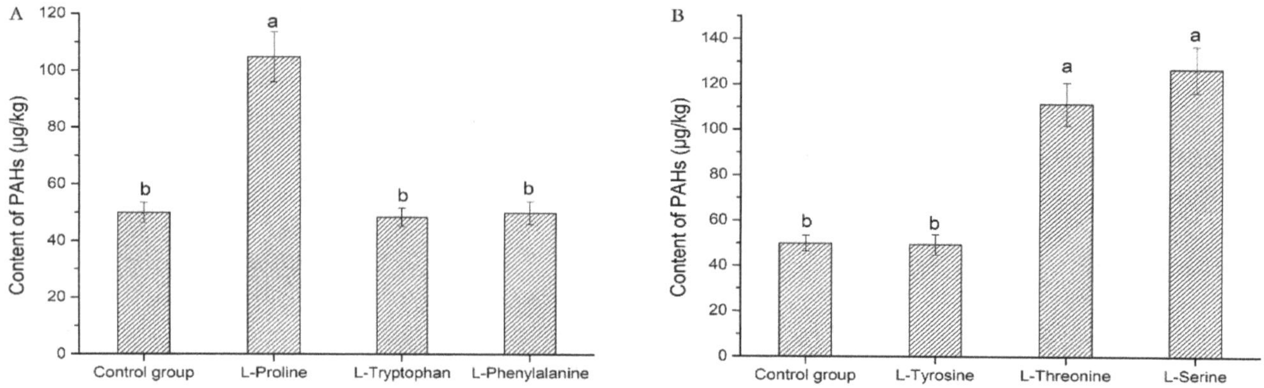

Figure 2. Effect of benzene ring in amino acid on polycyclic aromatic hydrocarbons (PAHs) content. (**A**) The effects of three non-polar amino acids on the content of PAHs in grilled pork sausages; (**B**) The effects of three polar amino acids on the content of PAHs in grilled pork sausages; a, b, different letters indicate significant differences in PAHs content between any two bar graphs in Figure 2A,B ($p < 0.05$).

3. Materials and Methods

3.1. Standards and Reagents

D-glucose (CAS number: 50-99-7), D-fructose (CAS number: 53188-23-1), 4-(α-D-Glucosido)-D-glucose (CAS number: 6363-53-7), cellulose (CAS number: 9004-34-6), Non-polar amino acids: L-proline (CAS number: 147-85-3), L-tryptophan (CAS number: 73-22-3) and L-phenylalanine (CAS number: 63-91-2); Polar amino acids: L-tyrosine (CAS number: 60-18-4), L-threonine (CAS number: 72-19-5) and L-serine (CAS number: 56-45-1); Basic amino acids: L-Lysine (CAS number: 56-87-1) and L-arginine (CAS number: 74-79-3); Acidic amino acids: L-glutamic acid (CAS number: 56-86-0) and L-aspartic acid (CAS No.: 56-84-8) (See Table 4 for specific information on the compounds studied above). The above reagents were all more than 99% pure and were purchased from Aladdin Reagent (Shanghai Aladdin Biochemical Technology Co., Ltd., Shanghai, China). Cyclohexane, dichloromethane (analytical pure) and acetonitrile (chromatographic pure) were purchased from Sinopharm Group. Twelve kinds of PAHs that were used as standards (Naphthalene (NA), Acenaphthene (Ac), Fluorene (FLU), Fluoranthene (FLT), Benzo[a]anthracene (BaA), Chrysene (CHR), Benzo[b]fluoranthene (BbFA), Benzo[k]fluoranthene (BkFA), Benzo[a]pyrene (BaP), Dibenzo[a,h]anthracene (DBahA), Benzo[g,h,i]perylene (BPE), and Indeno[1,2,3-c,d]pyrene (IPY)) were purchased at a purity of more than 99% from Toronto Research Chemicals, Brisbane, North York, ON, Canada.

3.2. Sample Preparation

Fresh hind legs and fat from pigs were purchased from a Carrefour supermarket in Hefei, China. After the visible connective tissues and fat had been removed, the meat was cut into small pieces and ground twice using an electric meat mincer (sieve plate aperture 4 mm; TJ12-H, Henglian, Guangdong, China). The fat was also ground twice using an electric meat mincer with a sieve plate aperture of 6 mm. Four thousand grams of ground meat, 1000 g of fat, 100 g of sodium chloride were thoroughly mixed, then divided into 14 parts, each about 360 g. One point eight grams of an amino acid or carbohydrate was added to each portion and then mixed well. After curing for 6 h at 6–10 °C, the two groups were stuffed into edible casings (Shengguan Casing Company, Guangxi, China) to form sausages. The diameter, length and weight of the sausages were 1.9 cm, 20 cm, and 100 ± 10 g, respectively. All experiments were carried out in triplicate.

Table 4. Molecular weight, pKa and logP of each study compound.

	Molecular Weight (g/mol)	pKa	logP
D-Glucose	180.16	12.43 (t = 18 °C)	−3.24
D-Fructose	180.16	12.06 (t = 18 °C)	−2.23
4-(α-D-Glucosido)-D-glucose	342.30	-	−5.03
Cellulose	>50000	-	-
L-Proline	115.13	10.64 (t = 25 °C)	−2.54
L-Tryptophan	204.22	7.38 (t = 25 °C)	−1.06
L-Phenylalanine	165.19	1.24 (t = 25 °C)	−1.38
L-Tyrosine	181.19	2.20 (t = 25 °C)	−2.26
L-Threonine	119.12	5.60 (t = 25 °C)	−2.94
L-Serine	105.09	2.21 (t = 25 °C)	−3.07
L-Lysine	146.19	3.12 (t = 0 °C)	−3.05
L-Arginine	174.20	2.24 (t = 0 °C)	−4.20
L-Glutamic acid	147.13	2.23 (t = 0 °C)	−3.69
L-Aspartic acid	133.10	2.01 (t = 0 °C)	−3.89

Explanations: pKa—The acid dissociation constant refers to a specific equilibrium constant to represent the ability of an acid to dissociate hydrogen ions; logP—The logP value refers to the logarithm of the partition coefficient of a substance in n-octane (oil) and water.

3.3. *Grilling Tools and Cooking Procedures*

In this study, the sausages were grilled using an electric oven (Electrolux, EOT5004K, Guang dong, China). The sausages were placed on a metal grid 12 cm below the heating source and baked for 20 min at 240 °C and flipped every 5 min.

3.4. *Extraction and Clean-Up*

The extraction and clean-up procedures were performed as described by Farhadian et al. [28]. After completion of the heat treatment, the sausage was removed and cooled to room temperature. Five grams of minced sausage was precisely weighed and placed into a 50 mL centrifuge tube containing 25 mL of cyclohexane. After shaking for 5 min with a quick mixer (SK-1, Jintan is Jairel Electric Co., Ltd., China), the centrifuge tube was subjected to ultrasonic treatment for 20 min (SB-5200D; Xinzhi Biotechnology Co., Ltd., Ningbo, China). After sonication, the solution was applied to a Florisil solid phase extraction (SPE) column previously treated with dichloromethane (3 mL) and cyclohexane (5 mL). The PAHs adsorbed by the SPE column were eluted with 9 mL of cyclohexane dichloromethane (3:1 v/v). The eluate was evaporated by rotary evaporation (40 °C, 30 rmp) until only 1–2 mL of concentrate remained. The concentrate was transferred into a 10 mL centrifuge tube and dried in a nitrogen atmosphere. A total of 2 mL of acetonitrile was then added to dissolve the extract, which was then filtered through a 0.22 μm membrane for analysis by HPLC.

3.5. *HPLC Analysis of PAHs*

The PAH analyses were performed based on the method described by Farhadian et al. [28]. PAH analysis was carried out using an HPLC apparatus (Agilent, Santa Clara, CA, USA) equipped with a 600 controller pump, fluorescence detector (G1312, Agilent, USA) and a 20 μL loop injector. A PAH column (250 mm × 4.6 mm, 5 μm particle size) (Agilent, Santa Clara, CA, USA) was used. The mobile phase was constituted of acetonitrile and water. The elution conditions applied were: 0–3 min, 60% of acetonitrile isocratic; 3–15 min, 60–100% of acetonitrile; 15–46 min, 100% of acetonitrile isocratic; 46–53, 100–60% of acetonitrile, gradient. The flow rate was 1.0 mL min^{-1}. A fluorescence detector operated at excitation/emission wavelength 265/327 nm for NAP, 285/320 nm for Ac, 256/300 for FLU, 275/450 nm for FLT, 274/382 nm for BaA, 260/360 for CHR, 283/430 for BbFA and BkFA, 285/410 for BaP, 284/395 for DBahA, 290/410 for BPE, and 301/480 for IPY. Separation was performed under isocratic conditions. Each PAH sample solution was passed through a 0.22 μm filter before injection into the HPLC system. The quantification of PAHs was performed using an external calibration curve method. The quantification of 12 PAHs was carried out through the external standard method.

3.6. Method Validation

The calibration curve, linearity (tested through square correlation coefficients R^2) and limits of detection and quantification (LOD and LOQ) were obtained using the signal-tonoise ratio of S/N = 3 and S/N = 10, respectively [29]. Accuracy was evaluated using spiked sausage samples at four levels of concentration (ranging from 0.25 to 20.0 $\mu g\ kg^{-1}$), and recoveries were calculated. The recoveries were calculated from the differences in total amounts of each PAHs between the spiked and unspiked samples. For precision, the sample was spiked with two levels (ranging from 0.25 to 8.0 $\mu g\ kg^{-1}$) on two different days, by a single analyst using the same equipment, and the relative standard deviation (RSD) was determined.

3.7. Statistical Analysis

The sample data are provided as the mean $\pm$ standard deviation. The data were subjected to analysis of variance (ANOVA) and Duncan's multiple range test for statistical significance ($p < 0.05$) using the SPSS 17.0 software system (SPSS Inc., Chicago, IL, USA).

4. Conclusions

This study tested the effects of molecule properties of carbohydrates and amino acids on PAHs formation in grilled pork sausage. The results showed that addition of aldehyde-based D-glucose could significantly increase the PAHs production when compared with that of the keto-based D-fructose group. The grilled sausage with smaller molecular weight, D-glucose, was determined to have more PAHs content as compared with the sausage with larger molecular weight, 4-(α-D-Glucosido)-D-glucose and cellulose. The addition of basic amino acids (L-lysine, L-arginine) was capable of producing more PAHs in grilled sausages than adding acidic amino acids (L-glutamic, L-aspartate). The addition of the amino acid containing a benzene ring produced fewer PAHs than other amino acids in the grilled sausages. Our study suggests that systematic consideration of the molecule properties is necessary when using food additives (amino acids and carbohydrates) for food processing.

Author Contributions: K.C., C.C., B.X. and W.N. conceived and designed the experiments. W.N., Y.L., Y.W., J.G. and S.Z. performed the experiments. W.N. wrote the manuscript. K.C. revised the manuscript.

Acknowledgments: Financial support of this research provided by the National Science Foundation of China [31501585], Major Science and Technology Program of Anhui, China [18030701159] and project of Yang sheng's student community. We also thank Xingyong Chen of Anhui Agricultural University for grammar correction of the manuscript.

References

1. Iwegbue, C.M.; Onyonyewoma, U.A.; Bassey, F.I.; Nwajei, G.E. Concentrations and health risk of polycyclic aromatic hydrocarbons in some brands of biscuits in the nigerian market. *Hum. Ecol. Risk Assess* **2015**, *21*, 338–357. [CrossRef]
2. Simon, R.; Gomez, J.A.; Von, H.C.; Wenzl, T.E. Results of a european inter-laboratory comparison study on the determination of eu priority polycyclic aromatic hydrocarbons (PAHs) in edible vegetable oils. *Anal. Bioanal. Chem.* **2008**, *391*, 1397–1408. [CrossRef]
3. Authority, E.F. Findings of the efsa data collection on polycyclic aromatic hydrocarbons in food. *Efsa J.* **2007**, *5*, 3–55. [CrossRef]
4. Alomirah, H.; Al-Zenki, S.; Al-Hooti, S.; Zaghloul, S.; Sawaya, W. Concentrations and dietary exposure to polycyclic aromatic hydrocarbons (PAHs) from grilled and smoked foods. *Food Control* **2011**, *22*, 2028–2035. [CrossRef]

5. Raj, A.; Prada, I.D.; Amer, A.A. A reaction mechanism for gasoline surrogate fuels for large polycyclic aromatic hydrocarbons. *Combust. Flame.* **2012**, *159*, 500–515. [CrossRef]
6. Richter, H.; Howard, J.B. Formation of polycyclic aromatic hydrocarbons and their growth to sootda review of chemical reaction pathways. *Prog. Energy Combust. Sci.* **2000**, *26*, 565–608. [CrossRef]
7. Mojica, M.; Francisco, M.A. The Diels-Alder cycloaddition reaction of substituted hemifullerenes with 1,3-butadiene: Effect of electron-donating and electron-withdrawing substituents. *Molecules* **2016**, *21*, 200. [CrossRef]
8. Farhadian, A.; Jinap, S.; Abas, F.S. Determination of polycyclic aromatic hydrocarbons in grilled meat. *Food Control* **2010**, *21*, 606–610. [CrossRef]
9. Saito, E.; Tanaka, N.; Miyazaki, A. Concentration and particle size distribution of polycyclic aromatic hydrocarbons formed by thermal cooking. *Food Chem.* **2014**, *153*, 285–291. [CrossRef]
10. Santos, C.; Gomes, A. Polycyclic aromatic hydrocarbons incidence in Portuguese traditional smoked meat products. *Food Chem. Toxicol.* **2011**, *49*, 2343–2347. [CrossRef]
11. Singh, L.; Agarwalauthor, T. Polycyclic aromatic hydrocarbons formation and occurrence in processed food. *Food Chem.* **2016**, *199*, 768–781. [CrossRef] [PubMed]
12. Zhou, C.; Li, J.; Tan, S. Effect of L-lysine on the physicochemical properties of pork sausage. *Food Sci. Biotechnol.* **2014**, *23*, 775–780. [CrossRef]
13. Zhou, C.; Li, J.; Tan, S. Effects of L-arginine on physicochemical and sensory characteristics of pork sausage. *Adv. J. Food Sci. Technol.* **2014**, *6*, 660–667. [CrossRef]
14. Phillip, F.; Britt, A.C.; Buchanan, C.V.; Owens, J.D.J. Glucose enhance the formation of nitrogen containing polycyclic aromatic compounds and polycyclic aromatic hydrocarbons in the pyrolysis of proline. *Fuel* **2004**, *83*, 1417–1432. [CrossRef]
15. Ramesh, K.S.; Geoffrey, C. Formation of low molecular weight heterocycles and polycyclic aromatic compounds(PACs) in the pyrolysis of a-amino acids. *J. Anal. Appl. Pyrolysis* **2003**, *66*, 97–121. [CrossRef]
16. Naknean, P.; Meenune, M. Factors affecting retention and release of flavour compounds in food carbohydrates. *Int. Food Res. J.* **2010**, *17*, 23–34. [CrossRef]
17. Goubet, I.; Le, Q.J.; Voilley, A.J. Retention of aroma compounds by carbohydrates: Influence of their physicochemical characteristics and of their physical state. A review. *J. Agric. Food. Chem.* **1998**, *46*, 1981–1990. [CrossRef]
18. Xiong, Y.L.; Noel, D.C.; Moody, W.G. Textural and sensory properties of low-fat beef sausages with added water and polysaccharides as affected by pH and salt. *J. Food Sci.* **2010**, *64*, 5. [CrossRef]
19. Appendix A to 40 CFR Part 423, 2010. United States Environmental Protection Agency (US-EPA). Available online: http://www.epa.gov/waterscience/methods/pollutants.htm (accessed on 1 July 2018).
20. COMMISSION REGULATION (EU) No 835 2011 of 19 August 2011 Amending Regulation (EC) No 18812006 as Regards Maximum Levels for Polycyclic Aromatic Hydrocarbons in Foodstuffs. Available online: http://www.docin.com/p-275090873.html (accessed on 19 August 2018).
21. Chen, B.H.; Lin, Y.S. Formation of polycyclic aromatic hydrocarbons during processing of duck meat. *J. Agric. Food Chem.* **1997**, *45*, 1394–1403. [CrossRef]
22. Bansal, V.; Kim, K.H. Review of PAH contamination in food products and their health hazards. *Environ. Inter.* **2015**, *84*, 26–38. [CrossRef]
23. Britt, P.F.; Buchanan, A.C.; Owens, C.V.; Todd, S.J. Formation of nitrogen containing polycyclic aromatic hydrocarbons from the co-pyrolysis of carbohydrates and amino acids. *Div. Fuel Chem.* **2002**, *47*, 400–403. [CrossRef]
24. Kandhro, A.; Sherazi, S.T.; Mahesar, S.A.; Bhanger, M.I.; Younis, T.M. GC-MS quantification of fatty acid profile including trans fa in the locally manufactured margarines of pakistan. *Food Chem.* **2008**, *109*, 207–211. [CrossRef]
25. Masuda, Y.; Mori, K.; Kuratsune, M. Polycyclic aromatic hydrocarbons formed by pyrolysis of carbohydrates, amino acids, and fatty acids. *Gan* **1967**, *58*, 69–74. [CrossRef] [PubMed]
26. Thomas, E.; McGrath, W.; Geoffrey, C. Low temperature mechanism for the formation of polycyclic aromatic hydrocarbons from the pyrolysis of cellulose. *J. Anal. Appl. Pyrolysis* **2003**, *66*, 51–70. [CrossRef]
27. Wanwisa, W.; Kanithaporn, V. Effects of oil types and pH on carcinogenic polycyclic aromatic hydrocarbons (PAHs) in grilled chicken. *Food Control* **2017**, *79*, 119–125. [CrossRef]

28. Farhadian, A.; Jinap, S.; Hanifah, H.N.; Zaidul, I.S. Effects of meat preheating and wrapping on the levels of polycyclic aromatic hydrocarbons in charcoal-grilled meat. *Food Chem.* **2010**, *124*, 141–146. [CrossRef]
29. Juan, C.; Zinedine, A.; Moltó, J.C.; Idrissi, L.; Mañes, J. Aflatoxins levels in dried fruits and nuts from rabat-salé area, morocco. *Food Control* **2008**, *19*, 849–853. [CrossRef]

7

Discrimination of Natural Mature Acacia Honey based on Multi-Physicochemical Parameters Combined with Chemometric Analysis

Tianchen Ma [1], Haoan Zhao [2], Caiyun Liu [3], Min Zhu [2], Hui Gao [1], Ni Cheng [1,3] and Wei Cao [1,3,*]

1 School of Food Science and Engineering, Northwest University, Xi'an 710069, China
2 School of Chemical Engineering, Northwest University, Xi'an 710069, China
3 Bee Product Research Center of Shaanxi Province, Xi'an 710065, China
* Correspondence: caowei@nwu.edu.cn

Abstract: Honey maturity is an important factor in evaluating the quality of honey. We established a method for the identification of natural mature acacia honey with eighteen physicochemical parameters combined with chemometric analysis. The analysis of variance showed significant differences between mature and immature acacia honey in physicochemical parameters. The principal component analysis explained 82.64% of the variance among samples, and indicated that total phenolic content, total protein content, and total sugar (glucose, fructose, sucrose) were the major variables. The cluster analysis and orthogonal partial least squares-discriminant analysis demonstrated that samples were grouped in relation to the maturity coinciding with the results of the principal component analysis. Meanwhile, the 35 test samples were classified with 100% accuracy with the method of multi-physicochemical parameters combined with chemometric analysis. All the results presented above proved the possibility of identifying mature acacia honey and immature acacia honey according to the chemometric analysis based on the multi-physicochemical parameters.

Keywords: natural mature honey; immature honey; chemometric analysis; multi-physicochemical parameters

---

1. Introduction

Acacia honey is the natural sweet substance produced by honeybees, which collect nectar from the flowers of *Robinia pesudoacacia* (Figure 1), transform and combine it with specific substances of their own, store it, and leave it in the honeycomb to ripen and mature [1]. In the process of maturation, honey properties and chemical compositions are also changed as a result of biotransformation, biodegradation, and bioaccumulation, which have a great influence on the quality and authenticity of natural honey [2–5]. Foraging bees collect nectar through their proboscis, and place it in the proventriculus (honey stomach) [4,6–9]. Meanwhile, proteins and salivary enzymes from the hypopharyngeal glands of bees begin breaking down sugars from the nectar [7,10–13]. Hydrogen peroxide and gluconic acid, formed by the degradation of glucose oxidase, are partly able to suppress bacterial growth and are responsible for increasing the acidity of honey [14,15]. As nectar collection is completed, the hive bees continually digest nectar and hydrolyze sucrose into glucose and fructose by using bee digestive enzymes [2,7,8,16–19]. Glucose and fructose indicate about 75% of the sugars found in honey, which play important roles in honey quality control and authenticity. The ratio between fructose and glucose, as well as their concentration, are commonly used for predicting honey crystallization and are a beneficial index for the classification of monofloral honeys [20,21]. Another major transformation that occurs during the maturity process is moisture evaporation. The hive bees store digested nectar in the honeycomb cells and transfer it from one cell to another. Hive bees flutter their wings continuously to circulate air and to evaporate moisture from the honey to about 18% and eventually cover the cells with wax to seal them [2,7,22]. Reduction of moisture content below 18% is deemed to be a secure level for retarding yeast activity, decreasing the rate of fermentation and avoiding the appearance of undesirable flavor [23–27]. According to all the above, honey with less than 18% moisture, a sugar concentration above the saturation point, and sealed honeycomb cells may be considered as natural mature honey.

Figure 1. Photograph of mature honey demonstration basement and acacia tree flower.

Because of the influence of components change during the honey maturation process, natural mature honey has its own characteristics, which differ from immature honey. Biologically speaking, immature honey lacks many of the positive properties of natural mature honey. Natural mature honey is a highly complex product with around 200 different substances, which cannot be artificially emulated [28,29]. Physicochemical parameters can provide useful and complete information for the composition and properties of honey and are effective to assess the honey quality and authenticity [30–33]. Chemometric methods can lessen the complexity of large data sets and offer better explication and construction of data sets, as well as identify the natural clustering pattern and group variables based on similarities between samples [34–36]. Combining the great deal of data acquired from physicochemical parameters with chemometrics may be an excellent measure for discrimination of natural mature honey.

The purpose of our research is to identify natural mature honey by establishing the method of multi-physicochemical parameters combined with chemometric analysis. Our study stands out as the first report comparing mature honey with immature honey using principal component analysis (PCA), cluster analysis (CA), and orthogonal partial least squares-discriminant analysis (OPLS-DA). Mature honey was differentiated by evaluating the similarities and discriminant features of honey samples qualitatively, a method that is expected to be applicable to quality control and authenticity identification of natural mature honey.

2. Materials and Methods

2.1. Honey Sample

A total of 85 acacia (*Robinia pseudoacacia*) honey samples were collected from several geographical areas of Shaanxi, China (Table 1) and were kept at 4 °C prior to analysis. The botanical origin of the samples was confirmed by the method of Lutier and Vassiere [37]. All honey samples were collected from a specific mature honey demonstration basement (Figure 1). The natural mature acacia honey (A1–A29) was collected from honey that was capped in the hive and brewed by bees for 7 to 10 days. Immature acacia honey (B30–B85) was collected by hive bees that brewed honey for one to three days.

In addition, all honey samples were extracted using a honey extractor and filtered to remove beeswax and other debris.

Table 1. Characterization of the analyzed acacia honey samples.

Samples	Type of Honey	Botanical Source	Production Region	Predominant Pollen (%)
A1–A29	Monofloral	*Robinia pseudoacacia*	Yan'an, Shaanxi	88.38 ± 2.56
B30–B45	Monofloral	*Robinia pseudoacacia*	Yan'an, Shaanxi	87.21 ± 1.78
B46–B55	Monofloral	*Robinia pseudoacacia*	Chunhua, Shaanxi	86.31 ± 3.01
B56–B60	Monofloral	*Robinia pseudoacacia*	Luochuan, Shaanxi	82.61 ± 1.89
B61–B65	Monofloral	*Robinia pseudoacacia*	Fufeng, Shaanxi	85.77 ± 3.21
B66–B70	Monofloral	*Robinia pseudoacacia*	Qianyang, Shaanxi	83.60 ± 2.18
B71–B75	Monofloral	*Robinia pseudoacacia*	Longxian, Shaanxi	84.34 ± 1.63
B76–B80	Monofloral	*Robinia pseudoacacia*	Yongshou, Shaanxi	88.63 ± 3.84
B81–B85	Monofloral	*Robinia pseudoacacia*	Tongchuan, Shaanxi	86.67 ± 1.67

The honey samples numbered A16–A29, B37–B45, B46–B53, B56–B58, B61–B63, B66–B68, B71–B73, B76–B78, and B81–B83 were randomly selected for the calibration set, and the remaining 35 samples were used as test samples to verify the accuracy of this method.

2.2. Pollen Analysis

The botanical origin of the samples was determined using the method of Lutier and Vassiere [37]. For floral identification, honey samples (5 g) were thoroughly mixed with distilled water (5 mL), and centrifuged at 3000 rpm for 10 min, to separate the pollens. Samples of separated pollen grains were spread with the help of a brush on a slide containing a drop of lactophenol. The slides were examined microscopically at 45× magnification, using a bright-field microscope (Olympus, Tokyo, Japan). According to the different volumes, contours, grooves, holes, and other characteristics of pollen morphology, as well as pictures of different varieties, pollen varieties were identified. A total of 40 horizons and a certain number of pollen grains were observed (the total number of pollen should be more than 100 grains).

$$Pollen\ Content\ (\%) = \frac{a\ certain\ pollen\ number\ of\ 40\ horizons}{a\ total\ pollen\ number\ of\ 40\ horizons} \times 100\%$$

2.3. Physicochemical Properties

The methods used for the quantitative analysis of physicochemical properties were determined mainly according to the Association of Official Analytical Chemists (AOAC) [38]. The details of the methods used in this study are summarized in Supplementary Material.

2.4. HPLC Conditions

The contents of fructose, glucose, and sucrose were determined by high-performance liquid chromatography (HPLC) and a refractive index detector (Shodex R1-201H, Shanghai, China). The column was a Waters carbohydrate high performance (4.6 × 250 mm, 4 μm; Waters). Honey samples (5 g) were thoroughly mixed with ultrapure water (60 mL), and the total volume of the mixture was adjusted to 100 mL with acetonitrile. The mobile phase was 78% acetonitrile and 22% ultrapure water (*v*/*v*), using an isocratic method. The solutions were filtered through a 0.45 μm membrane filter prior to use. The column was operated at 35 °C, the detector pool temperature was 35 °C, the flow-rate was 1.0 mL min^{-1}, and the injection volume was 15 μL.

2.5. Data Analysis

The multivariate statistical analysis was analyzed by SIMCA (Version 14.1, Umetrics, Umeå, Sweden). The main chemometrics methods used were principal component analysis (PCA), cluster analysis (CA), and orthogonal partial least squares-discriminant analysis (OPLS-DA). PCA simplifies multiple indexes into a small number of comprehensive indexes and uses as many variables as possible to reflect the information of the original variables [39]. CA is the aggregation of samples according to the similarity degree of quality characteristics, and the most similar priority polymerization [40,41]. OPLS-DA is a regression modeling method from multiple dependent variables to multiple independent variables [42]. CA and PCA were used to analyze the comprehensive change of honey, with mean-centered, UV scaled, and log-transformed data before building the PCA model. OPLS-DA was carried out to discriminate features with mean-centered, Pareto scaled and log-transformed data, and the validation of the model was tested using seven-fold internal cross-validation and permutation tests for 200 times. Significant differences were determined using Mann–Whitney U tests, and $p < 0.05$ was considered to be statistically significant. In order to avoid the influence of moisture on other variables, we preprocessed the variables and adopted the dry-weight value of each variable, that is, normalized the moisture content of the variables, and used the ratio of the actual variable value to the content of moisture as the dry weight value of the corresponding variables.

3. Results and Discussion

3.1. Pollen Analysis

Table 1 shows the floral origin of acacia honeys determined by microscopy pollen analysis. The data indicate that all the honey samples were monofloral. *Robinia pseudoacacia* pollen was detected in all samples more than 80.00%.

3.2. Physicochemical Parameters Analysis

Table 2 shows the mean values of the physicochemical parameters of natural mature acacia honey (NMH) and immature acacia honey (IMH) from different regions of the Shaanxi Province. Comparing the physicochemical parameters of NMH and IMH, we found significant differences between NMH and IMH in the mean values of moisture, sugars, protein content, total phenolic, and proline (Figure 2).

Moisture is an important standard for evaluating honey quality, as it can determine the shelf life of honey and its ability to resist fermentation deterioration [43]. The European and Codex standards established a limit of 20% in the case of honey, which can be kept for long periods of time without becoming spoiled [2,9,27]. The moisture of NMH ranged from 15.63% to 16.61%, and IMH was between 21.14% and 26.61%, exceeding the limit, making it difficult to store, producing acids and alcohols more easily, and seriously affecting the quality of honey [5,20].

Honey is a saturated solution of sugars, which accounts for about 70–75% of soluble sugar. Fructose and glucose account for the largest proportion of honey composition, but a small quantity of sucrose was also discovered [25,30]. They are the building blocks of more complex sugars such as disaccharides and maltose [33]. The content was within the limits of European and Codex standards of 65% minimum for glucose and fructose, where sucrose content should be not more than 5%. The sugar concentration (fructose and glucose) of NMH was approximately from 74.01% to 84.06%, which confirmed that samples were genuine honeys [30,32,44].

The total protein content of NMH was between 510.49 mg/kg and 622.29 mg/kg and that of IMH was lower (369.79–538.35 mg/kg). It is universally known that honey contains a trace amount of protein, usually formed by bees, which constantly swallow honey and flutter their wings [31]. The variability in the protein content of different maturity of types honey may be related to its brewing time and brewing degree [45].

During the ripening process, enzymes are proteins that help to speed up many chemical reactions in living organisms, and to convert certain substances into different products [46], especially diastase, invertase, and glucose oxidase. Owing to the high enzyme activity, NMH showed a richer nutritional value than IMH.

Table 2. Physicochemical property of samples (dry-weight basis).

Samples	L*	a*	b*	Conductivity (μS/cm)	pH	Free Acid (meq/kg Dry Matter)	Lacton (meq/kg Dry Matter)	Acid Value (meq/kg Dry Matter)	HMF (mg/kg Dry Matter)
A1–A29	47.60 ± 2.82 ^{a}	101.99 ± 4.30 ^{a}	23.49 ± 6.72	136.25 ± 3.38 ^{g}	3.17 ± 0.06 ^{c}	27.17 ± 1.72 ^{c}	5.98 ± 0.99 ^{a,b}	33.15 ± 1.58 ^{d}	N.D
B30–B45	70.99 ± 1.53 ^{b,c}	124.24 ± 2.29 ^{b}	8.79 ± 2.73 ^{a}	108.66 ± 5.58 ^{c,d}	2.88 ± 0.08 ^{a,b}	21.21 ± 1.09 ^{a,b}	4.94 ± 1.71 ^{a}	26.15 ± 2.38 ^{a,b}	N.D
B46–B55	70.71 ± 0.71 ^{b,c}	125.64 ± 1.71 ^{b}	3.62 ± 0.08 ^{a}	103.70 ± 1.58 ^{c}	2.89 ± 0.07 ^{a,b}	20.27 ± 0.74 ^{a}	3.78 ± 0.57 ^{a}	24.05 ± 0.94 ^{a}	N.D
B56–B60	72.95 ± 0.86 ^{d}	123.25 ± 2.41 ^{b}	5.92 ± 0.35 ^{a}	113.88 ± 0.67 ^{d,e}	2.94 ± 0.05 ^{a,b}	22.77 ± 0.25 ^{b}	4.76 ± 0.49 ^{a}	27.54 ± 0.25 ^{b}	N.D
B61–B65	72.25 ± 1.07 ^{c}	129.64 ± 1.88 ^{b}	4.01 ± 0.29 ^{a}	110.71 ± 3.93 ^{c,d}	2.89 ± 0.04 ^{a,b}	23.04 ± 0.86 ^{b}	4.86 ± 0.49 ^{a}	27.91 ± 1.31 ^{b,c}	N.D
B66–B70	71.45 ± 0.55 ^{b,c}	125.50 ± 1.65 ^{b}	3.75 ± 0.13 ^{a}	84.61 ± 2.25 ^{a}	3.09 ± 0.03 ^{c}	23.83 ± 0.49 ^{b}	8.79 ± 0.55 ^{c}	31.62 ± 1.04 ^{d}	N.D
B71–B75	69.48 ± 0.44 ^{b,c}	116.85 ± 0.58 ^{b}	6.29 ± 0.21 ^{a}	117.68 ± 4.80 ^{f}	2.83 ± 0.02 ^{a}	22.94 ± 0.83 ^{b}	7.98 ± 0.35 ^{b,c}	30.92 ± 0.84 ^{c,d}	N.D
B76–B80	71.65 ± 0.16 ^{b,c}	130.71 ± 1.11 ^{b}	2.56 ± 0.17 ^{a}	93.43 ± 2.15 ^{b}	2.99 ± 0.02 ^{b}	23.48 ± 0.43 ^{b}	7.35 ± 0.63 ^{b,c}	30.83 ± 0.98 ^{c,d}	N.D
B81–B85	68.34 ± 0.14 ^{b}	120.92 ± 1.44 ^{b}	3.73 ± 0.21 ^{a}	105.69 ± 1.09 ^{c}	2.86 ± 0.02 ^{a,b}	23.25 ± 0.59 ^{b}	7.75 ± 0.19 ^{b,c}	31.00 ± 0.44 ^{c,d}	N.D

Samples	Glucose (g/100 g Dry Matter)	Fructose (g/100 g Dry Matter)	Sucrose (g/100 g Dry Matter)	Total Sugar (g/100 g Dry Matter)	Total Phenolic (mg/kg Dry Matter)	Total Protein (mg/kg Dry Matter)	Amylase Activity (º Gothe)	Proline (mg/kg Dry Matter)	Glucose Oxidase (U/g Dry Matter)
A1–A29	26.04 ± 1.22 ^{b,c}	37.39 ± 1.12 ^{c}	1.97 ± 0.26 ^{d}	65.40 ± 1.71 ^{c}	126.59 ± 7.85 ^{c}	454.09 ± 11.48 ^{f}	39.15 ± 2.44 ^{f}	343.35 ± 11.42 ^{f}	1.26 ± 0.21 ^{a}
B30–B45	25.20 ± 0.36 ^{b}	34.07 ± 0.50 ^{b}	1.17 ± 0.17 ^{a,b}	60.44 ± 0.58 ^{b}	89.06 ± 1.21 ^{b}	361.14 = 10.13 ^{c,d}	33.86 ± 1.65 ^{d,e}	238.21 ± 11.65 ^{d}	2.36 ± 0.18 ^{b}
B46–B55	25.18 ± 0.24 ^{b}	33.93 ± 0.73 ^{b}	1.15 ± 0.11 ^{a,b}	60.26 ± 0.66 ^{b}	87.81 ± 2.46 ^{b}	388.65 ± 2.28 ^{d,e}	35.50 ± 1.23 ^{e}	232.51 ± 1.54 ^{d}	2.29 ± 0.30 ^{b}
B56–B60	25.61 ± 0.45 ^{b,c}	34.07 ± 0.43 ^{b}	1.43 ± 0.03 ^{b,c}	61.11 ± 0.10 ^{b}	88.71 ± 1.12 ^{b}	329.02 = 3.11 $^{a,b\,c}$	33.34 ± 1.03 ^{d,e}	266.43 ± 13.60 ^{e}	2.53 ± 0.16 ^{b}
B61–B65	24.67 ± 0.53 ^{b}	34.11 ± 0.87 ^{b}	1.14 ± 0.15 ^{a,b}	59.93 ± 0.59 ^{b}	79.35 ± 6.51 ^{a,b}	300.09 ± 11.81 ^{a}	29.62 ± 1.60 ^{b,c}	183.34 ± 0.41 ^{b}	3.25 ± 0.61 ^{c}
B66–B70	27.28 ± 1.14 ^{c}	35.61 ± 0.53 ^{b}	1.41 ± 0.15 ^{b,c}	64.29 ± 1.15 ^{c}	74.78 ± 4.65 ^{a}	416.62 ± 8.67 ^{e}	27.61 ± 1.38 ^{b}	133.24 ± 3.41 ^{a}	1.96 ± 0.58 ^{b}
B71–B75	24.89 ± 0.59 ^{b}	34.31 ± 0.83 ^{b}	0.89 ± 0.11 ^{a}	60.09 ± 1.48 ^{b}	89.39 ± 2.16 ^{c}	319.56 = 12.93 ^{a,b}	31.41 ± 0.68 ^{c,d}	213.51 ± 2.637 ^{c}	2.37 ± 0.86 ^{b}
B76–B80	24.88 ± 0.31 ^{b}	34.32 ± 0.13 ^{b}	1.68 ± 0.46 ^{c,d}	60.88 ± 0.68 ^{b}	74.97 ± 0.89 ^{a}	350.63 ± 12.13 $^{b,c\,d}$	21.40 ± 0.59 ^{a}	179.95 ± 0.32 ^{b}	1.22 ± 0.21 ^{a}
B81–B85	22.46 ± 1.86 ^{a}	30.52 ± 1.79 ^{a}	1.54 ± 0.07 ^{b,c}	54.52 ± 1.67 ^{a}	82.92 ± 2.77 ^{a,b}	366.10 ± 8.53 ^{c,d}	21.82 ± 1.03 ^{a}	199.24 ± 1.97 ^{b,c}	0.66 ± 0.15 ^{a}

A1–A29: mature honey; B30–B85: immature honey; N.D: not detected. Results presented in the table are expressed as the mean values ± standard deviation (SD). Different lower case letters correspond to significant differences at $p < 0.05$.

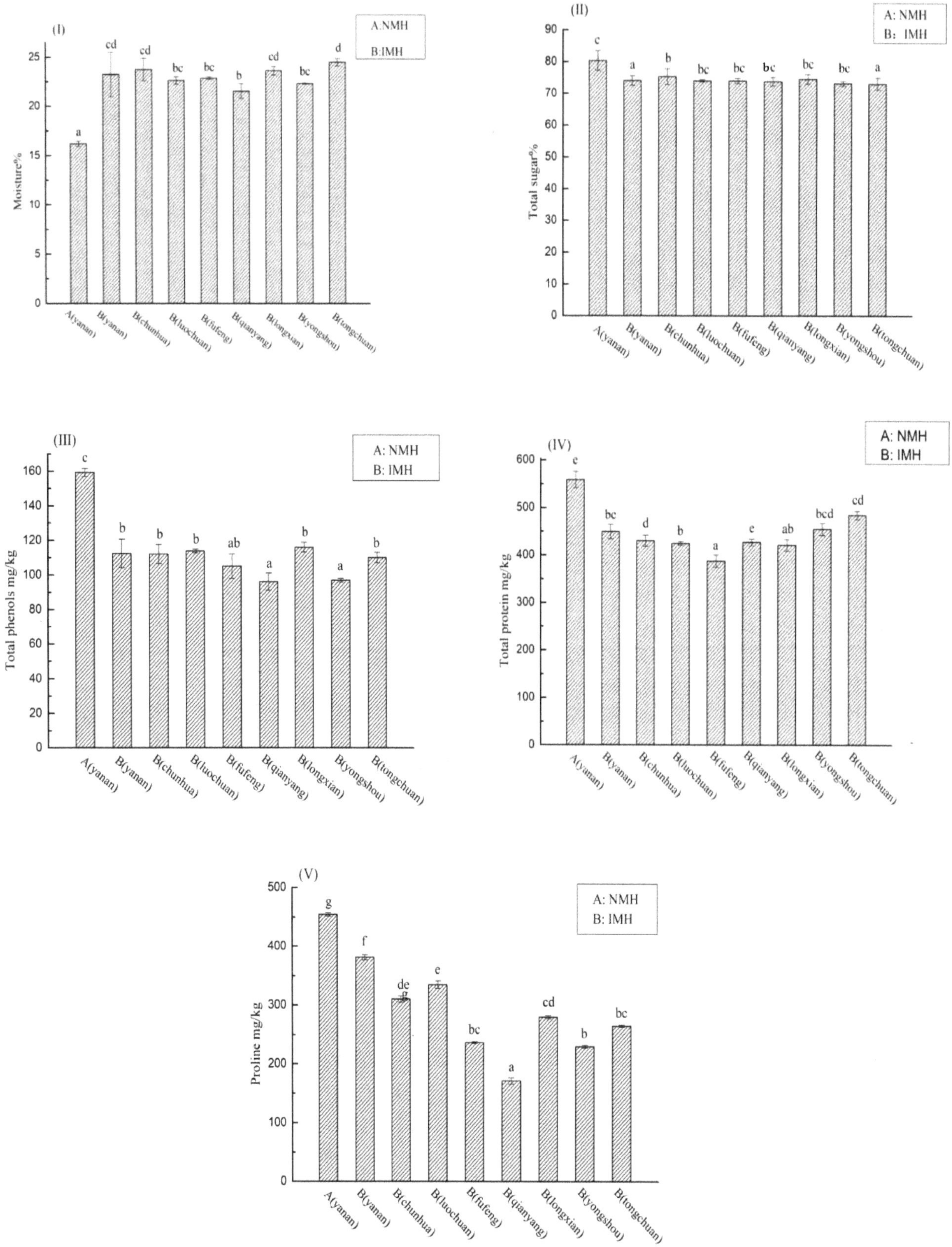

Figure 2. Significant physicochemical parameters of natural mature acacia honey (NMH) and immature acacia honey (IMH). Different lower case letters correspond to significant differences at $p < 0.05$. (I) represents moisture; (II) represents total sugar content; (III) represents total phenols content; (IV) represents total protein content; (V) represents proline content.

The electrical conductivity (EC) is based on the acid contents and ash of honey [27,31], and the free acidity is the content of free acids, which is determined by the equivalence point titration [47]. Furthermore, the pH is correlative with the stability and the shelf life of honey. The pH values of honey usually range from 3.5 to 5.5 [30,33]. The results of statistical analysis were not significantly different in electrical conductivity, pH, and free acidity. 5-hydroxymethylfurfural (HMF) is considered to be a useful indicator for heat treatment and long-term storage of honey [36]. An excessive amount of HMF has been considered evidence of overheating, characterized by a darkening of color and a loss of freshness of honey [23]. In this study, HMF was found to be below the detection limit, indicating that all samples were fresh and not overheating.

3.3. Principal Component Analysis (PCA)

PCA is a statistical method that simplifies several related indicators into a few comprehensive indicators [48]. PCA was combined with the physicochemical parameters in Table 2, in order to avoid the influence of moisture on other physicochemical parameters. The moisture was eliminated when the model was established, and the data of other parameters were normalized and used on a dry-weight basis. The principal component analysis results of the 17 indicators for 50 samples are shown in Table 3. There were three principal components, which are in agreement with the results in Table 3.

Table 3. Variance contribution rate and composition load matrix. PC, principal component.

Quality Index	PC1	PC2	PC3
L*	−0.963	0.029	0.153
a*	−0.870	0.014	0.175
b*	0.905	0.069	−0.134
Conductivity (μS/cm)	0.847	0.142	−0.333
pH	0.852	−0.142	0.218
Free acid (meq/kg)	0.829	−0.314	−0.034
Lacton (meq/kg)	0.106	−0.875	0.240
Acid value (meq/kg)	0.676	−0.651	0.088
Glucose %	0.425	0.195	0.785
Fructose %	0.817	0.164	0.338
Sucrose %	0.777	−0.252	−0.113
Total sugar %	0.814	0.154	0.533
Total phenolic content (mg/kg)	0.936	0.146	−0.160
Total Protein (mg/kg)	0.798	0.073	0.045
Amylase activity (°Gothe)	0.654	0.692	0.008
Proline (mg/kg)	0.872	0.294	−0.301
Glucose oxidase (U/g)	−0.538	0.516	0.234
Eigenvalues	10.208	2.343	1.497
Contribution rate %	60.045	13.785	8.806
Cumulative contribution%	60.045	73.830	82.636

We standardized data to ensure that all the elements had an equal influence over the results. The eigenvalues and the percentage variance, explained by principal components, are shown in Table 3. Three components, eigenvalues > 1, were extracted and used to examine the dataset. The first three components accounted for 82.64% of the total variance. Principal component 1 (PC1) expressed 60.05% of the variance, and the next principal components explained 13.79% and 8.81% of the variance, respectively. The loadings of each compound on the principal component analysis explicitly showed that the grouping of the different maturity honey was mainly influenced by certain compounds. PC1 and PC2 of all the samples explained 73.83% of the total variance at length. PC1 was directly relevant to L*, total phenolic content, and proline, and the dominant variables were mainly affected by protein, total phenol, and sucrose in PC2 (Figure 3B: loading). Moreover, honey samples were properly classified, NMH was classified into one category, and IMH was separated into another category

(Figure 3A: scores of sample). In summary, mature honey samples are significantly different from immature honey, which is in agreement with the result of the physicochemical parameters analysis.

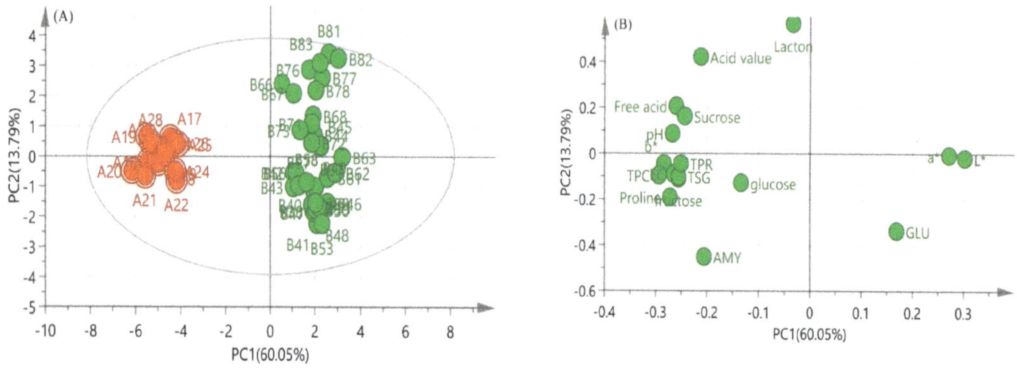

Figure 3. The principal component analysis (PCA) score plots (**A**) and PCA loading plots (**B**) of acacia honey samples. GLU: glucose oxidase; TSG: total sugar; AMY: amylase activity; EC: conductivity; TPC: total phenolic content; TPR: total protein content (for interpretation of the references to color in this figure legend, the reader is referred to the Web version of this article).

3.4. Cluster Analysis

Cluster analysis is a kind of statistical method used to classify the characteristics of multi-indicators and multi-objects [34]. It gradually aggregates according to the similarity of sample quality characteristics. The greatest degree of similarity is aggregated, and multiple varieties are integrated according to the comprehensive nature of the categories [41]. In this study, based on the physicochemical parameters, 50 samples were subjected to the Ward method. The cluster analysis of the pedigree chart is shown in Figure 4. Acacia honey samples were separated into two categories. The first category contained NMH and the second group, namely the remaining samples, contained IMH. The distance between these two categories was more than 100.

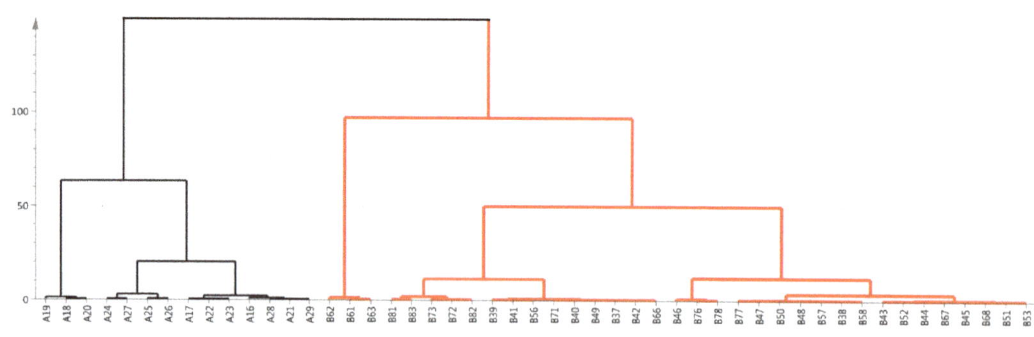

Figure 4. Results of hierarchical cluster analysis of samples.

According to the CA and the PCA, all honey samples were divided into two types, which were basically consistent with the analysis of the parameters mentioned above: color, electrical conductivity, pH, free acid, lactone, total acid, glucose, fructose, sucrose, total sugar content, total phenolic content, amylase activity, HMF, total protein content, glucose oxidase, and other basic physicochemical parameters.

3.5. Orthogonal Partial Least Squares-Discriminant Analysis

Orthogonal partial least squares-discriminant analysis (OPLS-DA) was developed to distinguish between NMH and IMH, and score plots of these models were applied for separating samples. The cross-validation method was used to verify the model. A total of three principal components were selected. The fitted model's index R^2X (cum) was 0.928, indicating that the four principal components explained 92.80% of the X variables, and the index of fitting the dependent variable R^2Y (cum) was 0.978, indicating that the four principal components interpreted 97.80% of the Y variable. The model prediction index Q^2 (cum) was 0.971, explaining that the model had a predictive power of 97.10% for mature honey and immature honey and that this model was stable and reliable. Permutation was conducted 200 times in order to further test the predictability of the OPLS-DA model. To show the predictability, all R^2 and Q^2 (Figure 5B) were > 0 and <-0.5, respectively.

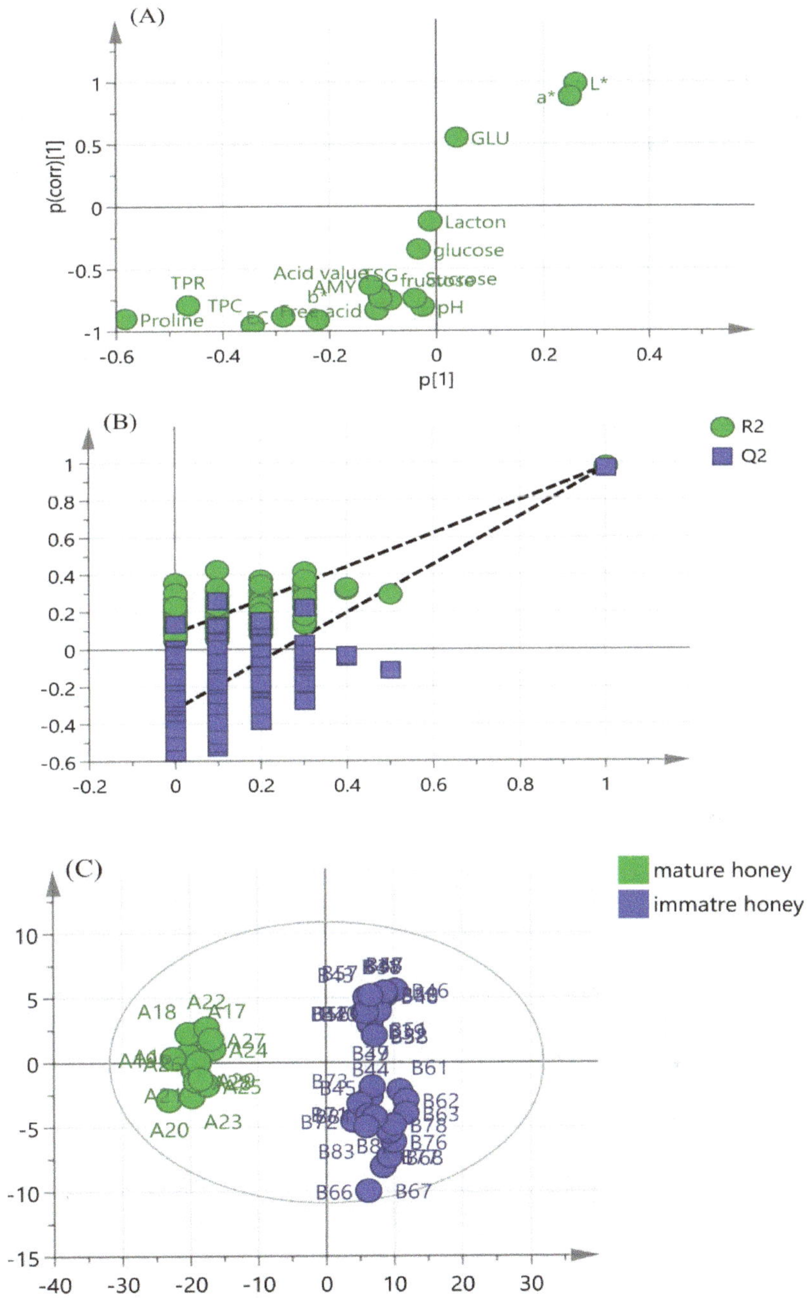

Figure 5. The orthogonal partial least squares-discriminant analysis (OPLS-DA) S-plots (**A**), validation plot (**B**), and score plots (**C**) of honey samples. P [1] is the loading vector of covariance in the first principal component. P (corr) [1] is loading vector of correlation in the first principal component. Variables with $|p| \geq 0.05$ and $|p\ (\text{corr})| \geq 0.5$ are considered statistically significant. R^2 is the fitted model's index; Q^2 is the model prediction index. GLU: glucose oxidase; TSG: total sugar; AMY: amylase activity; EC: conductivity; TPC: total phenolic content; TPR: total protein content (for interpretation of the references to color in this figure legend, the reader is referred to the Web version of this article).

An S-plot was adopted to visualize the influence of sensitive indices on the mature and immature honey (Figure 5A). Discriminant markers are located in the upper right and lower left corners of the S-plot, with higher absolute p [1] and p (corr) values to select the potential markers [49]. In this model, the sensitive markers were proline, total protein content, EC, L*, a*, b*, and total phenolic content. The variable importance to projection (VIP) values of these indicators are more important than 1. Figure 5 demonstrates a direct separation between NMH and IMH.

3.6. Test Samples Analysis

Thirty-five samples were tested. The results of the PCA score plots (Supplementary Figure S1) show that NMH samples were classified into one category and IMH samples into another. We took distance as a dependent variable and the test samples as independent variables for cluster analysis. Supplementary Figure S2 shows that 35 test samples were properly classified. The accuracy of both the method and cross-verification is 100%.

The score plots of the OPLS-DA models were developed for separating test samples. The R^2X of test honey supervised models was 0.901, R^2Y was 0.985, and Q^2 calculated from seven-fold cross-validation was 0.978. All R^2 and Q^2 were > 0 and <-0.5, respectively, showing a predictability. The classification results are shown in Supplementary Figure S3A,B. Thirty-five test samples were correctly classified on the basis of their maturity. The overall correct classification rate of the original and cross-validation methods is 100%.

Both the 50 honey samples and the 35 test samples achieved appropriate classification results according to the physicochemical parameters combined with chemometric methods. Therefore, physicochemical parameters combined with chemometric methods could be used in the classification of NMH and IMH.

4. Conclusions

In conclusion, the current results clearly show that there are significant differences in physicochemical parameters between natural mature acacia honey and immature acacia honey. Principal component analysis showed that total phenolic content, total protein content, and total sugar (glucose, fructose, sucrose) were the main parameters affected seriously by honey maturity. The cluster analysis and orthogonal partial least squares-discriminant analysis showed that samples were grouped in relation to the maturity and the overall correct classification rate reached 100%. The approach of multi-physicochemical parameters combined with chemometrics is effective in discriminating natural acacia mature honey. However, to confirm the applicability of the other monofloral honeys, this approach needs to be validated. We look forward to the development of this as a promising method for authenticity identification of natural mature honey in any quality control of businesses and government agencies.

Author Contributions: T.M. and W.C. designed the study and interpreted the results. C.L. and M.Z. collected test data. T. M., W.C., H.G., and N.C. drafted the manuscript.

Acknowledgments: We are grateful to Liming Wu and Xiaofeng Xue from the Institute of Apicultural Research, Chinese Academy of Agricultural Sciences for their assistance, and also thank the beekeepers for providing honey samples.

References

1. Council, E.U. Council Directive 2001/110/EC of 20 December 2001 relating to honey Official. *J. Eur. Communities L* **2002**, *10*, 47–52.
2. Ball, D.W. The chemical composition of honey. *J. Chem. Educ.* **2007**, *84*, 1647–1650. [CrossRef]
3. Chua, L.S.; Abdul-Rahaman, N.L.; Sarmidi, M.R.; Aziz, R. Multi-elemental composition and physical properties of honey samples from malaysia. *Food Chem.* **2012**, *135*, 880–887. [CrossRef]

4. Verma, L.R.; Partap, U. Foraging behaviour of apis cerana on cauliflower and cabbage and its impact on seed production. *J. Apic. Res.* **1994**, *33*, 231–236. [CrossRef]
5. Di Rosa, A.R.; Leone, F.; Cheli, F.; Chiofalo, V. Novel approach for the characterisation of Sicilian honeys based on the correlation of physicochemical parameters and artificial senses. *Ital. J. Anim. Sci.* **2018**, *18*, 389–397. [CrossRef]
6. Bailey, L. The action of the proventriculus of the honeybee (Apis Mellifera L.). *Bee World* **1952**, *32*, 92. [CrossRef]
7. Graham, J.M. The hive and the honey bee. *Bull. ESA* **1992**, *10*, 62.
8. Blatt, J.; Roces, F. The control of the proventriculus in the honeybee (Apis mellifera carnica L.) ii. Feedback mechanisms. *J. Insect Physiol.* **2002**, *48*, 683–691. [CrossRef]
9. Pita-Calvo, C.; Guerra-Rodriguez, M.E.; Vazquez, M. A review of the analytical methods used in the quality control of honey. *J. Agric. Food Chem.* **2017**, *65*, 690–703. [CrossRef]
10. Crailsheim, K. Intestinal transport of sugars in the honeybee (Apis mellifera L.). *J. Insect Physiol.* **1988**, *34*, 840–845. [CrossRef]
11. Almamary, M.; Almeeri, A.; Alhabori, M. Antioxidant activities and total phenolics of different types of honey. *Nutr. Res.* **2002**, *22*, 1041–1047. [CrossRef]
12. Cocker, L. The enzymic production of acid in honey. *J. Sci. Food Agric.* **2010**, *2*, 411–414. [CrossRef]
13. Oroian, M.; Ropciuc, S. Honey authentication based on physicochemical parameters and phenolic compounds. *Comput. Electron. Agric.* **2017**, *138*, 148–156. [CrossRef]
14. White, J.W.; Subers, M.H.; Schepartz, A.I. The identification of inhibine, the antibacterial factor in honey, as hydrogen peroxide and its origin in a honey glucose-oxidase system. *Biochim. Biophys. Acta* **1963**, *73*, 57–70. [CrossRef]
15. Zhu, W.; Hu, F.; Li, Y.; Zhan, Y. The antibacterial mechanism and the affected factors of honey. *Nat. Prod. Res. Dev.* **2004**, *16*, 372–374.
16. Free, J.B. Biology and behaviour of the honey bee Apis florea, and possibilities for beekeeping. *Bee World* **1981**, *62*, 46–59. [CrossRef]
17. Crane, E. Honey from honeybees and other insects. *Ethol. Ecol. Evol.* **1991**, *3*, 100–105. [CrossRef]
18. Peng, Y.S.; Marston, J.M. Filtering mechanism of the honey bee proventriculus. *Physiol. Entomol.* **2010**, *11*, 433–439. [CrossRef]
19. Bentabol Manzanares, A.; García, Z.H.; Galdón, B.R.; Rodríguez, E.R.; Romero, C.D. Differentiation of blossom and honeydew honeys using multivariate analysis on the physicochemical parameters and sugar composition. *Food Chem.* **2011**, *126*, 664–672. [CrossRef]
20. Silva, P.M.D.; Gauche, C.; Gonzaga, L.V. Honey chemical composition, stability and authenticity. *Food Chem.* **2016**, *196*, 309–323. [CrossRef]
21. Wu, L.; Du, B.; Vander Heyden, Y.; Chen, L.; Zhao, L.; Wang, M.; Xue, X. Recent advancements in detecting sugar-based adulterants in honey—A challenge. *TrAC Trends Anal. Chem.* **2017**, *86*, 25–38. [CrossRef]
22. Bogdanov, S.; Jurendic, T.; Sieber, R.; Gallmann, P. Honey for nutrition and health: A review. *J. Am. Coll. Nutr.* **2009**, *27*, 677–689. [CrossRef]
23. Subramanian, R.; Umesh Hebbar, H.; Rastogi, N.K. Processing of honey: A review. *Int. J. Food Prop.* **2007**, *10*, 127–143. [CrossRef]
24. Alvarez-Suarez, J.M.; Tulipani, S.; Romandini, S.; Bertoli, E.; Battino, M. Contribution of honey in nutrition and human health: A review. *Nutr. Metab.* **2010**, *3*, 15–23. [CrossRef]
25. Doner, L.W. The sugars of honey—A review. *J. Sci. Food Agric.* **2010**, *28*, 443–456. [CrossRef]
26. Gupta, R.K.; Reybroeck, W.; Veen, J.W.V.; Gupta, A. Beekeeping for poverty alleviation and livelihood security: II Quality control of honey and bee products. In *Beekeeping for Poverty Alleviation and Livelihood Security*; Springer: Berlin, Germany, 2014.
27. Pita-Calvo, C.; Vázquez, M. Differences between honeydew and blossom honeys: A review. *Trends Food Sci. Technol.* **2017**, *59*, 79–87. [CrossRef]
28. Grout, R.A. American Bee Journal Editor, M.G. Dadant. *Am. Bee J.* **2018**, *158*, 741–747.
29. Jibril, F.I.; Hilmi, A.B.M.; Manivannan, L. Isolation and characterization of polyphenols in natural honey for the treatment of human diseases. *Bull. Nat. Res. Cent.* **2019**, *43*, 4–5. [CrossRef]
30. Silvano, M.F.; Varela, M.S.; Palacio, M.A.; Ruffinengo, S.; Yamul, D.K. Physicochemical parameters and sensory properties of honeys from buenos aires region. *Food Chem.* **2014** *152*, 500–507. [CrossRef]

31. Sohaimy, S.A.E.; Masry, S.H.D.; Shehata, M.G. Physicochemical characteristics of honey from different origins. *Ann. Agric. Sci.* **2015**, *60*, 279–287. [CrossRef]
32. Siddiqui, A.J.; Musharraf, S.G.; Choudhary, M.I.; Rahman, A. Application of analytical methods in authentication and adulteration of honey. *Food Chem.* **2017**, *217*, 687–698. [CrossRef] [PubMed]
33. Popek, S.; Halagarda, M.; Kursa, K. A new model to identify botanical origin of polish honeys based on the physicochemical parameters and chemometric analysis. *LWT-Food Sci. Technol.* **2017**, *77*, 482–487. [CrossRef]
34. Yücel, Y.; Demir, C. Principal component analysis and cluster analysis for the characterisation of marbles by capillary electrophoresis. *Talanta* **2004**, *63*, 451–459. [CrossRef] [PubMed]
35. Yücel, Y.; Sultanoğlu, P. Characterization of hatay honeys according to their multi-element analysis using ICP-OES combined with chemometrics. *Food Chem.* **2013**, *140*, 231–237. [CrossRef] [PubMed]
36. Zhao, J.; Du, X.; Cheng, N.; Chen, L.; Xue, X.; Wu, L.; Cao, W. Identification of monofloral honeys using HPLC-ECD and chemometrics. *Food Chem.* **2016**, *194*, 167–174. [CrossRef] [PubMed]
37. Lutier, P.M.; Vaissière, B.E. An improved method for pollen analysis of honey. *Rev. Palaeobot. Palynol.* **1993**, *78*, 129–144. [CrossRef]
38. Association of Official Analytical Chemists (AOAC). Available online: www.aoac.org/aoac_prod_imis/AOAC/Publications/Official_Methods_of_Analysis/AOAC_Member/Pubs/OMA/AOAC_Official_Methods_of_Analysis.aspx (accessed on 21 July 2019).
39. Geladi, P.; Isaksson, H.; Lindqvist, L.; Wold, S.; Esbensen, K. Principal component analysis of multivariate images. *Chemom. Intell. Lab.* **1989**, *5*, 209–220. [CrossRef]
40. Everitt, B. *Cluster Analysis*; Quality & Quantity; Kluwer Academic Publishers: Norwell, MA, USA, 1980; Volume 14, pp. 75–100.
41. Everitt, B. *The Cambridge Dictionary of Statistics*; Cambridge University Press: Cambridge, UK, 2002.
42. Boccard, J.; Rutledge, D.N. A consensus orthogonal partial least squares discriminant analysis (OPLD-DA) strategy for multiblock omics data fusion. *Analytica Chimica Acta* **2013**, *769*, 30–39. [CrossRef]
43. Ouchemoukh, S.; Louaileche, H.; Schweitzer, P. Physicochemical characteristics and pollen spectrum of some algerian honeys. *Food Control* **2007**, *18*, 52–58. [CrossRef]
44. Lazarević, K.B.; Andrić, F.; Trifković, J.; Tešić, Ž.; Milojković-Opsenica, D. Characterisation of serbian unifloral honeys according to their physicochemical parameters. *Food Chem.* **2012**, *132*, 2060–2064. [CrossRef]
45. Azeredo, L.D.C.; Azeredo, M.A.A.; Souza, S.R.D.; Dutra, V.M.L. Protein contents and physicochemical properties in honey samples of apis mellifera of different floral origins. *Food Chem.* **2003**, *80*, 249–254. [CrossRef]
46. Weirich, G.F.; Collins, A.M.; Williams, V.P. Antioxidant enzymes in the honey bee, Apis mellifera. *Apidologie* **2002**, *33*, 3–14. [CrossRef]
47. Alqarni, A.S.; Owayss, A.A.; Mahmoud, A.A. Physicochemical characteristics, total phenols and pigments of national and international honeys in saudi arabia. *Arab. J. Chem.* **2016**, *9*, 114–120. [CrossRef]
48. Ares, G.; Jach, A. Applications of principal component analysis (PCA) in food science and technology. In *Mathematical and Statistical Methods in Food Science and Technology*; John Wiley & Sons, Ltd.: Hoboken, NJ, USA, 2014.
49. Zhao, H.; Cheng, N.; Zhang, Y.; Sun, Z.; Zhou, W.; Wang, Y.; Cao, W. The effects of different thermal treatments on amino acid contents and chemometric-based identification of overheated honey. *LWT-Food Sci. Technol.* **2018**, *96*, 133–139. [CrossRef]

8

Selection and Identification of Novel Aptamers Specific for Clenbuterol based on ssDNA Library Immobilized SELEX and Gold Nanoparticles Biosensor

Xixia Liu [1], Qi Lu [1], Sirui Chen [1], Fang Wang [1], Jianjun Hou [1,*], Zhenlin Xu [2,*], Chen Meng [1], Tianyuan Hu [1] and Yaoyao Hou [1]

1 Hubei Key Laboratory of Edible Wild Plants Conservation and Utilization, Hubei Normal University, Cihu Road, Huangshigang District, Huangshi 435002, China; liuxixia1@163.com (X.L.); 18271691835@163.com (Q.L.); 17371958850@163.com (S.C.); WF_0129@163.com (F.W.); 13329967633@163.com (C.M.); 18771808527@sina.cn (T.H.); dxhe@163.com (Y.H.)

2 Guangdong Provincial Key Laboratory of Food Quality and Safety, South China Agricultural University, Wushan Road, Tianhe District, Guangzhou 510642, China

* Correspondence: jjhou@mail.hzau.edu.cn (J.H.); jallent@163.com (Z.X.)

Abstract: We describe a multiple combined strategy to discover novel aptamers specific for clenbuterol (CBL). An immobilized ssDNA library was used for the selection of specific aptamers using the systematic evolution of ligands by exponential enrichment (SELEX). Progress was monitored using real-time quantitative PCR (Q-PCR), and the enriched library was sequenced by high-throughput sequencing. Candidate aptamers were picked and preliminarily identified using a gold nanoparticles (AuNPs) biosensor. Bioactive aptamers were characterized for affinity, circular dichroism (CD), specificity and sensitivity. The Q-PCR amplification curve increased and the retention rate was about 1% at the eighth round. Use of the AuNPs biosensor and CD analyses determined that six aptamers had binding activity. Affinity analysis showed that aptamer 47 had the highest affinity (Kd = 42.17 $\pm$ 8.98 nM) with no cross reactivity to CBL analogs. Indirect competitive enzyme linked aptamer assay (IC-ELAA) based on a 5′-biotin aptamer 47 indicated the limit of detection (LOD) was 0.18 $\pm$ 0.02 ng/L (n = 3), and it was used to detect pork samples with a mean recovery of 83.33–97.03%. This is the first report of a universal strategy including library fixation, Q-PCR monitoring, high-throughput sequencing, and AuNPs biosensor identification to select aptamers specific for small molecules.

Keywords: clenbuterol; systematic evolution of ligands by exponential enrichment; real-time quantitative PCR; high-throughput sequencing technology; aptamers; gold nanoparticles biosensor

---

1. Introduction

Clenbuterol (CBL) is a beta-adrenergic receptor agonist originally used to treat asthma [1]. After CBL enters the body of livestock and poultry, it promotes the growth of animals and increases the lean meat rate [2]. However, food poisoning can result from the human consumption of the edible parts of animals containing CBL. China's Ministry of Agriculture banned the use of adrenaline in animal production in March 1997. However, driven by economic interests, excess CBL in livestock and poultry products frequently causes human poisoning. Therefore, it is necessary to monitor the residue of CBL in livestock and poultry products over a long period. CBL residues are mainly found in livestock, and there is a high demand for domestic animal foods; thus, the detection of CBL residue-related products needs to be rapid. Currently, gold immunochromatography strips [3] and enzyme linked immunosorbent assay [4] are commonly used for this purpose. The key reagents of these two domain detection methods are antibodies. However, the production of antibodies is extremely complex and batch-to-batch variation occurs. Therefore, it is necessary to identify molecules that function similar to antibodies for the development of rapid detection products.

Aptamers are single-stranded DNA (ssDNA) or RNA selected and prepared by Systematic Evolution of Ligands by Exponential Enrichment (SELEX) [5]. Compared with antibodies, they have many advantages including different classes of target, easy modification, timesaving, strong specificity, and high affinity [6]. In the field of food detection, they have an extremely broad usage for the detection of small molecules, and they rival antibodies to some degree. Aptamers have been used to detect small molecules including patulin [7], ochratoxin [8], ractopamine [9] chloramphenicol [10], clenbuterol [11], aflatoxin B1 [12], and tetracycline [13]. The ochratoxin aptamer [14–16] is the most commonly used aptamer, also known as the star aptamer. Currently, there are insufficient numbers of aptamers to detect small molecules residual in food, because of their small molecular weight. How to quickly and efficiently select and identify novel aptamers specific for small molecules is an important question that needs to be answered urgently.

Many SELEX techniques are used to select aptamers, such as magnetic bead SELEX [17], graphene oxide (GO) SELEX [18], capillary electrophoresis SELEX [19], and affinity chromatography SELEX [20]. These methods mainly fix target molecules on the surface of carriers, and then enrich the specific aptamers in the library. Compared with biological macromolecules, the fixation of small molecules is difficult. Before the fixation, small molecules must be derivated, and a carboxyl group or amino active group generated. Then, they are coupled with a solid phase carrier. However, the selected aptamers often recognize the derivatives rather than the small molecules themselves, and therefore cannot be used for the selection of specific aptamers. To solve this problem, a previous study fixed the library on the surface of a solid phase carrier and small molecules were used to capture the aptamers on the solid phase carrier surface [21,22]. This method allowed the simultaneous selection of multiple targets and was also used to select aptamers against small molecules. The current method is to fix the library on the surface of magnetic beads [23] or graphene oxide (GO) [24]. Compared with GO, the magnetic bead method is based on the binding principle of streptavidin-biotin affinity. The library is more stable for fixation and elution compared with GO, and the number of selection rounds is reduced, which saves the selection cost and time. Duan [23] used self-made magnetic beads to select an aptamer specific for CBL. However, there were too many selection rounds (16 rounds). In addition, conventional cloning and sequencing were used to pick out candidate aptamers, resulting in few aptamer candidate sequences. Furthermore, fluorescence labeled aptamers and GO were used for affinity identification, although the cost of the fluorescence labeled aptamer was high. We propose that unmodified gold nanoparticles (AuNPs) based colorimetric biosensors could be used to identify active aptamers from mass enriched sequences. The principle is that AuNPs aggregate after an aptamer specifically combines with its target and the color changes from red to purple (blue). This method is simple, cheap, and quick, and it is suitable for picking out active aptamers from candidate aptamers. Although this method has been used to detect different kinds of small molecules [25,26], no report has identified active aptamers specific for small molecules from an enriched library using this method.

This study reports the first use of a multiple combination selection and identifying strategy, including ssDNA library fixation, Q-PCR monitoring, high-throughput sequencing and AuNPs biosensor identification. This universal technology allows the rapid selection and identification of novel aptamers specific for small molecules. This multiple combination selection and identifying strategy was used to select aptamers specific for CBL, and the specific aptamer successfully detected residual CBL in a pork sample.

2. Results

2.1. Magnetic Bead SELEX Selection of an Aptamer Specific for Clenbuterol

Q-PCR was used to monitor the selection process according to the Ct values and amplification curve for qualitative and quantitative analysis. The quantitative standard curve is shown in Figure 1 and the standard curve correlation coefficient R^2 was 0.9972, showing that the linear correlation between Ct value and logarithm DNA concentration of the template by Q-PCR was good. Therefore, this method can be used for the quantitative calculation of the enriched library amount for each round. The amplification curve and retention rate for all eight rounds of SELEX selection are shown in Figure 2. Figure 2A shows that the amplification curve changed with an increase in enriched rounds. The amplification curve was used to determine when to stop selection, and was related to the initial concentration of the eluent library, the counter selection conditions and specific or non-specific binding. For these reasons, the amplification curves were disordered, but this did not prevent us determining when to stop the selection. According to a report by Luo [27], the diversity of DNA libraries decreased during the SELEX selection, and this reduced mismatching of the DNA sequences. Consequently, the drop-in fluorescence disappeared and the change in fluorescence reached a plateau. Changes in the amplification curves directly represent the convergence of the aptamer species during the SELEX procedure. Therefore, in our study, by the end of the eighth round of selection, the amplification curve showed that the fluorescence had increased, which indicated that the enriched aptamer library was specific for CBL; thus, the selection was stopped. Figure 2B shows that the retention rate increased gradually from the first to fourth round of selection because the library was enriched for specific and non-specific binders. After adding counter selection, the retention rate decreased gradually from the fifth to seventh round of selection compared with the fourth round of selection because most of the non-specific binders were selected out. For the eighth round of selection, the retention rate was increased because this enriched library contained the most specific binders although counter selection was added. In summary, the amplification curve and retention rate indicated that the aptamer library was enriched specifically after eight rounds of selection.

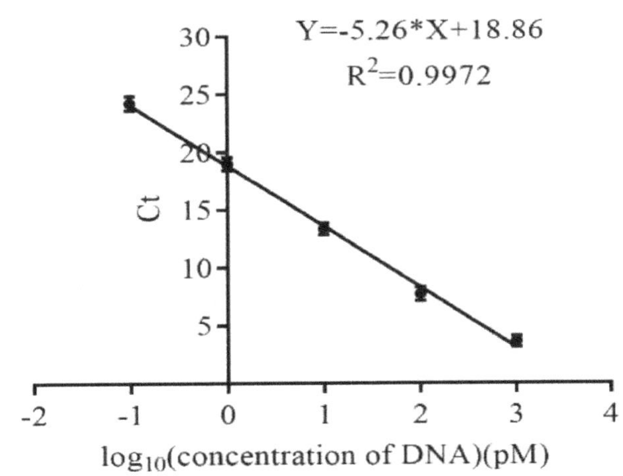

Figure 1. The quantitative standard curve of Q-PCR.

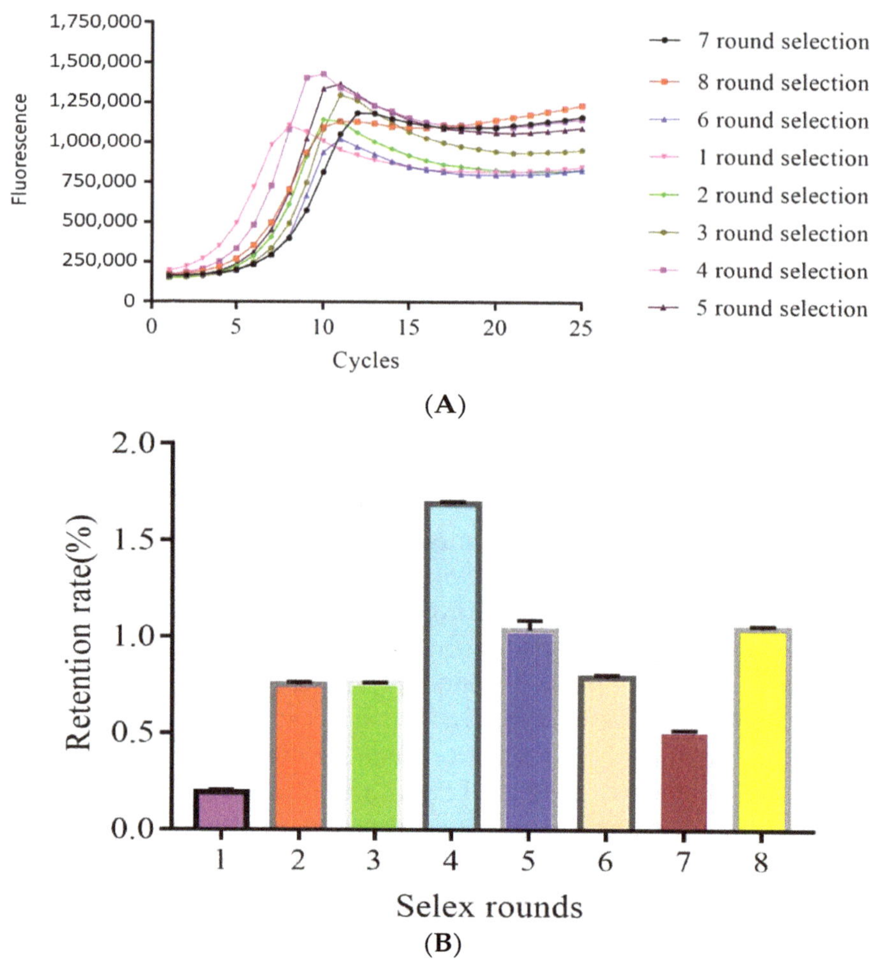

Figure 2. Q-PCR monitoring of the selection process: (**A**) amplification curve of Q-PCR for each round of selection; and (**B**) retention rate for each round of selection.

2.2. High-Throughput Sequencing and Sequence Analysis of the Enriched Library

High-throughput sequencing technology was used to sequence the enriched library of the eighth round. Overall, 105 sequences were obtained, and 172 sequences with a greater frequency of occurrence were selected, as shown in Figure S1. The 40 bp random area sequences underwent homologous alignment using clustal software, and the sequence homology rate was the highest when two sequences were nearest. Phylogenetic tree analysis of these sequences is shown in Figure S2. After removing the sequences that could not be clustered with other sequences, 104 sequences were selected. The mfold online analysis platform showed that 63 aptamers contained one ring region structure or more, and these were selected to identify binding activity. To save the cost of sequence synthesis, 11 bp were removed from the 5′ and 3′ constant regions of the aptamers. The truncated aptamers with 58 bp remaining had unchanged secondary structures, and these were used for binding activity analysis.

2.3. Establishment of a Gold Nanoparticles Biosensor to Identify Active Aptamers

The preliminarily selection results of the AuNPs biosensor are shown in Figure 3. The ratio between the experimental group and the blank group was close to 1 for most of the aptamers, indicating no binding activity between CBL and aptamers, and that the aptamer had no affinity. Interestingly, six aptamers had a ratio of >1.1 between the experimental group and the blank group. The results indicated that the $A_{620\ nm}/A_{520\ nm}$ value of the experimental group was greater than that of the blank group, which indicated that CBL and the aptamers produced a large number of combined complexes. Thus, these six aptamers had affinity with CBL. This method of identifying the binding activity of aptamers specific for a small molecule is the first to identify aptamers from an enriched library. The method is simple, rapid, and low-cost, and can be used to identify other active aptamers from the enriched library.

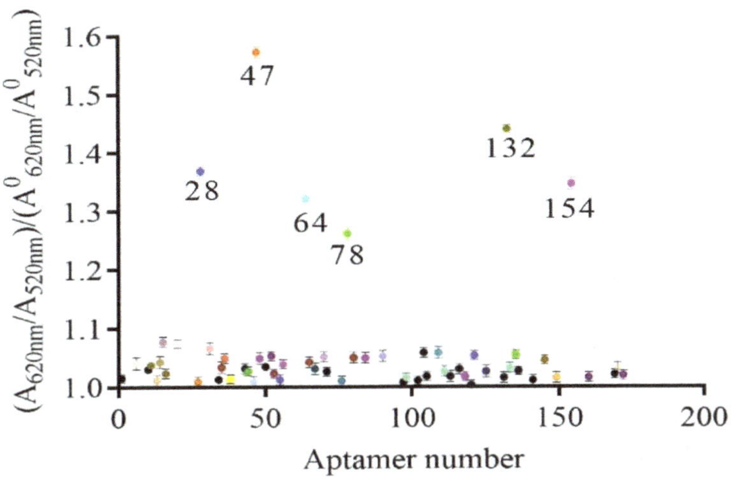

Figure 3. Preliminarily selection of active aptamers using the AuNPs biosensor.

2.4. *Characterization of Active Aptamers*

The sequence information of the six active aptamers are shown in Table 1. The secondary structures of these six aptamers were analyzed on the mfold platform, which showed that they contained two ring regions and Gibbs free energy (Table 1). The affinities of these six aptamers were analyzed, and the affinity constants were different. The affinity of aptamer 47 was the highest, with a Kd = 42.17 ± 8.98 nM. The secondary structure and affinity curve of aptamer 47 are shown in Figure 4.

Table 1. Sequence (5′−3′), dissociation constant (Kd) and dG values of aptamer candidates.

No.	Sequences	Kd (nM)	dG
28	5′ACGCATAGGATGCAACAACGCGAATGCCACCAATTCGTGCTTGTCGTGTCCTATGCGT3′	321 ± 104.5	−15.09
47	5′ACGCATAGGTCATGCCAAGCTGTACACCGTCCTGGCCTGGTTGGGATGTCCTATGCGT3′	42.17 ± 8.98	−13.76
64	5′ACGCATAGGTACACCACATCGGATCGTAATGCGTATGTGACTCAGTGTGCCTATGCGT3′	123.3 ± 14.28	−16.26
78	5′ACGCATAGGGTACGCCACAGATCGACCTCTGGTTGGTCCTGTTGTGTGGCCTATGCGT3′	201.1 ± 42.42	−16.00
132	5′ACGCATAGGGACCATGCGCAGTGAACTTGGTTCTTTGTGCTCATGTGTGCCTATGCGT3′	221.6 ± 41.87	−14.77
154	5′ACGCATAGGGCATCACACATGTCCTTGCCATTGCTGACTTGTTTGGTGTCCTATGCGT3′	268.8 ± 40.43	−15.40

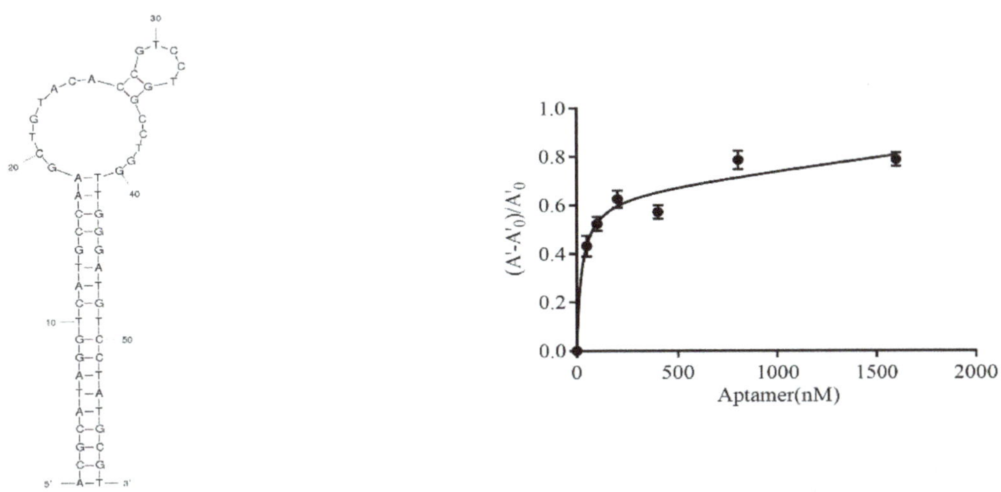

Figure 4. The secondary structure predicted by mfold and corresponding saturation curve of aptamer 47.

To confirm the binding activity between CBL and aptamers, conformation changes of the aptamer before and after CBL binding were studied by circular dichroism (CD) spectrum, a powerful and sensitive tool for studying the interaction between DNA and small molecules [28]. Figure 5 shows that CBL has an extremely weak CD signal, while the aptamer presents a positive and negative peak at about 275 nm and 250 nm, respectively, which was consistent with the reported CD spectrum of a single chain oligonucleotide [29]. When the ssDNA-CBL complex was formed by the binding of free CBL with the aptamer, the characteristic peak at 275 nm was significantly increased, while the characteristic peak at 250 nm was only slightly changed. This suggests that the binding of CBL to the aptamer embedded form led to the reduction of DNA base accumulation. Therefore, the aptamer conformation changed after binding with CBL, which confirmed the interaction between the six aptamers and CBL.

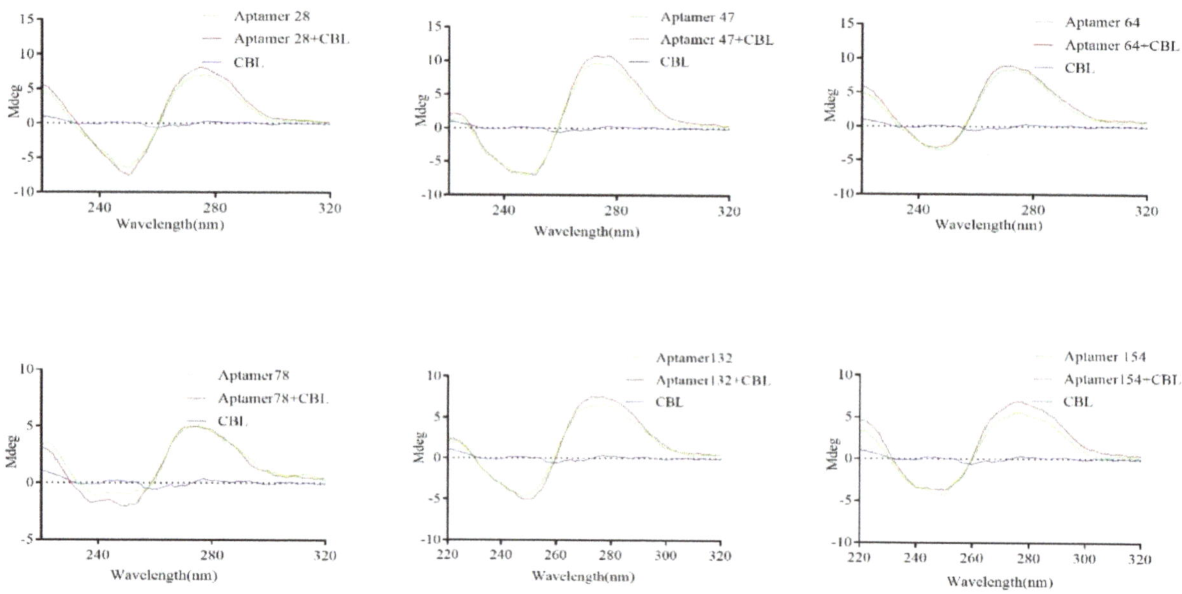

Figure 5. CD spectrum identification of binding activity between six aptamers and CBL.

The specific analysis results are shown in Figure 6. The response signals of the six aptamers binding with CBL exceeded that observed with the CBL analogs, which indicated that there was no cross reaction with the CBL analogs and that the six aptamers had high specificity.

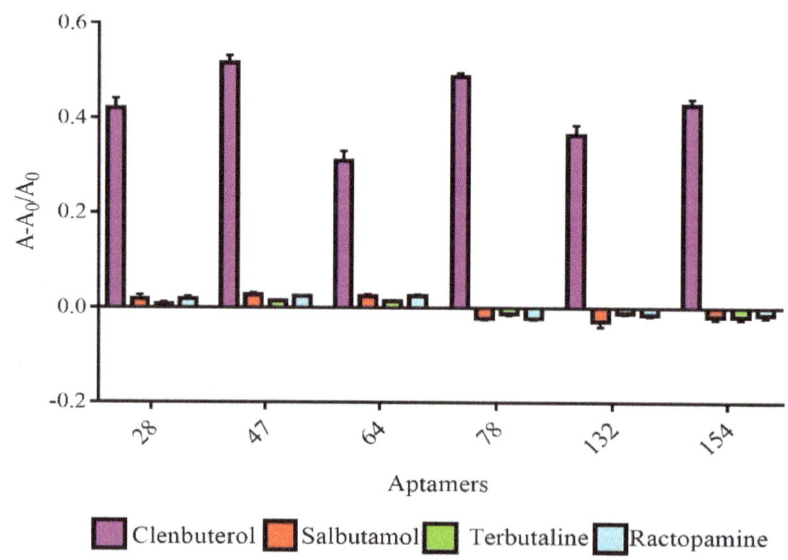

Figure 6. Identification of the specificity of six active aptamers against CBL.

2.5. Detection of CBL Using a Gold Nanoparticles Biosensor and Indirect Competitive Enzyme Linked Aptamer Assay

The sensitivity analysis of aptamer 47 and biotin-aptamer 47 are shown in Figure 7. The AuNPs biosensor results are shown in Figure 7A. The limit of detection (LOD) was 1.2 ± 0.005 ng/mL (n = 3), as estimated by the equation LOD = 3SD/slope, where SD represents the standard deviation of the blank, and the slope was obtained from the calibration curve. The linear detection range was 0.01–4 μg/mL. However, this method was affected by the salt concentration, and therefore was not suitable for the detection of real samples. The complex sample matrix in food may cause AuNPs to change color; therefore, it is necessary to establish another method for real sample detection.

The results of the indirect competitive enzyme-linked aptamer assay (ic-ELAA) are shown in Figure 7B. The mean concentration of CBL required for 50% inhibition of binding (IC_{50}) and LOD (LOD for 10% inhibition of binding) were 5.44 ± 0.01 ng/mL (n = 3) and 0.18 ± 0.02 ng/L (n = 3), respectively, and the linear response range extended from 0.61 to 92.20 ng/mL, showing that the 5′-biotin aptamer 47 could be used to develop ic-ELAA. We used the 5′-biotin aptamer 47 for pork analysis (Table 2). The mean recovery was 83.33–97.03%, indicating that the 5′-biotin aptamer 47 was suitable for the detection of residual CBL in real samples.

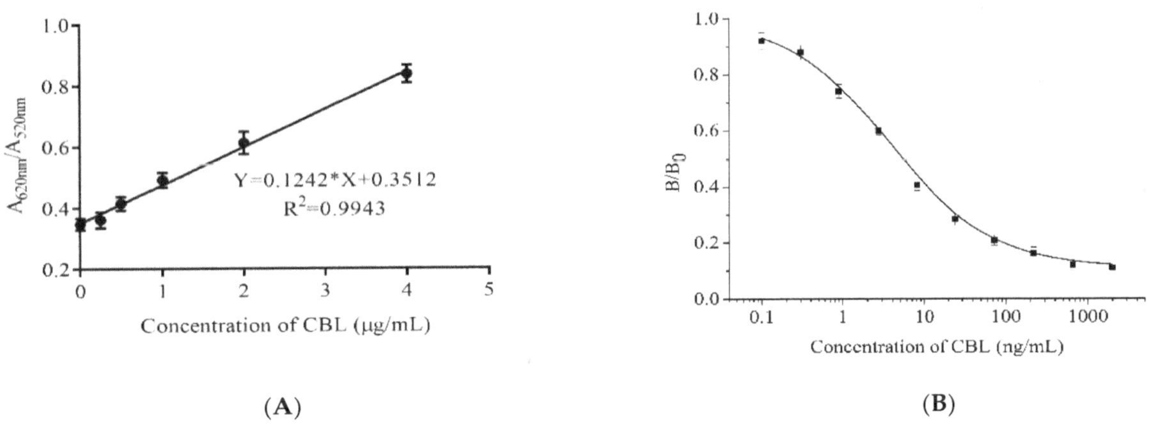

Figure 7. Standard curve for CBL detection: (**A**) standard curve for CBL detection with aptamer 47 using AuNPs biosensor; and (**B**) standard curve for CBL detection with 5′-biotin aptamer 47 using ic-ELAA.

Table 2. Recovery test results with pork samples (n = 3).

Samples	Spiked Concentration (ng/g)	Detection Concentration (ng/g)	Recovery (%)	RSD (%)
Pork	1	0.83 ± 0.02	83.33 ± 1.53	1.83
	5	4.66 ± 0.09	93.27 ± 1.70	1.82
	10	9.70 ± 0.05	97.03 ± 0.45	0.46

3. Discussion

SELEX selection generally has to undergo multiple rounds of library enrichment. It is very important to monitor whether the specific aptamer is enriched in the selection process. At present, the most widely used method is Q-PCR, and studies by Mencin [30] and Michiels [31] monitored the SELEX progress according to the melt curve of Q-PCR. However, Luo [27] found that a single melting curve analysis was insufficient to monitor a pool especially during small molecule aptamer selection, because the melting curve did not change during small molecule aptamer selection. They monitored the selection process using the Ct value and Q-PCR amplification curve. When the library presented with specific enrichment, the Ct value was significantly reduced and the amplification curve increased. This method was suitable to monitor the selection progress of aptamers specific for different classes of targets. However, it has not been used to monitor the selection of aptamers specific for CBL. In our study, we used this method to monitor SELEX progress, and after only eight rounds of selection,

SELEX was stopped. The number of selection rounds was lower than reported by a previous report for small molecule selection, which used 10–16 rounds. Therefore, this method might save time and cost, and aptamer selection was simple.

The preparation of a secondary ssDNA library is a key step of SELEX technology, and existing methods include asymmetric PCR [32], lambda exonuclease digestion [33], and unequal length primer PCR [34]. Among them, the asymmetric PCR method is often accompanied by the generation of double-stranded DNA during the preparation of ssDNA. The purity of the secondary library is not high and it is difficult to accurately determine concentrations. The lambda DNA enzyme method has high efficiency, but the enzyme cutting system is large, the cost of enzymes is high, and thus the selection cost is greatly increased. In the current study, we used unequal length primer PCR combined with fluorescence labeling technology to add a fluorescence label to short primers. Through PAGE degenerative electrophoresis separation, fluorescence bands were cut and prepared as for a secondary ssDNA library. This allows the secondary ssDNA library to be separated rapidly and at low cost. The main disadvantage of conventional PCR amplification during unequal length primer PCR is that overamplification increases nonspecific hybridization among different products and by-products, which may cause the loss of potential specific aptamers, inefficient selection, and even selection failure. It was reported [35] that emulsion PCR could overcome the shortcomings of conventional PCR. During emulsion PCR, different templates are separated by emulsion particles, allowing single-molecule PCR, and avoiding nonspecific hybridization. In our study, emulsion PCR was used for the preparation of a secondary ssDNA library.

After multiple rounds of SELEX selection, the active candidate aptamers were picked out. Currently, the conventional method for this is T cloning [36], which randomly picks a certain number of clones for sequencing and activity identification. The disadvantages of this method are that a limited number of clones were picked and some clones were duplicated. Therefore, in our study, high-throughput sequencing technology was used to analyze sequence information in the enriched library containing a large amount of sequence information, to avoid the multiple identification of repeated clones.

A method to identify the binding active aptamers from the enriched library is critical for aptamer selection. The AuNPs biosensor based on aptamers had the advantages of being simple, rapid, sensitive, and visible, which was convenient for the development of portable and on-site rapid detection [37,38]. Of note, the free-labeled aptamers could be identified, which saved cost and time. Therefore, this method was used to batch identify aptamers from the enriched library in our study. By using this convenient method, six novel aptamers were discovered, and aptamer 47 had the highest affinity. For the AuNPs biosensor based on aptamer 47, the LOD was lower than that of Han's patent [39] with a LOD = 4 μg/mL, and which used the ELAA method. Therefore, aptamer 47 has a greater potential for the development of a rapid detection method than the aptamer reported by Han. However, the AuNPs biosensor is not suitable for real sample detection. First, the sample matrix causes AuNPs to aggregate and change color, although this can be prevented by blocking the nanoparticle surface including adding DNA Spacer and changing aptamer graft density on AuNPs surface [40] to ensure greater stability in the conjugated colloids. However, this modification process is complicated. Second, the LOD of the AuNPs biosensor in our study was high and could not be used to detect samples with a low concentration of CBL residues. To solve this issue, we established an ic-ELAA based on the 5′-biotin aptamer 47. In the ic-ELAA analysis, the LOD was slightly higher than a previously reported aptamer with a LOD = 0.07 ng/mL [11] for CBL detection. The sequence of aptamer 47 is novel and the affinity constant is lower than a previously reported aptamer. The 5′-biotin aptamer 47 was also suitable for the development of a rapid detection kit to monitor low levels of residual CBL in pork samples. The sensitivity of the ic-ELAA was higher compared with the AuNPs biosensor, possibly because the reaction system and analysis principles were different. To avoid the interference of salt when using the AuNPs biosensor, aptamer 47 was reacted with CBL in water. For the ic-ELAA, the 5′-biotin aptamer 47 was reacted with CBL in binding buffer. Therefore, the sensitive ic-ELAA method is suitable for real sample detection.

4. Materials and Methods

4.1. Chemicals and Reagents

The clenbuterol and ractopamine standards were purchased from Shanghai Yuanye Biotechnology Ltd. (Shanghai, China). Terbutaline and salbutamol standards were purchased from Aladdin Co., Ltd. (Shanghai, China). Taq polymerase, dNTPs and 2× TBE-urea buffer were purchased from Sangon Biotech Ltd. (Shanghai, China). Streptavidin labeled magnetic beads were purchased from Thermo Fisher Scientific Ltd. (Shanghai, China). $HAuCl_4{\cdot}4H_2O$ was purchased from Sinopharm Chemical Reagent Co., Ltd. (Shanghai, China). Evagreen was purchased from Shanghai Open Biotechnology Ltd. (Shanghai, China). All other analytical pure chemical reagents were purchased from Sinopharm Chemical Reagents Ltd. (Shanghai, China). streptavidin-horseradish peroxidase was purchased from Boster Biological Technology Co. ltd. (Wuhan, China). Other solutions used in the experiments were treated with sterilization ultrapure water. The aptamer library was synthesized by Sangon Biotech Ltd. (Shanghai, China). The primers (Biotin-P, FAM-Forward, polyA-Reverse, Q-Forward, Q-Reverse, Table 3) were synthesized by Nanjing Genscript Biotechnology Ltd. (Nanjing, China). The aptamers and 5′-biotin aptamer 47 were synthesized by Suzhou Hongxun Biotechnology Ltd. (Suzhou, China). High-throughput sequencing was performed by Anhui Angputuomai Biotechnology Ltd. (Hefei, China). Secondary structures were calculated with mfold online bioinformatics platforms (http://unafold.rna.albany.edu/?q=mfold/DNA-Folding-Form).

Table 3. Primer sequences.

Name	Sequences (5′-3′)
Biotin-P	5′-CCTATGCGTGGAGTGCCAAT-3′-biotin
FAM-Forward	6-FAM-5′-ATTGGCACTCCACGCATAGG-3′
polyA-Reverse	5′-AAAAAAAAAAAAAAAAAAAAspacer18TTCACGGTAGCACGCATAGG-3′
Q-Forward	5′-ATTGGCACTCCACGCATAGG-3′
Q-Reverse	5′-TTCACGGTAGCACGCATAGG-3′

4.2. Selection of Aptamers Specific for CBL Based on ssDNA Library Immobilized SELEX

The procedure of ssDNA Library Immobilized SELEX for aptamers against CBL is shown in Figure 8. The synthetic ssDNA library (5′-ATTGGCACTCCACGCATAGG$(N)_{40}$CCTATGCGTGCTACCGTGAA-3′) was dissolved in binding buffer (0.1 g $CaCl_2$, 0.2 g KCl, 0.2 g KH_2PO_4, 0.1 g $MgCl_2.6H_2O$, 8 g NaCl, and 1.15 g Na_2HPO_4, 1 L). Biotin-P was mixed with the library at a ratio of 2:1, and slowly denatured and renatured (95 °C for 10 min, 60 °C for 1 min, 25 °C for 1 min). The above mixture was added to the streptavidin magnetic beads and incubated (the magnetic beads were washed 4 times with the binding buffer before use, and 700 μL magnetic beads was used in the first round of selection) for 45 min. The magnetic beads were washed 6 times, and incubated for 90 min with CBL in binding buffer (200 μL) to a final concentration of 100 μM at room temperature. The supernatant was collected by magnetic separation, the magnetic beads were washed with 200 μL binding buffer containing the CBL at a final concentration of 100 μM, and the supernatant was collected using magnetic separation. The supernatant from two elutions were pooled as the eluent library with a volume of 400 μL.

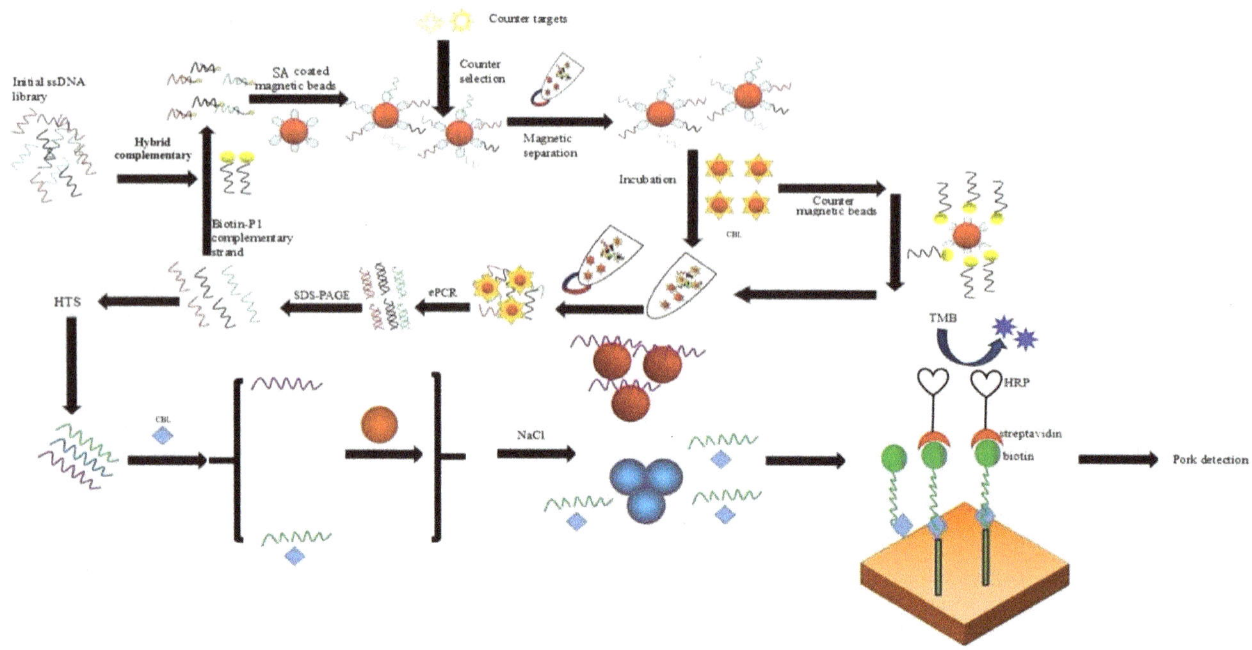

Figure 8. Graphical abstract.

4.3. Establishment of a Real-Time Quantitative PCR Method for the Monitoring Selection Process

A standard curve was prepared as follows: the initial ssDNA library was diluted to a concentration of 16,000, 1600, 160, 16, and 1.6 pM as a real-time quantitative PCR template. Then, 2 μL of template was mixed with 30 μL Q-PCR mix (1 μL Q-Forward at a concentration of 10 μM, 1 μL Q-Reverse at a concentration of 10 μM, 0.5 μL dNTP mix at a concentration of 10 mM, 3 μL 10× PCR buffer, 1 μL Taq DNA polymerase, 1 μL Evagreen, 15.5 μL sterile water, total volume of 30 μL), and the following Q-PCR program (StepOnePlus, ABI, Carlsbad, CA, USA) was used: 95 °C for 2 min, 95 °C for 0.5 min, 60 min at 0.5 °C, and 72 °C for 0.5 min with 25 cycles. The enriched libraries from each round were monitored using Q-PCR amplification according to the above method, and the template was the eluent library with a volume of 2 μL.

4.4. Preparation of Secondary Libraries

After each round of selection, all of the eluent library was added into the emulsion PCR mix (398 μL template DNA, 10 μL FAM-Forward at a concentration of 100 μM, 10 μL polyA-Reverse at a concentration of 100 μM, 40 μL dNTP mix at a concentration of 10 mM, 200 μL 10× PCR buffer, 8 μL Taq DNA polymerase, 1732 μL sterile water, total volume of 2 mL). After adding 8 mL emulsifier (1 mL EM 90, 25 μL triton X-100, 49 mL mineral oil), the emulsion was prepared by vortexing for 2 min, and left to stand for 5 min. The PCR program was as follows: (95 °C 2 min, 95 °C 1 min, 60 °C 1 min, 72 °C 1 min, 25 cycles). The PCR product was concentrated with n-butyl alcohol, then it was mixed with 2× TBE-urea buffer and boiled for 10 min. The sample was separated by denatured SDS-PAGE (400 V, 15 min). The fluorescent strip was cut off and boiled to separate the secondary ssDNA library, which was dialyzed overnight in binding buffer. After measuring of the secondary library concentration, the next round of selection was conducted. The first, second, third and fourth rounds of the selection procedures were consistent. The fifth, sixth and seventh rounds of selection were counter selection with analogs (salbutamol, ractopamine and terbutaline) and positive selection with CBL. The eighth round of selection was positive selection with CBL, and then biotin-P magnetic beads were added for counter selection. The amount of ssDNA library in the first round of selection was 1.3 nm, and in the later rounds it was 100 pM.

4.5. High-Throughput Sequencing and Sequence Analysis of the Enrichment Library

In the eighth round, the library was enriched and sent to Anhui Angputuomai Biotechnology Co., Ltd. for high-throughput sequencing by an Illumina high-throughput sequencing platform (HiSeq/MiSeq). Homologous comparison and evolutionary tree analysis were carried out by clustalX2 software (downloaded from http://www.clustal.org/download/current/) and then visualized using TreeView tool version 1.6.6 for 172 sequences, which were selected from high-throughput sequencing results according to the frequency of occurrence. Then, 104 sequences with a high homology rate were selected for secondary structure prediction using mfold online bioinformatics platforms (http://unafold.rna.albany.edu/?q=mfold/DNA-Folding-Form). Sixty-three sequences with several ring regions ≥ 2 were selected, and 11 bases were removed from the 5′ and 3′ constant regions, respectively. Each aptamer contained 58 bp and was synthesized.

4.6. Determination of Binding Activity between Aptamers and CBL Using a Gold Nanoparticles Biosensor

The binding activity of candidate aptamers were preliminarily determined using an AuNPs biosensor (Figure 8). AuNPs were prepared as previous report [41] by reducing chloroauric acid with sodium citrate. This was centrifuged at 12,000 r/min, 4 °C for 25 min, the supernatant was discarded, and the precipitate was dissolved in ultrapure water, as the concentration of AuNPs in ultrapure water was determined to be 8.76 nM using UV-vis spectroscopy according to the previously reported method [42]. First, 50 μL/well of aptamers were diluted with ultrapure water until the final concentration was 0.4 μM, and then they were incubated with 1 μg/mL CBL (50 μL/well) for 30 min at room temperature. AuNPs (50 μL/well, 8.76 nM) were added and incubated for 30 min at room temperature, and then 0.9 M NaCl (10 μL/well) was added. The absorbance values at 620 nm ($A_{620\,nm}$) and 520 nm ($A_{520\,nm}$) were measured on an automatic microplate reader (I3X, Molecular Devices, San Jose, CA, USA), and aptamers with binding activity were determined by the ratio of $A_{620\,nm}/A_{520\,nm}$ between the experimental group and the blank group.

4.7. Affinity Analysis of Active Aptamers

Aptamers with different concentrations (0, 50, 100, 200, 400, 800 and 1600 nM) were incubated with 1 μg/mL CBL for 30 min at room temperature. AuNPs were added and incubated for 30 min at room temperature, and then 0.9 M NaCl was added. The A_{520nm} was measured on an automatic microplate reader, $(A'-A'_0)/A'_0$ was used as the ordinate and the aptamer concentration was used as the abscissa. The affinity constant of aptamers was calculated by Graphpad Prism 7.0 software (GraphPad Software, La Jolla, CA, USA). A′ represents the A_{520nm} value at the concentration of each aptamer and A'_0 represents the value of A_{520nm} when the concentration of the aptamer is zero.

4.8. Circular Dichroism Spectrum Analysis of Active Aptamers

The aptamers were diluted to 1 μM with binding buffer, and incubated with 10 μM CBL for 30 min at room temperature. The circular dichroism of the mixture was scanned at a scanning wavelength of 220–320 nm. At the same time, the CD spectrum (J-810, Jasco, Tokyo, Japan) analysis of CBL or aptamers were performed under the same conditions.

4.9. Specificity Analysis of Active Aptamers

Clenbuterol, terbutaline, salbutamol, and ractopamine (1 μg/mL) were incubated with 0.4 μM of aptamers for 30 min at room temperature. AuNPs were added and incubated for 30 min at room temperature, and then 0.9 M NaCl was added. The $A_{620\,nm}$ and $A_{520\,nm}$ were measured on an automatic microplate reader. A: represents the value of ($A_{620\,nm}/A_{520\,nm}$) when the concentration of Clenbuterol, terbutaline, salbutamol, and ractopamine is 1 μg/mL and A_0: represents the value of (A_{620nm}/A_{520nm}) when the concentration of clenbuterol, terbutaline, salbutamol, and ractopamine is 0 μg/mL.

4.10. Detection of CBL Using a Gold Nanoparticles Biosensor and ic-ELAA

Aptamer 47 was diluted with ultrapure water until the final concentration was 0.4 μM, and it was then incubated with a series of CBL concentrations (0.01, 0.25, 0.50, 1.0, 2.0, and 4.0 μg/mL) for 30 min at room temperature. AuNPs were added and incubated for 30 min at room temperature, and then 0.9 M NaCl was added. The $A_{620\,nm}$ and $A_{520\,nm}$ was measured on an automatic microplate reader. ($A_{620\,nm}/A_{520\,nm}$) was used as the ordinate and the CBL concentration was used as the abscissa.

The sensitivity of the 5′-biotin aptamer 47 was also determined by ic-ELAA (Figure 8). A microtiter plate was coated with 5 μg/mL CBL-Bovine serum albumin (BSA) in coating buffer (15 mM Na_2CO_3 and 35 mM $NaHCO_3$, pH 9.6) at 37 °C overnight. The CBL-BSA was prepared as previously reported [43]. The plate was washed twice with washing buffer (6 g $Na_2HPO_4 \cdot 12H_2O$, 16 g NaCl, 1.2 mL Tween 20, pH 7.4, 2 L) and then blocked with 150 μL 3% (*w*/*v*) skim milk powder in PBS for 2 h. Subsequently, a series concentration of CBL standard (0, 0.1, 0.3, 0.9, 2.7, 8.1, 24.3, 72.9, 218.7, 656.1, and 1968.3 ng/mL, 50 μL) and 5′-biotin aptamer 47 (1 μM, 50 μL) were respectively added into each well, and the resulting solution was incubated for 1 h at 37 °C. After washing six times as described above, 100 μL of streptavidin-horseradish peroxidase (HRP) diluted 1:5000 was added, and the plate was incubated at 37 °C for 1 h. After washing the plate six times, 3,3′,5,5′-tetramethylbenzidine (TMB) was added to the reaction for 10 min. The reaction was stopped with 10% H_2SO_4, and A_{450nm} was measured on an automatic microplate reader.

The detection of CBL in real samples was also studied using ic-ELAA. Pork was purchased from a supermarket and pretreated according the following method. One hundred grams of pork was homogenized for 10 min. Next, 10 μL of CLB standard solutions at different concentrations (1, 5, and 10 ng/μL) were individually mixed with 10 g of pork paste and homogenized for 10 min. Then, 1 g of the above homogenized sample was mixed with 1 mL of 0.01 M HCl, vortexed for 2 min and centrifuged at 12,000 r/min for 10 min. The supernatant was adjusted to pH 7.4 and centrifuged at 12,000 r/min for 10 min. The supernatant was filtered through a 0.45 μm filtration membrane and collected for the next assay.

5. Conclusions

Aptamers are molecules that complement the function of antibodies for the development of rapid detection kits, where the biggest bottleneck is the selection of novel specific aptamers. For small molecules, it is difficult to select aptamers with specific binding activity. This is the first study to report the use of multiple combination selection and identification strategy including library fixation, Q-PCR monitoring, high-throughput sequencing, and AuNPs biosensor identification for the selection of novel specific aptamers. The selection, monitoring, and identification process was simple, intuitive, reliable, and high-throughput. After eight rounds of selection, six novel aptamers were discovered from the ssDNA aptamer library. Aptamer 47 had the highest affinity with a binding constant of 42.17 ± 8.98 nM. The ic-ELAA based on 5′-biotin aptamer 47 was used for pork sample detection with a mean recovery of 83.33–97.03%. Most importantly, the multiple combination selection and identification strategy proposed in this study is a universal technical method, which can be used for the selection and identification of other novel aptamers specific for small molecules.

6. Patents

Liu, X.X.; Hou, J.J.; Lu, Q.; Wang, F.; Hou, Y.Y.; Meng, C.; Hu, T.Y.; Chen, S.Y. An aptamer used to detect CBL of clenbuterol and its screening method and application. Patent 2018, Application number: 201810874140.X.

Author Contributions: Formal analysis, C.M., T.H. and Y.H.; Funding acquisition, X.L. and J.H.; Investigation, Z.X.; Methodology, Q.L., S.C. and F.W.; Project administration, X.L.; Supervision, X.L., J.H., and Z.X.; and Writing—original draft, X.L.

References

1. Torneke, K.; Larsson, C.I.; Appelgren, L.E. A comparison between clenbuterol, salbutamol and terbutaline in relation to receptor binding and in vitro relaxation of equine tracheal muscle. *J. Vet. Pharmacol. Ther.* **1998**, *21*, 388–392. [CrossRef] [PubMed]
2. Li, Y.; Qi, P.; Ma, X.; Zhong, J.G. Quick detection technique for clenbuterol hydrochloride by using surface plasmon resonance biosensor. *Eur. Food Res. Technol.* **2014**, *239*, 195–201. [CrossRef]
3. Wang, J.Y.; Zhang, L.; Huang, Y.J.; Dandapat, A.; Dai, L.W.; Zhang, G.G.; Lu, X.F.; Zhang, J.W.; Lai, W.H.; Chen, T. Hollow Au-Ag Nanoparticles labeled immunochromatography strip for highly sensitive detection of clenbuterol. *Sci. Rep.-UK* **2017**, *7*, 41419. [CrossRef] [PubMed]
4. Ma, L.Y.; Nilghaz, A.; Choi, J.R.; Liu, X.Q.; Lu, X.N. Rapid detection of clenbuterol in milk using microfluidic paper-based ELISA. *Food Chem.* **2018**, *246*, 437–441. [CrossRef] [PubMed]
5. Tian, R.Y.; Lin, C.; Yu, S.Y.; Gong, S.; Hu, P.; Li, Y.S.; Wu, Z.C.; Gao, Y.; Zhou, Y.; Liu, Z.S.; et al. Preparation of a specific ssDNA aptamer for brevetoxin-2 using SELEX. *J. Anal. Methods Chem.* **2016**, *2016*, 9241860. [CrossRef] [PubMed]
6. Wang, Y.; Li, J.; Qiao, P.; Jing, L.; Song, Y.Z.; Zhang, J.Y.; Chen, Q.; Han, Q.Q. Screening and application of a new aptamer for the rapid detection of Sudan Dye III. *Eur. J. Lipid Sci. Tech.* **2018**, *120*, 1700112. [CrossRef]
7. Wu, Z.Z.; Xu, E.; Jin, Z.Y.; Irudayaraj, J. An ultrasensitive aptasensor based on fluorescent resonant energy transfer and exonuclease-assisted target recycling for patulin detection. *Food Chem.* **2018**, *249*, 136–142. [CrossRef] [PubMed]
8. Zhu, X.; Kou, F.X.; Xu, H.F.; Han, Y.J.; Yang, G.D.; Huang, X.J.; Chen, W.; Chi, Y.W.; Lin, Z.Y. Label-free ochratoxin A electrochemical aptasensor based on target-induced noncovalent assembly of peroxidase-like graphitic carbon nitride nanosheet. *Sensor. Actuat. B-Chem.* **2018**, *270*, 263–269. [CrossRef]
9. Zhu, C.; Zhang, G.L.; Huang, Y.F.; Yan, J.; Chen, A.L. Aptamer based ultrasensitive determination of the beta-adrenergic agonist ractopamine using PicoGreen as a fluorescent DNA probe. *Microchim. Acta* **2017**, *184*, 439–444. [CrossRef]
10. Duan, Y.; Wang, L.H.; Gao, Z.Q.; Wang, H.S.; Zhang, H.X.; Li, H. An aptamer-based effective method for highly sensitive detection of chloramphenicol residues in animal-sourced food using real-time fluorescent quantitative PCR. *Talanta* **2017**, *165*, 671–676. [CrossRef] [PubMed]
11. Chen, D.; Yang, M.; Zheng, N.J.; Xie, N.; Liu, D.L.; Xie, C.F.; Yao, D.S. A novel aptasensor for electrochemical detection of ractopamine, clenbuterol, salbutamol, phenylethanolamine and procaterol. *Biosens.Bioelectron.* **2016**, *80*, 525–531. [CrossRef] [PubMed]
12. Peng, G.; Li, X.Y.; Cui, F.; Qiu, Q.Y.; Chen, X.J.; Huang, H. Aflatoxin B1 electrochemical aptasensor based on tetrahedral DNA nanostructures functionalized three dimensionally ordered macroporous MoS2-AuNPs film. *ACS Appl. Mater. Int.* **2018**, *10*, 17551–17559. [CrossRef] [PubMed]
13. Qi, M.Y.; Tu, C.Y.; Dai, Y.Y.; Wang, W.P.; Wang, A.J.; Chen, J.R. A simple colorimetric analytical assay using gold nanoparticles for specific detection of tetracycline in environmental water samples. *Anal. Methods-UK* **2018**, *10*, 3402–3407. [CrossRef]
14. Chen, Y.Q.; Chen, M.L.; Chi, J.X.; Yu, X.; Chen, Y.X.; Lin, X.C.; Xie, Z.H. Aptamer-based polyhedral oligomeric silsesquioxane (POSS)-containing hybrid affinity monolith prepared via a "one-pot" process for selective extraction of ochratoxin A. *J. Chromatogr. A* **2018**, *1563*, 37–46. [CrossRef] [PubMed]
15. Shen, P.; Li, W.; Ding, Z.; Deng, Y.; Liu, Y.; Zhu, X.R.; Cai, T.T.; Li, J.L.; Zheng, T.S. A competitive aptamer chemiluminescence assay for ochratoxin A using a single silica photonic crystal microsphere. *Anal. Biochem.* **2018**, *554*, 28–33. [CrossRef] [PubMed]
16. Al Rubaye, A.; Nabok, A.; Catanante, G.; Marty, J.L.; Takacs, E.; Szekacs, A. Detection of ochratoxin A in aptamer assay using total internal reflection ellipsometry. *Sensor. Actuat. B-Chem.* **2018**, *263*, 248–251. [CrossRef]
17. Lin, C.; Liu, Z.S.; Wang, D.X.; Li, L.; Hu, P.; Gong, S.; Li, Y.S.; Cui, C.; Wu, Z.C.; Gao, Y.; et al. Generation of internal-image functional aptamers of okadaic acid via magnetic-bead SELEX. *Mar. Drugs* **2015**, *13*, 7433–7445. [CrossRef] [PubMed]

18. Nguyen, V.T.; Kwon, Y.S.; Kim, J.H.; Gu, M.B. Multiple GO-SELEX for efficient screening of flexible aptamers. *Chem. Commun.* **2014**, *50*, 10513–10516. [CrossRef] [PubMed]
19. Luo, Z.F.; Zhou, H.M.; Jiang, H.; Ou, H.C.; Li, X.; Zhang, L.Y. Development of a fraction collection approach in capillary electrophoresis SELEX for aptamer selection. *Analyst* **2015**, *140*, 2664–2670. [CrossRef] [PubMed]
20. Setlem, K.; Monde, B.; Ramlal, S.; Kingston, J. Immuno affinity SELEX for simple, rapid, and cost-effective aptamer enrichment and identification against aflatoxin B_1. *Front. Microbiol.* **2016**, *7*, 1909. [CrossRef] [PubMed]
21. Wu, Y.G.; Zhan, S.S.; Wang, L.M.; Zhou, P. Selection of a DNA aptamer for cadmium detection based on cationic polymer mediated aggregation of gold nanoparticles. *Analyst* **2014**, *139*, 1550–1561. [CrossRef] [PubMed]
22. He, J.A.; Liu, Y.A.; Fan, M.T.; Liu, X.J. Isolation and identification of the DNA aptamer target to acetamiprid. *J. Agr. Food Chem.* **2011**, *59*, 1582–1586. [CrossRef] [PubMed]
23. Duan, N.; Gong, W.H.; Wu, S.J.; Wang, Z.P. Selection and application of ssDNA aptamers against clenbuterol hydrochloride based on ssDNA library immobilized SELEX. *J. Agric. Food Chem.* **2017**, *65*, 1771–1777. [CrossRef] [PubMed]
24. Duan, N.; Gong, W.H.; Wu, S.J.; Wang, Z.P. An ssDNA library immobilized SELEX technique for selection of an aptamer against ractopamine. *Anal. Chim. Acta.* **2017**, *961*, 100–105. [CrossRef] [PubMed]
25. Zhang, D.W.; Yang, J.Y.; Ye, J.; Xu, L.R.; Xu, H.C.; Zhan, S.S. Colorimetric detection of bisphenol A based on unmodified aptamer and cationic polymer aggregated gold nanoparticles. *Anal. Biochem.* **2016**, *499*, 51–56. [CrossRef] [PubMed]
26. Yang, C.; Wang, Y.; Martyc, J.L.; Yang, X.R. Aptamer-based colorimetric biosensing of Ochratoxin A using unmodified gold nanoparticles indicator. *Biosens. Bioelectron.* **2011**, *26*, 2724–2727. [CrossRef] [PubMed]
27. Luo, Z.F.; He, L.; Wang, J.J.; Fang, X.N.; Zhang, L.Y. Developing a combined strategy for monitoring the progress of aptamer selection. *Analyst* **2017**, *142*, 3136–3139. [CrossRef] [PubMed]
28. Ramana, M.M.V.; Betkar, R.; Nimkar, A.; Ranade, P.; Mundhe, B.; Pardeshi, S. In vitro DNA binding studies of antiretroviral drug nelfinavir using ethidium bromide as fluorescence probe. *J. Photoch. Photobio. B* **2015**, *151*, 194–200. [CrossRef] [PubMed]
29. Zhang, A.M.; Jia, L.P.; Zhang, J.J. Study on the interaction between amikacin and DNA by molecular spectroscopy. *J. Liaocheng Univer.* **2008**, *21*, 67–70.
30. Mencin, N.; Smuc, T.; Vranicar, M.; Mavri, J.; Hren, M.; Galesa, K.; Krkoc, P.; Ulrich, H.; Solar, B. Optimization of SELEX: Comparison of different methods for monitoring the progress of in vitro selection of aptamers. *J. Pharmaceyt Biomed.* **2014**, *91*, 151–159. [CrossRef] [PubMed]
31. Vanbrabant, J.; Leirs, K.; Vanschoenbeek, K.; Lammertyn, J.; Michiels, L. Remelting curve analysis as a tool for enrichment monitoring in the SELEX process. *Analyst* **2014**, *139*, 589–595. [CrossRef] [PubMed]
32. Heiat, M.; Ranjbar, R.; Latifi, A.M.; Rasaee, M.J.; Farnoosh, G. Essential strategies to optimize asymmetric PCR conditions as a reliable method to generate large amount of ssDNA aptamers. *Biotechnol. Appl. Bioc.* **2017**, *64*, 541–548. [CrossRef] [PubMed]
33. Citartan, M.; Tang, T.H.; Tan, S.C.; Gopinath, S.C.B. Conditions optimized for the preparation of single-stranded DNA (ssDNA) employing lambda exonuclease digestion in generating DNA aptamer. *World J. Microb. Biot.* **2011**, *27*, 1167–1173. [CrossRef]
34. Williams, K.P.; Bartel, D.P. PCR product with strands of unequal length. *Nucleic. Acids Res.* **1995**, *23*, 4220–4221. [CrossRef] [PubMed]
35. Shao, K.K.; Shi, X.H.; Zhu, X.J.; Cui, L.L.; Shao, Q.X.; Ma, D. Construction and optimization of an efficient amplification method of a random ssDNA library by asymmetric emulsion PCR. *Biotechnol. Appl. Bioc.* **2017**, *64*, 239–243. [CrossRef] [PubMed]
36. Wu, S.J.; Li, Q.; Duan, N.; Ma, H.L.; Wang, Z.P. DNA aptamer selection and aptamer-based fluorometric displacement assay for the hepatotoxinmicrocystin-RR. *Microchim. Acta* **2016**, *183*, 2555–2562. [CrossRef]
37. Zhou, X.T.; Wang, L.M.; Shen, G.Q.; Zhang, D.W.; Xie, J.L.; Mamut, A.; Huang, W.W.; Zhou, S.S. Colorimetric determination of ofloxacin using unmodified aptamers and the aggregation of goldnanoparticles. *Microchim. Acta* **2018**, *185*, 355. [CrossRef] [PubMed]
38. Shi, Q.A.; Shi, Y.P.; Pan, Y.; Yue, Z.F.; Zhang, H.; Yi, C.Q. Colorimetric and bare eye determination of urinary methylamphetamine based on the use of aptamers and the salt-induced aggregation of unmodified gold nanoparticles. *Microchim. Acta* **2015**, *182*, 505–511. [CrossRef]

39. Han, Q.Q.; Qiao, P.; Song, Y.Z.; Zhang, J.Y.; Chen, Q. A Aptamer Specific CBL and Its Application. Patent 201510485760.0, 23 December 2015.
40. Zhao, W.A.; Chiuman, W.; Lam, J.C.F.; McManus, S.A.; Chen, W.; Cui, Y.G.; Pelton, R.; Brook, M.A.; Li, Y.F. DNA aptamer folding on gold nanoparticles: From colloid chemistry to biosensors. *J. Am. Chem. Soc.* **2008**, *130*, 3610–3618. [CrossRef] [PubMed]
41. Yao, X.; Ma, X.D.; Ding, C.; Jia, L. Colorimetric determination of lysozyme based on the aggregation of gold nanoparticlescontrolled by a cationic polymer and an aptamer. *Microchim. Acta* **2016**, *183*, 2353–2359. [CrossRef]
42. Haiss, W.; Thanh, N.T.K.; Aveyard, J.; Fernig, D.G. Determination of size and concentration of nanoparticles from UV-Vis spectra. *Anal. Chem.* **2007**, *79*, 4215–4221. [CrossRef] [PubMed]
43. Kim, Y.H.; Kim, Y.S. Effects of active immunization against clenbuterol on the growth-promoting effect of clenbuterolin rats. *J. Anim. Sci.* **1997**, *75*, 446–453. [CrossRef] [PubMed]

9

Synthesis and Structural Identification of a Biaryl Ether-Linked Zearalenone Dimer

Julia Keller [1], Luisa Hantschke [1], Hajo Haase [2] and Matthias Koch [1,*]

[1] Department Analytical Chemistry, Reference Materials, Bundesanstalt für Materialforschung und -prüfung (BAM), Richard-Willstätter-Straße 11, 12489 Berlin, Germany; juliakeller19@yahoo.de (J.K.); luisa.hantschke@web.de (L.H.)

[2] Department of Food Chemistry and Toxicology, Technische Universität Berlin, Gustav-Meyer-Allee 25, 13355 Berlin, Germany; haase@tu-berlin.de

* Correspondence: matthias.koch@bam.de

Abstract: A new dimer of the food-relevant mycotoxin zearalenone was isolated after electrochemical and chemical oxidation. The structure was determined as a 16-*O*-15′-biaryl ether-linked dimer based on spectroscopic analyses (^{1}H- and ^{13}C-NMR, COSY, HMBC, and HSQCAD) and high-resolution mass spectrometry analysis (Q-TOF).

Keywords: mycotoxin; dimerization; HRMS; NMR

---

1. Introduction

The fungal secondary metabolite zearalenone (ZEN) is found worldwide and is primarily produced by *Fusarium* species [1–3]. Often found in common crops like corn, wheat, rice, soybeans, sorghum, spices, or walnuts, it poses a health risk to human and animals [4–6]. As mycoestrogen, it causes swelling of the uterus and vulva, infertility, and atrophy of ovaries reported in swine and cattle [7,8]. Several metabolites derived from plants, fungi, and mammalian metabolism are part of ongoing research due to unknown toxic effects and occurrence [9–12].

Oxidative reactions of ZEN lead to hydroxylated species obtained from in vitro assays with liver microsomes of rodent and non-rodent liver cells [13,14]. A recent study proposed the production of numerous hydroxylated as well as new dimeric species of ZEN by using electrochemistry coupled to mass spectrometry [15]. The production, isolation, and structural elucidation of the predominant dimeric species is now achieved.

2. Results and Discussion

Compound **1** was obtained as a pale orange solid substance after electrochemical and chemical oxidation of ZEN with a molecular formula of $C_{36}H_{42}O_{10}$, which was measured by high-resolution mass spectrometry with *m/z* 633.2658 $[M - H]^-$ in an ESI negative ionization mode (theoretical exact mass *m/z* 633.2702), as previously described [15]. The MS/MS measurements of compound 1 revealed a fragment with *m/z* 589.2756, which is due to the loss of CO_2, and a fragment with *m/z* 565.2746 because of a loss of C_3O_2. The signal with *m/z* 491.1658, corresponds to a loss of $C_8H_4O_2$, which led to the fragment *m/z* 447.1763 after the loss of CO_2 (Supplement Figure S1). The assumed fragments and their chemical formulas with theoretical exact masses are shown in Figure 1.

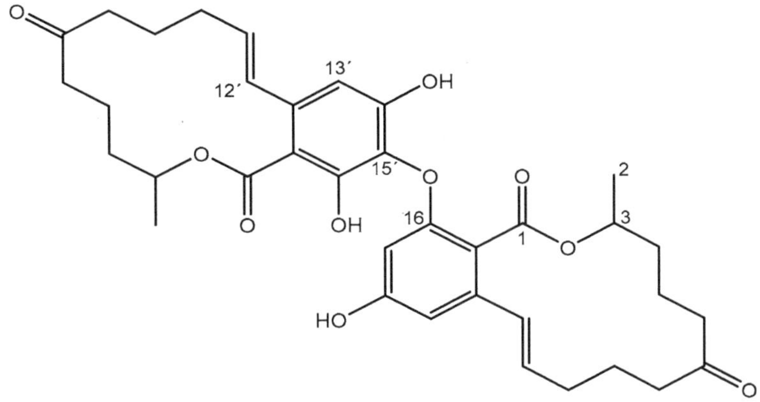

Figure 1. Structure of compound **1**.

The ^{1}H-NMR spectrum of compound 1 measured at 400 MHz in MeOH-d_4 gave the following information with 1.00–2.85 (m, H, J = 6.8 Hz), 1.34 (d, 3H, J = 6.2 Hz), 1.39 (d, 3H, J = 6.4 Hz), 5.06–5.16 (m, 1H), 5.20–5.36 (m, 1H), 5.71 (ddd, 1H, J = 4.0, 9.8, 15.6 Hz), 5.77 (d, 1H, J = 2.1 Hz), 6.03 (d, 1H, J = 4.4, 9.6, 14.6 Hz), 6.34–6.41 (m, 2H), 6.42 (s, 1H), and 6.58 (d, 1H, J = 2.1 Hz). The ^{13}C-NMR measurements with 100 MHz revealed the following chemical shifts with 20.2, 21.0, 22.1, 22.6, 23.4, 32.4, 35.9, 37.6, 38.3, 43.9, 44.7, 73.7, 74.3, 100.8, 104.3, 106.7, 126.4, 130.2, 134.4, 137.2, 139.4, 158.0, 161.0, 172.1, 213.8, and 214.1 ppm (Supplement Figures S2 and S3). The assignments of the carbons and protons are summarized in Table 1.

Table 1. ^{1}H-NMR and ^{13}C-NMR data of **1** (400/100 MHz, Methanol-d_4).

Position	^{1}H (ppm)	^{13}C (ppm)
1, 1′	-	213.8, 214.1
2	1.39 (d, J = 6.4 Hz, 3H)	35.9
3	5.20–5.36 (m, 1H)	73.7
10, 10′	-	32.4, 32.4
11	6.03 (ddd, J = 4.4, 9.6, 14.6 Hz, 1H)	134.4
12	6.38–6.41 (m, 1H)	130.2
13	6.58 (d, J = 2.1 Hz, 1H)	106.7
14	-	161.0
15	5.77 (d, J = 2.1 Hz, 1H)	100.8
2′	1.34 (d, J = 6.2 Hz, 3H)	35.9
3′	5.06–5.16 (m, 1H)	74.3
11′	5.71 (ddd, J = 4.0, 9.8, 15.6 Hz, 1H)	137.2
12′	6.34–6.37 (m, 1H)	126.4
13′	6.42 (s, 1H)	104.3

The ^{1}H-NMR and COSY spectra revealed two aromatic systems (Supplement Figures S2 and S4). One of these aromatic systems contained two protons (d 6.58 ppm, 1H, J = 2.1 Hz, and d 5.77 ppm, 1H, J = 2.1 Hz) while the second aromatic system had only one proton (s 6.42 ppm, 1H). From this observation, the structures connected over the C-C linkages 15, 15′, 13, 13′, and 13, 15′ can be excluded since these dimers would have two aromatic systems with only one proton. Thus, only dimers having an C-O-C ether bridge between the monomers are possible. The COSY spectrum showed nearly identical chemical structures for the spectral part of the molecule (Supplement Figure S4). As a result, the dimerization of two ZEN molecules is only likely over an ether-link between 14-O-13′, 14-O-15′, 16-O-13′, or 16-O-15′. The HMBC spectrum (8 Hz) indicates a common coupling partner of the single aromatic proton and 12′-H, which is located in the aromatic-olefinic region (Supplement Figure S5). Due to the spatial proximity, it should be at the singlet (6.42 ppm) and, as a result, act around the 13′-H position. Consequently, an O-15′-linkage is conceivable.

The observed significant difference of the chemical shifts of 3-H and 3′-H indicated a different chemical environment. For a C-14 link, the closer chemical environment of 3-H and 3′-H would be very similar. A C-16 link, on the other hand, would be a greater steric influence and inductive effects would occur. Thus, the compound 16-O-15′-biaryl ether bond is the most likely structure, which is shown in Figure 2. Whether this dimer can be found naturally in food or feed remains to be analyzed in detail. Especially in plants, lichen, bacteria, and fungi regio-selective and stereo-selective biaryl C-C and biaryl ether C-O-C linkages are often found and it is conceivable that dimers of ZEN might not be uncommon in nature [16].

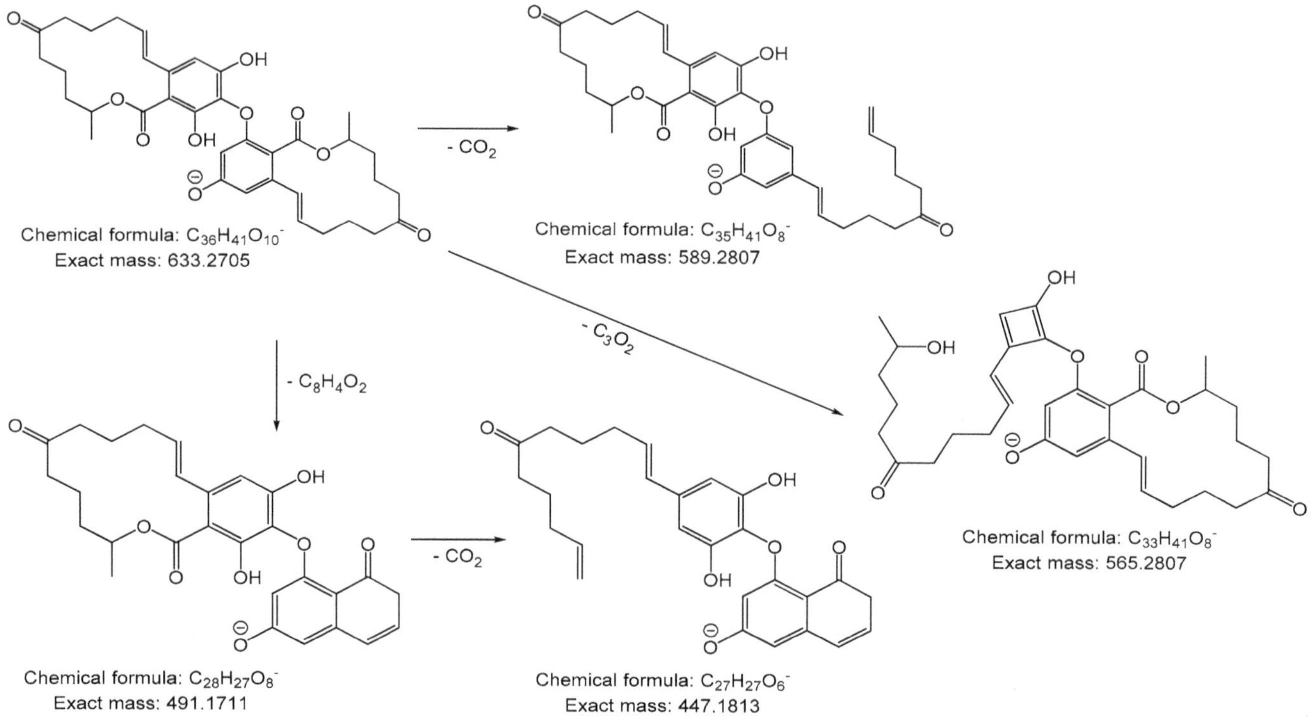

Figure 2. Postulated fragments of the ZEN-dimer, according to the MS/MS measurements with a molecular structure and formula along with the theoretical exact masses.

3. Materials and Methods

3.1. Chemicals and General Experimental Procedures

Zearalenone with a purity over 98% was obtained from Fermentek (Jerusalem, Israel) and Cerium(IV)sulfate was purchased from Sigma-Aldrich (Steinheim, Germany). Ultrapure water was generated by a Seralpur PRO 90 CN system (Ransbach-Baumbach, Germany). All standard chemicals were of p.a. grade and all solvents were of an HPLC grade. Electrochemical oxidation was achieved by using the Roxy® system synthesis cell (Antec, Zoeterwoude, The Netherlands) equipped with a platinum working electrode. The HPLC system used for fractionation consisted of an Agilent 1200 series autosampler, a 1260 series pump, a 1200 series diode array detector, and a column oven. A Macherey-Nagel Nucleosil C18 100-5 150 × 4.6 mm column (Düren, Germany) was used. The TripleTOF® 6600 Quadrupole Time-Of-Flight (QTOF) mass analyzer (Sciex, Darmstadt, Germany) was operated in negative ionization mode and 10 μM of the dimer sample were dissolved in methanol with 0.1% of formic acid. The used parameters were as follows:

Gas temperature 350 °C, ion source gas 1 (nitrogen) 20 L/min, ion source gas 2 (nitrogen) 15 L/min, curtain gas (nitrogen) 20 L/min, and ion spray voltage floating −4500 V. The MS/MS-spectrum was recorded in the targeted MS/MS mode with the following parameters: De-clustering potential (−15 V), collision energy (−40 V), and TOF Masses 100–640 Da. Confirmation of the 16-O-15′-biaryl ether-linked dimer structure was conducted by nuclear magnetic resonance spectroscopy (NMR).

The NMR spectra were recorded in methanol-d_4 on an Agilent 400-MR NMR spectrometer (Agilent Technologies, Waldbronn, Germany) at 30 °C. For the measurements, the ATB 5mm-probe head was operated at 399.8 MHz for ^{1}H and 100.5 MHz for ^{13}C data.

3.2. Electrochemical and Chemical Production of ZEN Dimer

Electrochemical: The optimal potential was tested by taking aliquots after different time points and using potentials of 0, 1.0, 1.4, and 1.8 V vs. Pd/H_2. For the electrochemical oxidation, 80 mL of 250 μM ZEN in acetonitrile/water (50/50, *v*/*v*) was stirred for 48 h using 1.4 V vs. Pd/H_2. The solution was subsequently evaporated to dryness by using a rotary evaporator and dissolved in water/acetonitrile (65/35, *v*/*v*) for HPLC fractionation.

Chemical: For the oxidative production of ZEN dimers, Ce(IV)sulfate was used. About 100 mg of ZEN and 350 mg of Ce(IV)sulfate were dissolved in 200 mL of acetonitrile/water (50/50, *v*/*v*) and stirred for two hours at 70 °C. Subsequently, the sample was stirred for 24 h at room temperature and a white precipitate was formed. The yellow solvent mixture was extracted three times with 20 mL of ethyl acetate. After the extraction, the ethyl acetate was colored yellow and the acetonitrile water mixture was colorless. After evaporation, a deep orange and highly viscous fluid was obtained. After freeze-drying, a pale orange solid was formed with a yield of 10%.

3.3. Purification of ZEN Dimer

For the separation of ZEN dimers, an already described HPLC method was adapted [15] by using a flowrate of 1.2 mL/min and an isocratic eluent consisting of water/acetonitrile, 65/35, *v*/*v* without modifiers. The ZEN dimer was isolated by collecting the fraction between 15.4 min and 16.2 min of the retention time by using a Foxy® R1 fraction collector (Teledyne ISCO, Nebraska, NE, USA). The purity of the dimer was determined to be 92% based on DAD spectra by using a wavelength at λ = 254 nm (Supplement Figure S7).

4. Conclusions

A new dimeric species of the food-relevant mycotoxin zearalenone was synthesized electrochemically and chemically with Ce(IV)sulfate and structurally identified. Among other possible dimers, the occurrence of the 16-*O*-15′-biaryl ether-linked dimer in food and feed is conceivable because the dimerization of phenolic compounds is often observed in plants, fungi, or lichen.

Supplementary Materials:
Figure S1. MS/MS fragmentation spectrum of the ZEN dimer obtained with Q-TOF in negative ionization mode. Figure S2. ^{1}H-NMR spectrum of the 16-*O*-15′-biaryl ether-linked zearalenone dimer. Figure S3. ^{13}C-NMR spectrum of the 16-*O*-15′-biaryl ether-linked zearalenone dimer. Figure S4. COSY spectrum of the 16-*O*-15′-biaryl ether-linked zearalenone dimer. Figure S5. HMBC spectrum of the 16-*O*-15′-biaryl ether-linked zearalenone dimer. Figure S6. HSQC spectrum of the 16-*O*-15′-biaryl ether-linked zearalenone dimer. Figure S7. HPLC-DAD chromatograms (λ = 254 nm) of the zearalenone dimer reaction mixture before fractionation (top) and after fractionation (bottom).

Author Contributions: Conceptualization, J.K. and L.H. Methodology, J.K., L.H., and M.K. Investigation, L.H. Data Curation, J.K. and M.K. Writing-Original Draft Preparation, J.K. Writing-Review & Editing, H.H. and M.K. Supervision, M.K and H.H.

Acknowledgments: The authors would like to thank René Kudick (ASCA GmbH, Berlin, Germany) for the NMR measurements and analytical advice.

References

1. Caldwell, R.W.; Tuite, J.; Stob, M.; Baldwin, R. Zearalenone production by fusarium species. *Appl. Microbiol.* **1970**, *20*, 31–34. [PubMed]
2. Fraeyman, S.; Croubels, S.; Devreese, M.; Antonissen, G. Emerging fusarium and alternaria mycotoxins: Occurrence, toxicity and toxicokinetics. *Toxins* **2017**, *9*, 228. [CrossRef] [PubMed]
3. Placinta, C.M.; D'Mello, J.P.F.; Macdonald, A.M.C. A review of worldwide contamination of cereal grains and animal feed with fusarium mycotoxins. *Anim. Feed Sci. Technol.* **1999**, *78*, 21–37. [CrossRef]
4. Alexander, J.; Benford, D.; Boobis, A.; Ceccatelli, S.; Cottrill, B.; Cravedi, J.P.; Di Domenico, A.; Doerge, D.; Dogliotti, E.; Edler, L.; et al. Scientific opinion on the risks for public health related to the presence of zearalenone in food efsa panel on contaminants in the food chain. *EFSA J.* **2011**, *9*, 2197.
5. Hetmanski, M.T.; Scudamore, K.A. Detection of zearalenone in cereal extracts using high-performance liquid chromatography with post-column derivatization. *J. Chromatogr.* **1991**, *588*, 47–52. [CrossRef]
6. Koppen, R.; Riedel, J.; Proske, M.; Drzymala, S.; Rasenko, T.; Durmaz, V.; Weber, M.; Koch, M. Photochemical trans-/cis-isomerization and quantitation of zearalenone in edible oils. *J. Agric. Food Chem.* **2012**, *60*, 11733–11740. [CrossRef] [PubMed]
7. Drzymala, S.S.; Binder, J.; Brodehl, A.; Penkert, M.; Rosowski, M.; Garbe, L.-A.; Koch, M. Estrogenicity of novel phase i and phase ii metabolites of zearalenone and cis-zearalenone. *Toxicon* **2015**, *105*, 10–12. [CrossRef] [PubMed]
8. Pfohlleszkowicz, A.; Chekirghedira, L.; Bacha, H. Genotoxicity of zearalenone, an estrogenic mycotoxin—DNA adduct formation in female mouse-tissues. *Carcinogenesis* **1995**, *16*, 2315–2320. [CrossRef]
9. Binder, S.B.; Schwartz-Zimmermann, H.E.; Varga, E.; Bichl, G.; Michlmayr, H.; Adam, G.; Berthiller, F. Metabolism of zearalenone and its major modified forms in pigs. *Toxins* **2017**, *9*, 56. [CrossRef] [PubMed]
10. Borzekowski, A.; Drewitz, T.; Keller, J.; Pfeifer, D.; Kunte, H.-J.; Koch, M.; Rohn, S.; Maul, R. Biosynthesis and characterization of zearalenone-14-sulfate, zearalenone-14-glucoside and zearalenone-16-glucoside using common fungal strains. *Toxins* **2018**, *10*, 104. [CrossRef] [PubMed]
11. Brodehl, A.; Moller, A.; Kunte, H.J.; Koch, M.; Maul, R. Biotransformation of the mycotoxin zearalenone by fungi of the genera rhizopus and aspergillus. *FEMS Microbiol. Lett.* **2014**, *359*, 124–130. [CrossRef] [PubMed]
12. Kiessling, K.H.; Pettersson, H. Metabolism of zearalenone in rat liver. *Acta Pharmacol. Toxicol.* **1978**, *43*, 285–290. [CrossRef]
13. Bravin, F.; Duca, R.C.; Balaguer, P.; Delaforge, M. In vitro cytochrome p450 formation of a mono-hydroxylated metabolite of zearalenone exhibiting estrogenic activities: Possible occurrence of this metabolite in vivo. *Int. J. Mol. Sci.* **2009**, *10*, 1824–1837. [CrossRef] [PubMed]
14. Hildebrand, A.A.; Pfeiffer, E.; Rapp, A.; Metzler, M. Hydroxylation of the mycotoxin zearalenone at aliphatic positions: Novel mammalian metabolites. *Mycotoxin Res.* **2012**, *28*, 1–8. [CrossRef] [PubMed]
15. Keller, J.; Haase, H.; Koch, M. Hydroxylation and dimerization of zearalenone: Comparison of chemical, enzymatic and electrochemical oxidation methods. *World Mycotoxin J.* **2017**, *10*, 297–307. [CrossRef]
16. Wezeman, T.; Bräse, S.; Masters, K.-S. Xanthone dimers: A compound family which is both common and privileged. *Nat. Prod. Rep.* **2015**, *32*, 6–28. [CrossRef] [PubMed]

10

The Content of Biogenic Amines in Croatian Wines of Different Geographical Origins

Ivana Mitar [1], Ivica Ljubenkov [1,*], Nikolina Rohtek [2], Ante Prkić [3], Ivana Anđelić [1] and Nenad Vuletić [1]

1 Department of Chemistry, Faculty of Science, University of Split, Ruđera Boškovića 33, 21000 Split, Croatia; imitar@pmfst.hr (I.M.); angel@pmfst.hr (I.A.); nenov@pmfst.hr (N.V.)

2 University Department for Forensic Sciences, University of Split, Ruđera Boškovića 33, 21000 Split, Croatia; rohtekn@yahoo.com

3 Department of Analytical Chemistry, Faculty of Chemistry and Technology, University of Split, Ruđera Boškovića 35, 21000 Split, Croatia; prkic@ktf-split.hr

* Correspondence: iljubenk@pmfst.hr

Abstract: Samples of white and red wines produced in two different wine-growing regions, coastal (Dalmatia) and continental (Hrvatsko zagorje) of Croatia, were analysed for biogenic amines content. Biogenic amines content was determined, and its concentration levels were associated with the geographical origin of the wine. Due to its high sensitivity, HPLC method with ultraviolet detector was used, including the derivatisation step with dansyl chloride. The method was applied to detect and quantify 11 biogenic amines in 48 red and white wines. It was found that both Dalmatian red and white wines are characterised by tryptamine (0.23–1.22 mg L^{-1}), putrescine (0.41–7.5 mg L^{-1}) and ethanolamine (2.87–24.32 mg L^{-1}). White wines from the Hrvatsko zagorje region are characterised by content of isopentylamine (0.31–1.47 mg L^{-1}), putrescine (0.27–1.49 mg L^{-1}) and ethanolamine (3.80–17.96 mg L^{-1}). In contrast to white wines from the Hrvatsko zagorje region, in the red wines, all biogenic amines except ethylamine, were found and equally presented.

Keywords: Croatian wines; biogenic amines; HPLC; geographical origin

1. Introduction

Recently, there has been a great interest of scientists to find a way to control the global wine market. There are numerous parameters determining the quality of the wine. These parameters can be classified as chemical and sensory parameters. Wine is a beverage wherein the quality depends on many factors, among which grape variety, origin, vintage, grape growing conditions, winemaking practice and maturation process, physical conditions of production and way of storage, are the most significant, and they also influence sensory characteristics. Over the past century, chemists have played a significant role in the determination of wine chemical composition and its association with wine flavour and sensory attributes. In the global wine market, wine identity (brand, type, vintage and origin) is extremely important and all those characteristics are crucial for the determination of its price. In the past century, chemists had developed powerful tools for detecting adulteration of wine, such as the addition of water, glycerol, alcohol, dyes, sweeteners, flavour substances and a non-authorised addition of sugar or acidity adjustment [1]. Therefore, in the last few years, there has been a great interest from scientists as well as consumers and commercial wine producers, on the geographical origin and authenticity of wines in terms of quality and price determination. In the 19th and early 20th centuries, the focus was on detecting fraud, while more recently the emphasis has been on quantifying trace compounds, especially those that may be related to a grape variety [2]. There is a large number of studies dealing with the classification of wines according to geographical origin, e.g., assuming that the concentrations or ratios of some chemical parameters in wine depend on its geographical origin [1]. According to some studies, volatile components [3], polyphenols [3–7], as well as elemental composition [8–11], can be linked to the geographical origin or grape variety.

Biogenic amines (BAs) are in the focus of wine quality analysis, and generally of food quality control of many studies. BAs can be found in different fermented foods, such as milk, cheese or beer as well as in wine. They are low molecular weight organic compounds. In wine they can originate from grapes, or can be produced during fermentations (alcoholic and malolactic). In most wines, the level of BAs is low after alcoholic fermentation, while after malolactic fermentation their level increases [12]. Also, they can be formed during wine ageing or storage processes, especially if wine is exposed to the microorganisms' activity or free amino acids are present in it. The formations of BAs from free amino acids can be through reaction of decarboxylation (histamine, tyramine, putrescine, cadaverine), transamination, reductive amination or degradation of their amino acids precursors [12]. The conditions such as storage temperature, pH, and presence of oxygen, sulphur dioxide or sodium chloride content in wine are important factors that affect the concentration of BAs in the final product [13–18]. All mentioned conditions act synergistically. Obviously, with some combinations, BAs may be increased through progressive fermentation process, e.g., elevated storage temperature, pH and O_2 presence. On the other hand, elevated SO_2 and NaCl concentrations diminish fermentation process and decrease BAs content in wines.

Although BAs are considered essential for many physiological functions, such as: body temperature regulation, stomach pH value and brain activity, the frequent and prolonged intake of BAs through wine consummation causes various health problems, such as headaches, flushing, itching, skin irritation, hypertension, etc. [13,15]. BAs are found to be toxic in cases of the intake of foods or beverages that contain them in large amounts. BAs suspected of having toxicological effects are histamine, tryptamine and phenethylamine. Therefore, the level of BAs in wine can be a safety indicator as well as an important parameter for grading wine quality [15]. Polyamines are usually associated with deficient sanitary conditions, especially putrescine and cadaverine. There is a large variability in content and distribution of BAs in wine.

A large number of studies on BAs in commercially available wines have been carried out in the world's most important wine producing countries: Turkish red wines [19], Greek red wines [20,21], Chinese red wines [22], Chilean wines [23], Brazilian wines [16], Portuguese [24] and Spanish red and white wine samples [25–27], purchased from local stores with the aim of monitoring or determining the content of biogenic amines, the high content of which disrupts the quality or proves bad hygiene conditions of wine. More than 20 biogenic amines have been identified in the mentioned studies.

However, there are only a few studies on BAs content with the purpose of determining wine geographical origin or grape variety. BAs are naturally present in grapes (putrescine, spermidine, histamine and tyramine) [28] and in initial musts (ethanolamine, tyramine, putrescine, cadaverine, phenylethylamine and spermidine) [29]. Del Prete et al. detected ethanolamine, ethylamine and putrescine in grapes [30], while Ladente et al. found that differences in putrescine concentration may be attributed to certain grape varieties [31]. According to those studies, the concentrations of BAs that are naturally present in must are directly connected to a grape variety or a soil type, and finally, to the grape's geographical origin. Marques et al., explained a correlation between BAs and grape variety on the samples of red wines produced in three different Portuguese regions [32], while a group of Italian researchers in their study on red wines concluded that BAs composition is a feature of a particular geographic region [4]. Red wines are richer in the content of biogenic amines [15], which can be related to the fact that secondary fermentation (malolactic fermentation) is less usual in white wines [21,27]. Therefore, significantly smaller number of studies are conducted on white wine samples.

Croatia, like many other Mediterranean countries such as Spain, France, Italy, Greece and Turkey, has a long-standing tradition of wine production. In Croatia, there is a large number of small manufacturers who produce small quantities of wine, so the Croatian market is also known as a market of a large number of monovarietal wines [33]. As the authors know, there are only a few studies on BAs content in Croatian wines. Kovačević Ganić et al. investigated BAs content in red wine samples from Slavonia wine region and changes in BAs content during the winemaking and maturation processes. They reported the presence of 10 BAs: tryptamine, hydroxylamine, phenylethylamine, putrescine, cadaverine, histamine, tyramine, serotonin, spermine and spermidine in investigated samples [14]. Jeromel et al. investigated the concentration of BAs in red wines from northwest Croatia with the aim of comparison of BAs levels in wines produced by classic and cold maceration. In their study, the most abundant BAs were histamine, tryptamine and 2-phenylethylamine, while tyramine, putrescine, cadaverine, spermidine, spermine and serotonin were also detected in significantly lower concentrations [34]. There is also a lack of studies on BAs in Croatian white wine samples.

Although numerous analytical methods have been reported for the determination of the BAs in wines and other beverages or food samples [24,35,36], high-performance liquid chromatography (HPLC) is preferred by most researchers [27,30,37–39]. There are differences in the derivatisation procedures among studies, but the most commonly used reagents are *o*-phthalalhehyde (OPA), dabsyl chloride (DABS-Cl) and dansyl chloride (DNS-Cl).

Although, researchers are usually using OPA as a derivatisation agent, we used dansyl chloride because of its stability when exposed to UV-Vis spectra (in detection system). Also, OPA reacts only with primary amines, which prevents the determination of polyamines such as spermine and spermidine whose presence has been previously reported in Croatian wines.

In this study, 48 samples of Croatian white and red wines were analysed. The investigated wines were produced from native and introduced grape varieties, characteristic of two Croatian wine regions; Dalmatia (coastal wine region) and Hrvatsko zagorje (continental wine region). Also, this research was conducted with the aim of comparison of the BAs concentration in red and white wines from the same wine regions.

2. Results

2.1. Samples

A total of 48 samples of wines from two different Croatian wine regions are listed in Table 1 (samples from coastal wine region of Dalmatia) and Table 2 (samples from continental wine region of Hrvatsko zagorje) with corresponding grape variety and origin. All wine samples were obtained directly from small farmers of the corresponding region.

Table 1. The investigated red and white wine samples from Croatian coastal wine region of Dalmatia.

Number	Sample Mark	Grape Variety	Origin	Type of Wine
1	1DB	Debit	Drniš	white
2	2DB	Pošip	Bol, Brač	white
3	3DB	Pošip barrique	Bol, Brač	white
4	4DB	Kujundžuša	Rogoznica	white
5	5DB	Maraština	Kaštela	white
6	6DB	Maraština	Bol, Brač	white
7	7DB	Chardonnay	Kaštela	white
8	8DB	Pošip barrique	Kaštela	white
9	9DB	Kujundžuša + Graševina	Imotski	white
10	10DB	Debit	Sinj	white
11	11DB	Debit	Šibenik	white
12	1DC	Plavina + Lasin + Shiraz	Drniš	red
13	2DC	Bogondon	Bol, Brač	red
14	3DC	Tribidrag	Bol, Brač	red
15	4DC	Plavac	Rogoznica	red
16	5DC	Plavac	Kaštela	red
17	6DC	Plavac	Bol, Brač	red
18	7DC	Crljenak	Kaštela	red
19	8DC	Plavac	Kaštela	red
20	9DC	Crljenak	Bol, Brač	red
21	10DC	Crljenak	Imotski	red
22	11DC	Plavac	Sinj	red
23	12DC	Plavac	Šibenik	red
24	13DC	Plavac	Knin	red

Table 2. The investigated red and white wine samples from Croatian continental wine region of Hrvatsko zagorje.

Number	Sample Mark	Grape Variety	Origin	Type of Wine
1	1ZB	Graševina	Beretinec	white
2	2ZB	Pinot	Moslavec	white
3	3ZB	Rajnski rizling	Varaždin	white
4	4ZB	Graševina	Varaždin	white
5	5ZB	Muškat žuti	Varaždin	white
6	6ZB	Sivi pinot	Varaždin	white
7	7ZB	Manzoni	Beretinec	white
8	8ZB	Graševina	Varaždin	white
9	9ZB	Traminac	Beretinec	white
10	14ZB	Sauvignon	Cestica	white
11	1ZC	Frankovka	Beretinec	red
12	2ZC	Isabella + Farber	Kneginec	red
13	3ZC	Shiraz	Varaždin	red
14	4ZC	Frankovka + Plavac	Ludbreg	red
15	5ZC	Frankovka	Cestica	red
16	6ZC	Frankovka	Ledinec	red
17	7ZC	Merlot + Frankovka	Beretinec	red
18	8ZC	Portugizac	Ivanec	red
19	10ZC	Isabella	Ledinec	red
20	11ZC	Frankovka	Beretinec	red
21	12ZC	Frankovka + Merlot	Beretinec	red
22	13ZC	Isabella	Beretinec	red
23	14ZC	Isabella	Ledinec	red
24	15ZC	Frankovka	Beretinec	red

2.2. Biogenic Amines Determinations

The content of biogenic amines was determined by HPLC method, as described by Manetta et al. [40], with slight modifications.

The derivatisation was performed without pre-treatment of the samples as follows: 0.25 mL of BAs standard solution or wine sample was mixed with 70 μL of a saturated sodium hydrogen carbonate solution and 65 μL 0.1 M potassium hydroxide solution. Then, 1 mL of dansyl chloride (0.5% *w/v* in acetone) was added and the mixture was incubated for 45 min at 40 °C in thermoblock with occasional stirring. After that, 100 μL of ammonia solution (25% *w/w*) was added and after strong stirring by vortex, the reaction mixtures were left in dark for 30 min. The volume of samples was made up to 5 mL with acetonitrile, and after the shaking they were filtrated and ready for the analysis. The control samples (blank) was prepared by the same procedure, but instead of standards or wine samples, ultrapure water was used.

All samples were prepared and analysed in triplicate, and the data are presented as a mean value ± standard deviation.

Gradient elution was conducted using acetonitrile (solvent A), and ultrapure water (solvent B) according to the program presented in Table 3.

Table 3. The HPLC gradient elution program used for the analysis of biogenic amines.

Time (min)	Solvent A (%)	Solvent B (%)
0.5	40	60
25	80	20
30	95	5
34	95	5
35	40	60
43	40	60

The applied flow rate was 1 mL min^{-1}, column temperature was 25 °C, the sample injection volume was 10 μL and the detection wavelength was 254 nm. The identification of BAs was carried out by comparing their retention times, individually or in a mixture of standards as is shown at Figure 1. Peaks appearing in the chromatogram that are not assigned to any standard are secondary products of the derivatisation process. The quantification was done using linear calibration curves that were created for every standard compound of BAs.

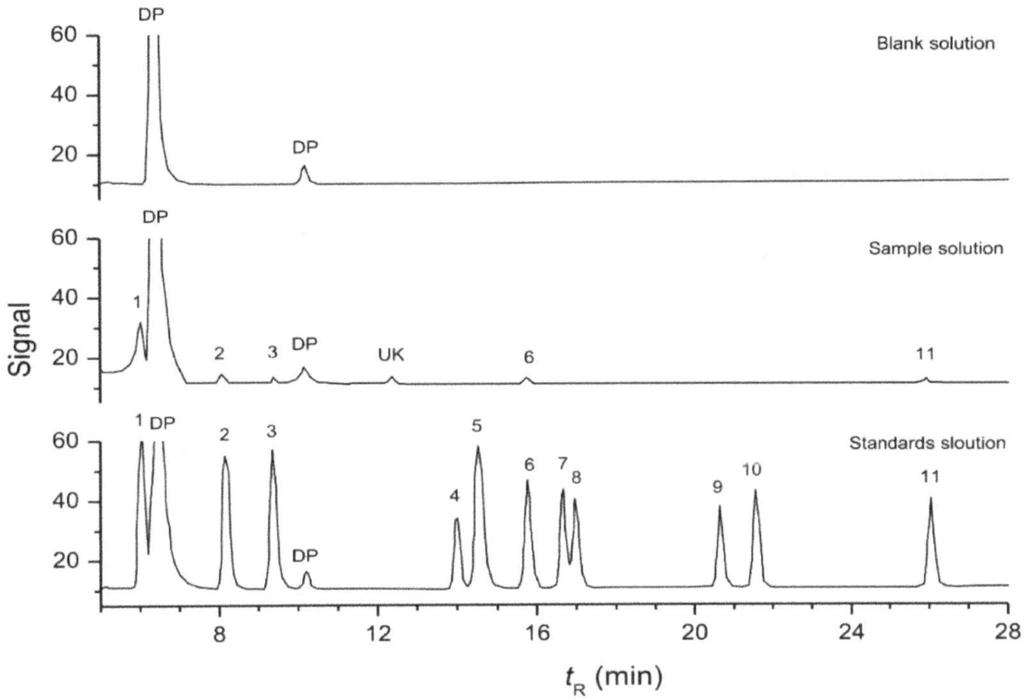

Figure 1. HPLC chromatogram of the blank sample, wine sample and standards' solution of biogenic amines. The numbers correspond to amines reported in Table 4, DP—derivation peak, UK—unknown peak.

As ethanolamine, ethylamine and methylamine were obtained as hydrochloride salts, their concentrations in standard solutions were corrected as for a free base.

The limits of detection (LODs) and the limits of quantification (LOQs) were determined using signal-to-noise-ratio (S/N) of 3 and 10 respectively for all standards. Table 4 is showing calculated analytical parameters for the applied method.

The detected concentrations of 11 investigated biogenic amines (tryptamine, putrescine, cadaverine, histamine, tyramine, spermidine, spermine, isopentylamine, ethanolamine, methylamine and ethylamine) in 48 samples of Croatian red and white wines from Hrvatsko zagorje and Dalmatia wine regions are given in Table 5. The samples can be divided into four groups: Dalmatian white wines (11 samples), Dalmatian red wines (13 samples), Hrvatsko zagorje white wines (10 samples) and Hrvatsko zagorje red wines (14 samples). The results are presented in Table 5 as the average value ± relative standard deviation (RSD) (%) of all samples.

Table 4. The analytical parameters of a chromatographic method.

Number	Amine Name (Short Name)	t_R (min) ± RSD (n = 3)	LOD (mg L^{-1})	LOQ (mg L^{-1})	R^2	Recovery White Wines (%)	Recovery Red Wines (%)	Linear Range (mg L^{-1})
1	Ethanolamine (ETHA)	5.79 ± 0.04	0.85	2.50	0.998	102	106	2.50–300.00
2	Methylamine (MA)	7.92 ± 0.05	0.14	0.41	0.999	123	88	0.41–100.00
3	Ethylamine (ETHYL)	9.07 ± 0.04	0.20	0.59	0.999	104	90	0.59–100.00
4	Tryptamine (TRP)	14.05 ± 0.04	0.03	0.06	0.998	103	82	0.06–215.00
5	Isopentylamine (IPA)	14.49 ± 0.04	0.22	0.66	0.999	107	93	0.66–200.00
6	Putrescine (PUT)	15.81 ± 0.05	0.03	0.09	0.998	97	88	0.09–96.70
7	Cadaverine (CAD)	16.71 ± 0.04	0.03	0.10	0.998	117	109	0.10–107.50
8	Histamine (HIS)	16.97 ± 0.04	0.03	0.10	0.998	99	91	0.10–107.50
9	Tyramine (TYR)	20.66 ± 0.04	0.04	0.13	0.999	107	92	0.13–107.50
10	Spermidine (SPD)	21.60 ± 0.03	0.03	0.09	0.998	107	101	0.09–96.70
11	Spermine (SPM)	26.12 ± 0.02	0.03	0.10	0.998	108	93	0.10–107.50

t_R—retention time, RSD—relative standard deviation, LOD—limit of detection, LOQ—limit of quantification, R—correlation coefficient.

Table 5. The content of biogenic amines (mg L^{-1}) in 48 investigated wine samples.

Sample	TRP	IPA	PUT	CAD	HIS	TYR	SPD	SPM	ETHA	MA	ETHYL
1DB	<0.03	<0.22	1.21 ± 0.04	<0.03	<0.03	<0.04	<0.03	<0.03	24.32 ± 8.05	<0.14	<0.59
2DB	<0.03	0.29 ± 0.21	1.19 ± 0.02	<0.03	<0.03	<0.04	<0.03	0.19 ± 0.09	10.42 ± 0.09	<0.14	<0.20
3DB	<0.03	<0.22	0.99 ± 0.04	<0.03	<0.10	0.39 ± 0.04	<0.03	0.39 ± 0.21	12.08 ± 0.83	1.40 ± 0.01	<0.20
4DB	0.49 ± 0.38	<0.22	0.41 ± 0.03	<0.03	<0.03	<0.04	<0.09	<0.10	9.09 ± 0.24	<0.14	<0.20
5DB	<0.03	<0.22	1.06 ± 0.00	<0.03	<0.03	<0.04	<0.09	<0.03	6.71 ± 0.00	<0.14	<0.20
6DB	0.42 ± 0.13	<0.22	0.86 ± 0.53	<0.03	<0.03	<0.04	<0.03	<0.03	2.87 ± 0.45	<0.14	<0.20
7DB	1.22 ± 1.12	<0.22	1.15 ± 0.01	<0.03	<0.03	<0.04	<0.03	0.11 ± 0.01	7.83 ± 0.33	<0.14	<0.20
8DB	0.68 ± 0.04	<0.22	1.25 ± 0.01	<0.03	0.56 ± 0.03	0.40 ± 0.01	<0.03	<0.10	4.22 ± 0.91	<0.14	<0.20
9DB	0.27 ± 0.18	<0.22	0.72 ± 0.02	<0.03	<0.03	<0.04	<0.03	<0.03	<2.5	<0.14	<0.20
10DB	0.38 ± 0.25	<0.22	2.10 ± 0.06	<0.10	<0.03	<0.04	<0.09	<0.03	<0.85	<0.14	<0.59
11DB	<0.03	<0.22	1.58 ± 0.02	<0.10	<0.03	<0.04	<0.03	<0.10	9.68 ± 0.04	<0.14	<0.59
1DC	<0.03	<0.22	1.58 ± 0.03	<0.03	<0.03	<0.04	<0.03	<0.03	4.85 ± 0.25	<0.14	<0.20
2DC	<0.03	1.13 ± 0.16	7.50 ± 0.17	<0.03	2.33 ± 0.08	2.00 ± 0.00	0.18 ± 0.04	0.31 ± 0.15	13.24 ± 0.13	<0.14	<0.20
3DC	0.54 ± 0.15	<0.22	1.76 ± 0.04	<0.10	0.89 ± 0.05	1.57 ± 0.01	<0.03	0.27 ± 0.13	14.00 ± 0.38	<0.14	<0.20
4DC	1.20 ± 0.01	<0.22	1.31 ± 0.00	<0.03	<0.03	<0.04	<0.03	<0.10	<0.85	<0.14	<0.20
5DC	0.43 ± 0.21	<0.22	1.35 ± 0.02	<0.03	<0.03	<0.04	0.09 ± 0.00	<0.03	<2.5	<0.14	<0.20
6DC	0.43 ± 0.18	<0.22	3.47 ± 0.09	<0.03	1.64 ± 0.01	1.88 ± 0.04	0.14 ± 0.05	0.24 ± 0.20	8.77 ± 0.50	<0.14	<0.20
7DC	0.23 ± 0.04	<0.22	0.78 ± 0.04	1.75 ± 0.03	0.89 ± 0.02	0.32 ± 0.03	<0.03	0.14 ± 0.03	5.11 ± 0.07	<0.14	<0.20
8DC	0.71 ± 0.04	<0.22	1.45 ± 0.02	<0.03	0.52 ± 0.01	<0.13	<0.03	<0.03	<2.5	<0.14	<0.20

Table 5. *Cont.*

Sample	TRP	IPA	PUT	CAD	HIS	TYR	SPD	SPM	ETHA	MA	ETHYL
9DC	0.34 ± 0.15	<0.22	0.78 ± 0.00	1.98 ± 0.09	1.17 ± 0.01	<0.13	0.46 ± 0.25	<0.10	9.22 ± 0.06	<0.14	<0.20
10DC	0.43 ± 0.35	<0.22	1.91 ± 0.11	<0.03	<0.03	<0.04	<0.03	0.17 ± 0.08	<0.85	<0.14	<0.20
11DC	<0.03	<0.22	1.77 ± 0.01	0.12 ± 0.06	<0.03	<0.04	0.12 ± 0.01	<0.03	10.75 ± 0.45	<0.14	<0.20
12DC	0.46 ± 0.30	<0.22	2.77 ± 0.07	<0.03	<0.03	<0.13	<0.09	<0.10	6.58 ± 0.17	<0.14	<0.20
13DC	0.66 ± 0.04	<0.22	0.89 ± 0.03	1.01 ± 0.14	0.36 ± 0.04	<0.04	<0.03	<0.03	6.36 ± 0.03	<0.14	<0.59
1ZB	0.68 ± 0.02	1.29 ± 0.04	0.27 ± 0.01	0.40 ± 0.00	<0.03	<0.04	<0.03	<0.10	4.07 ± 0.28	<0.14	<0.20
2ZB	<0.03	0.87 ± 0.49	1.49 ± 0.67	0.82 ± 0.31	<0.03	<0.04	<0.03	<0.03	17.96 ± 0.01	<0.14	<0.20
3ZB	<0.03	1.32 ± 0.00	0.62 ± 0.01	<0.03	<0.03	<0.04	<0.03	<0.03	4.80 ± 0.51	<0.14	<0.20
4ZB	<0.03	1.47 ± 0.32	1.19 ± 0.01	<0.03	<0.03	<0.04	<0.03	<0.10	8.61 ± 1.19	<0.14	<0.20
5ZB	<0.03	0.50 ± 0.09	1.08 ± 0.01	<0.03	<0.03	<0.04	<0.03	<0.03	<0.85	<0.14	<0.20
6ZB	<0.03	0.82 ± 0.21	<0.09	<0.03	<0.03	<0.04	<0.03	<0.03	3.80 ± 0.00	<0.14	<0.20
7ZB	<0.03	0.42 ± 0.46	1.42 ± 0.01	<0.03	<0.03	<0.04	<0.03	<0.03	4.61 ± 0.01	<0.14	<0.20
8ZB	<0.03	0.31 ± 0.22	0.83 ± 0.04	<0.03	<0.03	<0.04	0.12 ± 0.02	<0.10	4.16 ± 0.14	<0.14	<0.20
9ZB	0.33 ± 0.00	1.00 ± 0.80	1.16 ± 0.02	<0.03	<0.10	<0.04	<0.03	<0.10	7.19 ± 0.08	<0.14	<0.20
14ZB	<0.03	<0.22	0.94 ± 0.01	<0.10	<0.03	<0.04	<0.09	<0.03	<0.85	<0.14	<0.59
1ZC	0.75 ± 0.87	<0.22	0.17 ± 0.01	<0.03	<0.03	<0.13	0.21 ± 0.05	<0.03	<2.5	<0.14	<0.20
2ZC	1.82 ± 0.06	0.79 ± 0.32	2.06 ± 0.04	0.47 ± 0.05	1.06 ± 0.04	<0.13	0.75 ± 0.04	3.55 ± 0.04	15.41 ± 0.02	<0.14	<0.59
3ZC	3.31 ± 0.33	1.19 ± 0.01	3.75 ± 0.06	<0.10	9.63 ± 0.13	<0.13	0.26 ± 0.13	0.74 ± 0.04	4.10 ± 0.23	1.28 ± 0.01	<0.59
4ZC	4.89 ± 0.27	0.50 ± 0.06	0.39 ± 0.04	<0.10	0.35 ± 0.08	0.45 ± 0.01	0.15 ± 0.01	0.67 ± 0.07	8.04 ± 0.16	<0.14	<0.20
5ZC	3.61 ± 0.45	0.87 ± 0.04	3.17 ± 0.11	0.10 ± 0.00	0.50 ± 0.01	0.50 ± 0.04	0.22 ± 0.02	1.81 ± 0.01	16.80 ± 0.23	1.57 ± 0.04	<0.59
6ZC	0.88 ± 0.22	3.71 ± 0.14	<0.03	<0.10	<0.03	0.31 ± 0.08	2.51 ± 0.21	<0.03	95.84 ± 0.20	0.99 ± 0.55	<0.20
7ZC	0.39 ± 0.31	2.44 ± 0.18	0.66 ± 0.78	0.21 ± 0.01	0.98 ± 0.02	2.10 ± 0.14	0.57 ± 0.07	<0.03	63.96 ± 0.20	<0.14	<0.20
8ZC	0.09 ± 0.01	2.97 ± 2.44	<0.09	<0.03	<0.10	0.78 ± 0.71	0.33 ± 0.00	<0.03	31.14 ± 0.17	0.65 ± 0.03	<0.20
10ZC	0.67 ± 0.37	1.81 ± 0.02	0.16 ± 0.09	0.44 ± 0.20	<0.03	2.31 ± 0.05	0.15 ± 0.00	<0.03	51.79 ± 3.21	<0.14	<0.20
11ZC	9.18 ± 0.07	2.14 ± 0.48	<0.09	<0.10	<0.03	1.45 ± 0.22	6.05 ± 1.22	<0.03	35.19 ± 10.47	0.56 ± 0.15	<0.59
12ZC	0.87 ± 0.61	3.27 ± 0.35	<0.09	0.27 ± 0.32	2.01 ± 0.14	2.97 ± 0.09	1.97 ± 0.20	<0.03	46.03 ± 12.33	<0.41	1.17 ± 0.21
13ZC	<0.03	1.55 ± 1.27	1.00 ± 0.04	<0.03	0.35 ± 0.06	0.13 ± 0.03	<0.03	<0.10	23.78 ± 1.50	<0.14	<0.20
14ZC	<0.03	0.33 ± 0.32	0.24 ± 0.02	<0.10	<0.03	<0.04	<0.09	<0.10	5.22 ± 0.28	<0.14	<0.20
15ZC	<0.03	0.70 ± 0.10	<0.03	<0.03	<0.03	<0.04	<0.03	<0.10	4.37 ± 0.20	<0.14	<0.20

The table contains abbreviations for biogenic amines, and the corresponding full names are reported in Table 4. DB = Dalmatian white samples, DC = Dalmatian red samples, ZB = Hrvatsko zagorje white samples and ZC = Hrvatsko zagorje red samples.

3. Discussion

In the case of polyamines (putrescine, cadaverine, spermidine and spermine), their high concentrations are usually associated with unsanitary conditions.

If we look at the detected concentrations of those polyamines in the investigated samples, in most white wines cadaverine, spermidine and spermine were not detected, while in red wines the highest concentrations detected were 1.98 mg L^{-1} of cadaverine, 2.51 mg L^{-1} of spermidine and 3.55 mg L^{-1} of spermine. Therefore, it can be concluded that the sanitary conditions of all the samples were satisfactory.

Histamine level also plays a special role as indicator amine. Histamine is the most toxic amine, although the toxicity is caused by histamine and the total content of amines, ethanol and acetaldehyde [41]. The allowed concentrations of histamine in wines are different across countries. According to available references, the highest histamine concentration of 10 mg L^{-1} is allowed in Switzerland [42]. According to the reported results, the highest concentration of histamine among tested samples was observed in the 3ZC sample (9.63 mg L^{-1}), while its content in other red wine samples was generally low (from 0.35 to 2.01 mg L^{-1}), and in white wines even below the detection limit, except for sample 8 DB, where it was found at concentration of 0.56 mg L^{-1}.

Putrescine and ethanolamine were the most prominent amines in all of the samples, regardless of the type of wine or its origin.

According to the Bover-Cid et al. [25], Glória et al. [43] and Kiss et al. [44] putrescine, spermine and spermidine are naturally present in grapes and their presence could be an indicator of the wine's geographical region or grape variety.

Putrescine concentrations were slightly higher in red wines, which can be explained by the fact that it can be formed during malolactic fermentation that usually occurs in the process of red wine production [25]. The highest concentration of putrescine in red wine samples was found in sample 2DC (7.5 mg L^{-1}). In comparison to results reported in other studies, these concentrations were relatively low [15,16,19,21–24,27,30,37,45–47]. The putrescine content in samples was investigated by Landete et al. ranged from 30 to 50 mg L^{-1} [31], while in our study they ranged from 0.16 to 3.75 mg L^{-1}. Spermidine was not detected in white wines, except in the sample 8ZB where its concentration was 0.12 mg L^{-1}. In red wine samples, spermidine was found in the range from 0.09 to 0.46 mg L^{-1} in Dalmatian red wines and from 0.15 to 6.05 mg L^{-1} in wine samples from Hrvatsko zagorje region.

Spermine was found in low concentrations in four samples of white wines from Dalmatian region (ranged from 0.11 to 0.39 mg L^{-1}), while in white wines from Hrvatsko zagorje region it was not detected. If we compare the results for spermine in red wines, slightly higher concentrations were detected in samples from Hrvatsko zagorje region than in Dalmatian wines but final concentrations correspond to those that are reported in the literature [20,43].

Del Prete et al. confirmed the presence of ethanolamine, ethylamine and putrescine in grapes [30].

Ethanolamine is an amine that was detected in almost all samples at significant concentrations and especially in red wines from Hrvatsko zagorje region. Its concentrations correspond to those reported in other studies of Mediterranean wines, such as in samples from Italy [4,30], Portugal [47], and Greece [20]. A large number of studies didn't research the ethanolamine content, but in our study, its concentration was found to be significant in almost all of the samples.

Ethylamine was detected only in sample 12ZC, and thus in a very low concentration of 1.17 mg L^{-1}, as well as methylamine, which was found in sample 3DB and in few red wines from Hrvatsko zagorje wine region at concentration range from 0.99 to 1.57 mg L^{-1}.

Jeromel et al. studied BAs in Croatian red wines, but in their study the content of ethylamine and methylamine were not investigated [34], while histamine and tryptamine were found to be the most abundant biogenic amines.

In this study, tryptamine was detected in the range from 0.2 to 1.2 mg L^{-1} in the wines from Dalmatia, while in the red samples from Hrvatsko zagorje region it was found in higher concentrations (from 0.09 to 9.18 mg L^{-1}).

It is interesting to see the distribution of isopentylamine in the samples. Its concentrations in the samples from Hrvatsko zagorje ranged from 0.3 to 1.5 mg L^{-1} in white wines and from 0.33 to 3.71 mg L^{-1} in red wines, while among Dalmatian samples, only samples 2DB and 2DC contained this compound in concentration from 0.29 and 1.13 mg L^{-1}, respectively.

Marques et al. in their research proved the connection between the content of tyramine and type of wine especially as its concentration is higher in red wines after malolactic fermentation [32].

Tyramine was found in very low concentrations in two white wines from Dalmatian region, while it was not detected at all in white wines from Hrvatsko zagorje region. As expected, red samples contained higher amounts of tyramine, but still significantly lower than those reported in the literature [32,45,46]. A very wide range of total BAs content has been reported, from not-detected to 130 mg L^{-1} with the main amines e.g., putrescine, histamine, tyramine and cadaverine [48].

A scree plot in Figure 2 suggests involving four principal components in the model. Those four PC (columns PC1–PC4) explain 70% of the total variance in the data, (Table 6). Since graphical presentation only allows for using two columns, the cut off point for loading values was >0.30 and it is marked throughout Table 6 in boldface type only for PC1 and PC2.

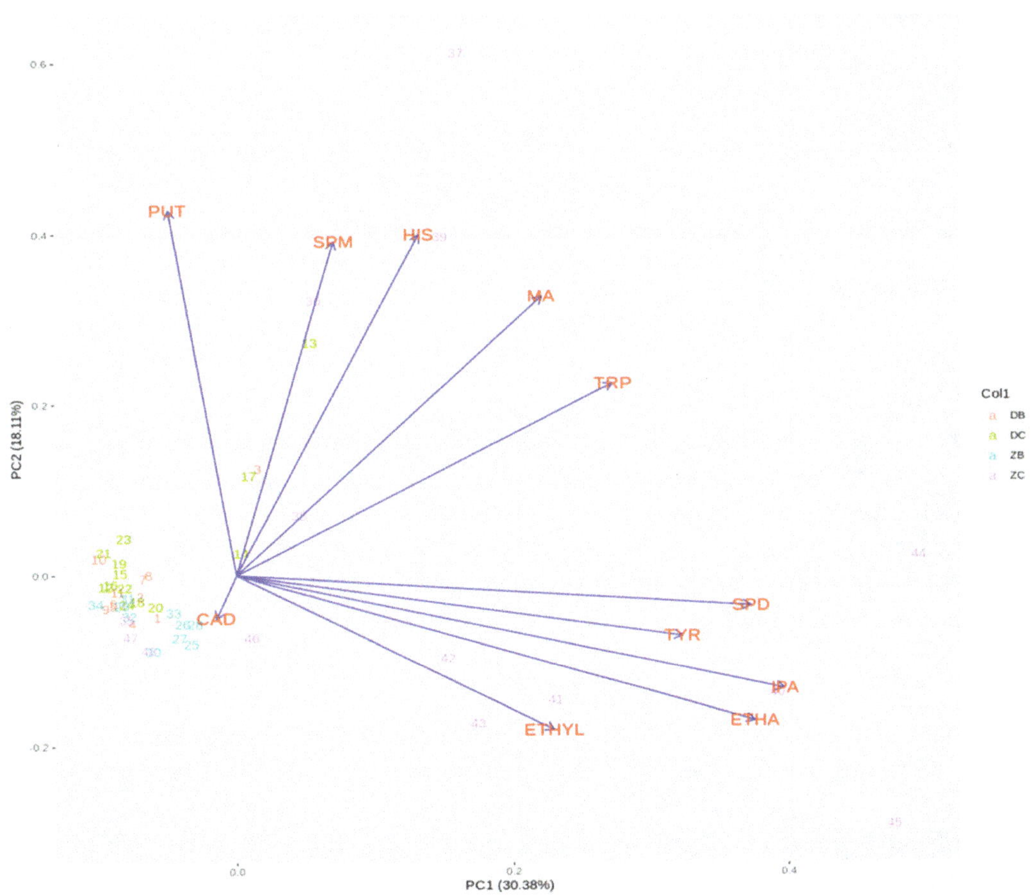

Figure 2. PCA plot. Table 4 contains abbreviations for biogenic amines. Numbers correspond to samples' in Tables 1 and 2. DB = Dalmatian white samples, DC = Dalmatian red samples, ZB = Hrvatsko zagorje white samples and ZC = Hrvatsko zagorje red samples.

Table 6. Loading values for PCA.

	PC1	PC2	PC3	PC4
TRP	**0.3160**	0.2615	−0.4331	0.1254
IPA	**0.4612**	−0.1534	0.0624	−0.1102
PUT	−0.0565	**0.4970**	0.3967	−0.1822
CAD	−0.0167	−0.0588	0.1155	0.9085
HIS	0.1533	**0.4641**	0.3395	0.0760
TYR	**0.3744**	−0.0824	0.4237	−0.0500
SPD	**0.4325**	−0.0409	−0.3308	0.0847
SPM	**0.3160**	0.2615	−0.4331	0.1254
ETHA	**0.4612**	−0.1534	0.0624	−0.1102
MA	−0.0565	**0.4970**	0.3967	−0.1822
ETHYL	−0.0167	−0.0588	0.1155	0.9085

Short amine names correspond to amines' name in Table 4.

The loading values express how well the new PCs correlate with old variables. The first PC, which explains 30.38% of the total variance correlates positively with Tryptamine (TRP), Isopentylamine (IPA), Tyramine (TYR), Spermidine (SPD) and Ethanolamine (ETHA). The second PC (18.11% of the total variance) correlates positively with Putrescine (PUT), Histamine (HIS), Spermine (SPM) and Methylamine (MA). On the other hand, it can be seen that CAD has small values. All data are given in Tables 6 and 7.

Table 7. Explanation of variance in statistical analysis.

	PC1	PC2	PC3	PC4
Standard deviation	1.8279	1.4114	1.1870	1.03047
Proportion of variance	0.3038	0.1811	0.1281	0.09653
Cumulative proportion	0.3038	0.4849	0.6130	0.70949

4. Materials and Methods

4.1. Chemical and Reagents

All used reagents were of analytical grade.

Dansyl chloride and amine standards (isopenthylamine, ethanolamine, methylamine, ethylamine, spermidine, spermine, putrescine, tyramine, histamine, cadaverine and tryptamine) were purchased from Sigma-Aldrich (Steinheim, Germany).

Hydrochloric acid (37%, *w/w*), ammonia solution (25%, *w/w*), sodium hydrogen carbonate, acetonitrile (HPLC grade) were also purchased from Sigma-Aldrich.

Ultrapure water was obtained from ELGA Purelab flex.

4.2. Apparatus and Software

A Perkin Elmer Series 200 HPLC system, equipped with an autosampler, binary pump and UV/Vis detector (all of Series 200), was applied with a TotalChrom Workstation software (PerkinElmer, Waltham, MA, USA).

Chromatographic separations were performed on a Restek Ultra IBD C18 column (5 μm particle size, 250 × 4.6 mm i.d.) with Ultra IBD guard column (5 μm particle size, 10 × 4 mm i.d.), Restek, Bellefonte, PA, USA.

4.3. Statistical Analysis

For exposing the underlying patterns in the data, principal component analysis (PCA) was used with the intention of showing which biogenic amines (wine samples) carry comparable information, and which of them are unique. The statistical analysis was carried out using the RStudio ver. 1.1.383 [49] while PCA analysis was done using 'prcomp' method by using singular value decomposition (SVD). Since SVD has slightly better numerical accuracy, therefore, 'prcomp' is the preferred function.

Biogenic amines were taken as variables (columns of the input matrix) and the various wines as cases (rows of the matrix). The underlying patterns, 'components' are represented by new variables called principal components.

5. Conclusions

This work shows that biogenic amines content can be a differentiation factor for a grape variety and geographical origin for red wines. It can be stated that Dalmatian white wines are characterised by tryptamine, putrescine and ethanolamine. Their content is in the following ranges: tryptamine from 0.23 to 1.22 mg L^{-1}; putrescine from 0.41 to 7.5 mg L^{-1} and ethanolamine from 2.87 to 24.32 mg L^{-1}. White wines from the Hrvatsko zagorje region are characterised by content of isopentylamine (from 0.31 to 1.47 mg L^{-1}), putrescine (from 0.27 to 1.49 mg L^{-1}) and ethanolamine (from 3.80 to 17.96 mg L^{-1}). On the other hand, in red wines from the Hrvatsko zagorje region all BAs, except ethylamine, were found. According to the PCA, the wines of the Hrvatsko zagorje red group samples are the most distinguished. Wines from the Hrvatsko zagorje red group marked as 36, 37 and 39 have a higher concentration of spermine, histamine, methylamine and putrescine, as well as lower concentration of spermidine, tyramine, isopentylamine, ethanolamine and ethylamine than wines from the same group in lines 41–45. All other wines are mostly concentrated around similar values, with the exceptions of Dalmatian red wines marked as 13 and 17, and Dalmatian white wine marked 3, respectively.

Author Contributions: Conceptualisation, I.M. and I.L.; methodology, I.L.; software, I.L. and A.P.; validation, I.M. and I.L.; formal analysis, A.P.; investigation, I.M. and N.R.; resources, N.R. and I.L.; data curation, I.M. and I.L.; writing—original draft preparation, I.M.; writing—review and editing; visualization, I.M.; supervision, I.L.; funding acquisition, I.A. and N.V.

Acknowledgments: The authors would like to thank all companies that provided their samples to this research.

References

1. Schlesier, K.; Fauhl-Hassek, C.; Forina, M.; Cotea, V.; Kocsi, E.; Schoula, R.; van Jaarsveld, F.; Wittkowski, R. Characterisation and determination of the geographical origin of wines. Part I: Overview. *Eur. Food Res. Technol.* **2009**, *230*, 1–13. [CrossRef]
2. Ebeler, S.E.; Thorngate, J.H. Wine Chemistry and Flavor: Looking into the Crystal Glass. *J. Agric. Food Chem.* **2009**, *57*, 8098–8108. [CrossRef] [PubMed]
3. Marquez, R.; Castro, R.; Natera, R.; Garcia-Barroso, C. Characterisation of the volatile fraction of Andalusian sweet wines. *Eur. Food Res. Technol.* **2008**, *226*, 1479–1484. [CrossRef]
4. Galgano, F.; Caruso, M.; Perretti, G.; Favati, F. Authentication of Italian red wines on the basis of the polyphenols and biogenic amines. *Eur. Food Res. Technol.* **2011**, *232*, 889–897. [CrossRef]
5. Jaitz, L.; Siegl, K.; Eder, R.; Rak, G.; Abranko, L.; Koellensperger, G.; Hann, S. LC–MS/MS analysis of phenols for classification of red wine according to geographic origin, grape variety and vintage. *Food Chem.* **2010**, *122*, 366–372. [CrossRef]
6. Agatonovic-Kustrin, S.; Hettiarachchi, C.G.; Morton, D.W.; Razic, S. Analysis of phenolics in wine by high performance thin-layer chromatography with gradient elution and high resolution plate imaging. *J. Pharm. Biomed. Anal.* **2015**, *102*, 93–99. [CrossRef] [PubMed]
7. Marković, M.; Martinović Bevanda, A.; Talić, S. Antioxidant activity and total phenol content of white wine Žilavka. *Bull. Chem. Technol. Bosn. Herceg.* **2015**, *44*, 1–4.
8. Galgano, F.; Favati, F.; Camso, M.; Scarpa, T.; Palma, A. Analysis of trace elements in southern Italian wines and their classification according to provenance. *LWT-Food Sci. Technol.* **2008**, *41*, 1808–1815. [CrossRef]
9. Geana, I.; Iordache, A.; Ionete, R.; Marinescu, A.; Ranca, A.; Culea, M. Geographical origin identification of Romanian wines by ICP-MS elemental analysis. *Food Chem.* **2013**, *138*, 1125–1134. [CrossRef] [PubMed]

10. Ražić, S.; Čokeša, D.; Sremac, S. Multivariate data visualization methods based on elemental analysis of wines by atomic absorption spectrometry. *J. Serb. Chem. Soc.* **2007**, *72*, 1487–1492. [CrossRef]
11. Ražić, S.; Onjia, A. Trace Element Analysis and Pattern Recognition Techniques in Classification of Wine from Central Balkan Countries. *Am. J. Enol. Viticult.* **2010**, *61*, 506–511. [CrossRef]
12. Mo Dugo, G.; Vilasi, F.; La Torre, G.; Pellicano, T. Reverse phase HPLC/DAD determination of biogenic amines as dansyl derivatives in experimental red wines. *Food Chem.* **2006**, *95*, 672–676. [CrossRef]
13. Guo, Y.Y.; Yang, Y.P.; Peng, Q.; Han, Y. Biogenic amines in wine: A review. *Int. J. Food Sci. Technol.* **2015**, *50*, 1523–1532. [CrossRef]
14. Kovačević Ganić, K.; Gracin, L.; Komes, D.; Ćurko, N.; Lovrić, T. Changes of the content of biogenic amines during winemaking of Sauvignon wines. *Croat. J. Food Sci. Technol.* **2009**, *1*, 21–27.
15. Anlı, R.E.; Bayram, M. Biogenic Amines in Wines. *Food Rev. Int.* **2008**, *25*, 86–102. [CrossRef]
16. Souza, S.C.; Theodoro, K.H.; Souza, E.R.; da Motta, S.; Beatriz, M.; Gloria, A. Bioactive Amines in Brazilian Wines: Types, Levels and Correlation with Physico-Chemical Parameters. *Braz. Arch. Biol. Technol.* **2005**, *48*, 53–62. [CrossRef]
17. Košmerl, T.; Šućur, S.; Prosen, H. Biogenic amines in red wine: The impact of technological processing of grape and wine. *Acta Agric. Slov.* **2013**, *101*, 249–261. [CrossRef]
18. Cecchini, F.; Morassut, M. Effect of grape storage time on biogenic amines content in must. *Food Chem.* **2010**, *123*, 263–268. [CrossRef]
19. Anlı, R.E.; Vural, N.; Yılmaz, S.; Vural, Ỳ.H. The determination of biogenic amines in Turkish red wines. *J. Food Compos. Anal.* **2004**, *17*, 53–62. [CrossRef]
20. Soufleros, E.H.; Bouloumpasi, E.; Zotou, A.; Loukou, Z. Determination of biogenic amines in Greek wines by HPLC and ultraviolet detection after dansylation and examination of factors affecting their presence and concentration. *Food Chem.* **2007**, *101*, 704–716. [CrossRef]
21. Proestos, C.; Loukatos, P.; Komaitis, M. Determination of biogenic amines in wines by HPLC with precolumn dansylation and fluorimetric detection. *Food Chem.* **2008**, *106*, 1218–1224. [CrossRef]
22. Li, Z.; Wu, Y.; Zhang, G.; Zhao, Y.; Xue, C. A survey of biogenic amines in Chinese red wines. *Food Chem.* **2007**, *105*, 1530–1535. [CrossRef]
23. Pineda, A.; Carrasco, J.; Pena-Farfal, C.; Henriquez-Aedo, K.; Aranda, M. Preliminary evaluation of biogenic amines content in Chilean young varietal wines by HPLC. *Food Control* **2012**, *23*, 251–257. [CrossRef]
24. Fernandes, J.; Ferreira, M. Combined ion-pair extraction and gas chromatography–mass spectrometry for the simultaneous determination of diamines, polyamines and aromatic amines in Port wine and grape juice. *J. Chromatogr. A* **2000**, *886*, 183–195. [CrossRef]
25. Bover-Cid, S.; Iquierdo-Pulido, M.; Marine-Font, A.; Vidal-Carou, M.C. Biogenic mono-, di- and polyamine contents in Spanish wines and influence of a limited irrigation. *Food Chem.* **2006**, *96*, 43–47. [CrossRef]
26. Romero, R.; Gázquez, D.; Bagur, M.; Sánchez-Viñas, M. Optimization of chromatographic parameters for the determination of biogenic amines in wines by reversed-phase high-performance liquid chromatography. *J. Chromatogr. A* **2000**, *871*, 75–83. [CrossRef]
27. Romero, R.; Sanchez-Vinas, M.; Gazquez, D.; Bagur, M.G. Characterization of selected spanish table wine samples according to their biogenic amine content from liquid chromatographic determination. *J. Agric. Food Chem.* **2002**, *50*, 4713–4717. [CrossRef] [PubMed]
28. Moreno-Arribas, M.V.; Polo, M.C. *Wine Chemistry and Biochemistry*; Springer: New York, NY, USA, 2009; Volume 735.
29. Wang, Y.-Q.; Ye, D.-Q.; Zhu, B.-Q.; Wu, G.-F.; Duan, C.-Q. Rapid HPLC analysis of amino acids and biogenic amines in wines during fermentation and evaluation of matrix effect. *Food Chem.* **2014**, *163*, 6–15. [CrossRef] [PubMed]
30. Del Prete, V.; Costantini, A.; Cecchini, F.; Morassut, M.; Garcia-Moruno, E. Occurrence of biogenic amines in wine: The role of grapes. *Food Chem.* **2009**, *112*, 474–481. [CrossRef]
31. Landete, J.M.; Ferrer, S.; Polo, L.; Pardo, I. Biogenic amines in wines from three spanish regions. *J. Agric. Food Chem.* **2005**, *53*, 1119–1124. [CrossRef] [PubMed]
32. Marques, A.P.; Leitao, M.C.; Romao, M.V.S. Biogenic amines in wines: Influence of oenological factors. *Food Chem.* **2008**, *107*, 853–860. [CrossRef]
33. Alpeza, I.; Prša, I.; Mihaljević, B. Vinogradarstvo i vinarstvo Republike Hrvatske u okviru svijeta. *Glasnik Zaštite Bilja* **2014**, *37*, 6–13.

34. Jeromel, A.; Kovačević Ganić, K.; Herjavec, S.; Mihaljević, M.; Korenika Jagatić, A.M.; Rendulić, I.; Čolić, M. Concentration of Biogenic Amines in 'Pinot Noir' Wines Produced in Croatia. *Agric. Conspec. Sci.* **2012**, *77*, 37–40.
35. Almeida, C.; Fernandes, J.; Cunha, S. A novel dispersive liquid–liquid microextraction (DLLME) gas chromatography-mass spectrometry (GC–MS) method for the determination of eighteen biogenic amines in beer. *Food Control* **2012**, *25*, 380–388. [CrossRef]
36. Onal, A. A review: Current analytical methods for the determination of biogenic amines in foods. *Food Chem.* **2007**, *103*, 1475–1486. [CrossRef]
37. Busto, O.; Miracle, M.; Guasch, J.; Borrull, F. Determination of biogenic amines in wines by high-performance liquid chromatography with on-column fluorescence derivatization. *J. Chromatogr. A* **1997**, *757*, 311–318. [CrossRef]
38. Ganić, K.K.; Ćurko, N.; Kosić, U.; Komes, D.; Gracin, L. Determination of Biogenic Amines in Red Croatian Wines. In *32nd World Congress of Vine and Wine, Zagreb, Croatia 2009*; Kubanović, V., Ed.; Ministry of Agriculture, Fisheries and Rural Development: Zagreb, Croatia, 2009.
39. Galgano, F.; Caruso, M.; Favati, F.; Romano, P.; Caruso, M. HPLC determination on agmatine and other amines in wine. *J. Int. Sci. Vigne. Vin.* **2003**, *37*, 237–242. [CrossRef]
40. Manetta, A.C.; Di Giuseppe, L.; Tofalo, R.; Martuscelli, M.; Schirone, M.; Giammarco, M.; Suzzi, G. Evaluation of biogenic amines in wine: Determination by an improved HPLC-PDA method. *Food Control* **2016**, *62*, 351–356. [CrossRef]
41. Garcia-Marino, M.; Trigueros, A.; Escribano-Bailon, T. Influence of oenological practices on the formation of biogenic amines in quality red wines. *J. Food Compos. Anal.* **2010**, *23*, 455–462. [CrossRef]
42. Lehtonen, P. Determination of amines and amino acids in wine—A review. *Am. J. Enol. Vitic.* **1996**, *47*, 127–133.
43. Gloria, M.B.A.; Watson, B.T.; Simon-Sarkadi, L.; Daeschel, M.A. A survey of biogenic amines in Oregon Pinot noir and Cabernet Sauvignon wines. *Am. J. Enol. Viticult.* **1998**, *49*, 279–282.
44. Sass-Kiss, A.; Kiss, J.; Havadi, B.; Adanyi, N. Multivariate statistical analysis of botrytised wines of different origin. *Food Chem.* **2008**, *110*, 742–750. [CrossRef]
45. Lehtonen, P.; Saarinen, M.; Vesanto, M.; Riekkola, M.-L. Determination of wine amines by HPLC using automated precolumn derivatisation with *o*-phthalaldehyde and fluorescence detection. *Z. Lebensm.-Unters. Forsch.* **1992**, *194*, 434–437. [CrossRef]
46. Herbert, P.; Santos, L.; Alves, A. Simultaneous Quantification of Primary, Secondary Amino Acids, and Biogenic Amines in Musts and Wines Using OPA/3-MPA/FMOC-Cl Fluorescent Derivatives. *J. Food Sci.* **2001**, *66*, 1319–1325. [CrossRef]
47. Mafra, I.; Herbert, P.; Santos, L.; Barros, P.; Alves, A. Evaluation of Biogenic Amines in Some Portuguese Quality Wines by HPLC Fluorescence Detection of OPA Derivatives. *Am. J. Enol. Viticult.* **1999**, *50*, 128–132.
48. Restuccia, D.; Loizzio, M.R.; Spizzirri U., G. Accumulation of Biogenic Amines in Wine: Role of Alcocholic and Malolactic Fermentation. *Fermentation* **2018**, *4*, 6. [CrossRef]
49. R Development Core Team. *Integrated Development for R. Rstudio*; R Development Core Team: Vienna, Austria, 2016.

11

A UHPLC-UV Method Development and Validation for Determining Kavalactones and Flavokavains in *Piper methysticum* (Kava)

Yijin Tang * and Christine Fields *

Applied Food Sciences, Inc., 2500 Crosspark Road, Coralville, IA 52241, USA
* Correspondence: ytang@appliedfoods.com (Y.T.); cfields@appliedfoods.com (C.F.)

Academic Editors: Alessandra Gentili and Chiara Fanali

Abstract: An ultra-high-performance liquid chromatographic (UHPLC) separation was developed for six kava pyrones (methysticin, dihydromethysticin (DHM), kavain, dihydrokavain (DHK), desmethoxyyangonin (DMY), and yangonin), two unidentified components, and three Flavokavains (Flavokavain A, B, and C) in *Piper methysticum* (kava). The six major kavalactones and three flavokavains are completely separated ($R_s > 1.5$) within 15 min using a HSS T3 column and a mobile phase at 60 °C. All the peaks in the LC chromatogram of kava extract or standard solutions were structurally confirmed by LC-UV-MS/MS. The degradations of yangonin and flavokavains were observed among the method development. The degradation products were identified as cis-isomerization by MS/MS spectra. The isomerization was prevented or limited by sample preparation in a non-alcoholic solvent or with no water. The method uses the six kava pyrones and three flavokavains as external standards. The quantitative calibration curves are linear, covering a range of 0.5–75 μg/mL for the six kava pyrones and 0.05–7.5 μg/mL for the three flavokavains. The quantitation limits for methysticin, DHM, kavain, DHK, DMY, and yangonin are approximately 0.454, 0.480, 0.277, 0.686, 0.189, and 0.422 μg/mL. The limit of quantification (LOQs) of the three flavokavains are about 0.270, 0.062, and 0.303 μg/mL for flavokavain C (FKC), flavokavain A (FKA), and flavokavain B (FKB). The average recoveries at three different levels are 99.0–102.3% for kavalactones (KLs) and 98.1–102.9% for flavokavains (FKs). This study demonstrates that the method of analysis offers convenience and adequate sensitivity for determining methysticin, DHM, kavain, DHK, yangonin, DMY, FKA, FKB, and FKC in kava raw materials (root and CO_2 extract) and finished products (dry-filled capsule and tablet).

Keywords: *Piper methysticum* (kava); kavalactones; flavokavains; UHPLC-UV; mass spectra; isomerization; single-laboratory validation; quality control

1. Introduction

Kava (*Piper methysticum*) has been used for a traditional beverage in the Pacific islands, from ancient times, for its relaxant and anxiolytic effects [1,2].

The active constituents in kava root have been reported as a group of structurally related lipophilic lactone derivatives, kavalactones (KLs), with an arylethylene-α-pyrone skeleton. More than 18 kavalactones have been isolated from kava [3–5], including six major KLs (Figure 1A) as follows: Kavain, 5,5-dihydrokavain (DHK), methysticin, dihydromethysticin (DHM), yangonin, and desmethoxy-yangonin (DMY). Other types of compounds identified in kava include alkaloids, chalcones (flavokavains A, B, and C, Figure 1B), avanones (pinostrobin, 5,7-dimethoxy avanone), cinnamic acid derivatives (bornyl ester of 3,4-methylene dioxycinnamic acid, cinnamic acid bornyl ester), long-chain fatty acids and alcohols, and sterols [6–8].

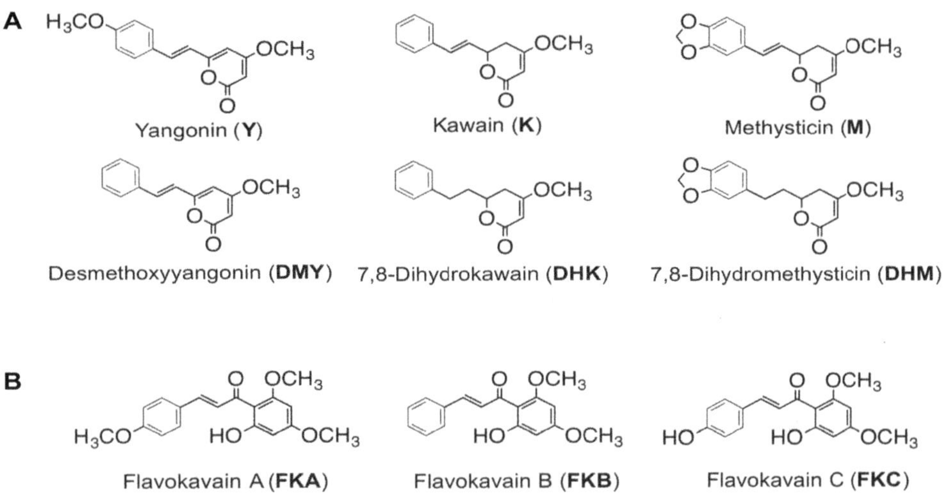

Figure 1. Chemical structures of (**A**) six major kavalactones and (**B**) three flavokavains.

Over a decade ago, potential safety issues with kava applications arose when several cases of liver damage were associated with kava consumption [9–11]. Given the concerns around potential liver toxicity with kava usage, the possible mechanisms for hepatotoxicity were investigated [12–14]. However, the cause of hepatotoxicity from kava was never clearly associated with one key factor. Kava continues to be consumed on a regular basis within local cultures, with no recent cases of hepatotoxicity reported. Recently, the World Health Organization revisited many of the reported kava-associated hepatotoxicity cases, evaluating the potential causes and providing guidance for safe use as well as recommendations for further studies to identify the potentially hepatotoxic compounds. The chalcone-based flavokavains A (FKA), B (FKB) and C (FKC) were studied with their potential risks and benefits. The studies suggested that the chalcones (FKA, FKB, and FKC), especially FKB, caused hepatotoxicity [15–17]. The possible toxicities of flavokavains (FKs) indicated that FKs should be limited to a very low level in the finished products being consumed.

As the main psychoactive components of kava, the contents and chemotypes of six major KLs, are the main criteria of kava beverage quality. Considering the possible hepatotoxicity of FKs, the reliable qualitative and/or quantitative determination of kava compounds is important for the best usage and purification techniques of kava. The RP-HPLC method was considered as an effective method with a good separation and a high accuracy, among many existing analytical methods for determining KLs [18–22]. Most recently, the HPLC method from Meissner and Häberlein [18] and the rapid HPLC method from Brown [22] were proposed for quantifying FKs, along with KLs. The two methods either had a long run time (50 min) or sacrificed the resolution among some key compounds. The UPLC-MS/MS method [23] was also reported for analysis of KLs, however, the linear range of UPLC-MS/MS was limited.

The main goal of this study was to develop and validate a more effective (U)HPLC method for quick and reliable quantification of the six major KLs and three FKs in kava raw roots and rhizomes, as well as finished products, based on the most prevalent instrumentation and standards available. During the method development of this study, it was observed that improper sample preparation of (standard) samples could lead the degradation of yangonin, as well the possible degradations of FKs, utilizing LC-UV-MS. The (U)HPLC methods were fully validated based on the AOAC Guideline [24] for Single-Laboratory Validation of Chemical Methods and ICH Q2 Guideline [25].

2. Results and Discussions

2.1. *Identification of Isomerization of Yangonin and Flavokavains (A, B, and C), along with Sample Preparation*

(U)HPLC-UV is still a more applicable method for quantitation of KLs and FKs in most of kava products. In this study, both standards and kava CO_2 extract samples were applied for the analysis and method development on LC-UV. The chromatographic separation was extended from the previous HPLC study [26] and achieved under a gradient separation at 60 °C. Optimum separation of KLs and FKs was achieved using an UHPLC column (Acquity HSS T3, 100 mm × 2.1 mm, 1.8 μm). Gradient elution was performed using water (100%, no addition of buffer or acids) as solvent A and isopropanol (100%, no addition of buffer or acids) as solvent B, with the gradient program listed in Table 1. The structures of KL or FK compounds were confirmed by the MS/MS spectra (Support Figure S1).

Table 1. Gradient UHPLC elution profile.

Time/min	A (H_2O, %)	B (IPA, %)	Flow Rate (mL/min)
0.00	95	5	0.50
0.35	90	10	0.50
0.39	78	22	
2.00	78	22	0.43
7.00	78	22	0.41
7.50	71	29	0.41
10.50	25	75	0.41
11.50	0	100	0.41
12.80	0	100	0.41
13.00	95	5	0.41
15.00	95	5	0.44
15.50	End		0.50

The two unknown compounds (U1 and U2) reported in a previous study [26] were also presented in the kava CO_2 extract (Figure 2). The $(M + H)^+$ was 261.1 Da for U1 and 263.1 Da for U2, from MS/MS spectra. The MS spectra suggest that the two compounds are very likely to be 5,6-dihydroyangonin (DHY) for U1 and 5,6,7,8-tetrahydroyangonin (THY) for U2, as the literature reported [2]. The structures of THY and DHY are associated with the mass fragmentations of the $(M + H)^+$ as 145.1 Da for U1 and 147.1 and 121.1 Da for U2. The two compounds are found as minor KLs in all-natural kava products. The quantitation of the major KLs could be interfered by the two minor components under a low-resolution LC.

The individual stock standard solutions were initially prepared in methanol. Then, the mixed standard solution was diluted with 50% H_2O/MeOH for LC analysis. Four unknown compounds were observed for the mixed standard solution over several weeks after the preparation. The unknown compounds were with the same molecular weight as both yangonin and flavokavains, but eluted significantly earlier (Figure 3). The fragmentation of the $(M + H)^+$ (parent ion) gives an identical product ion spectrum as that of yangonin and flavokavains. The isomerization from yangonin to cis-yangonin in kava standards or extract solutions was reported, especially if aqueous or alcoholic solutions were used [27,28]. MS/MS spectra suggested one of the unknown compounds as cis-yangonin, an isomer of yangonin. MS/MS spectra also indicated the other three compounds related to the isomerization of flavokavains A, B, and C. Those compounds are very likely to be the cis-isomer of flavokavains A, B, and C. For practical reasons, it was assumed that cis products could be formed with an alcoholic solvent or water [27,28]. Many LC methods were attempted for a good resolution ($R_s > 1.5$) for all KLs, especially for the separation of M and DHM. It is very difficult to achieve a good chromatograph separation ($R_s > 1.5$) for all KLs and FKs with the additions of cis-yangonin and cis-FKs. It is suggested that during validation of the proposed method, the stock standard solutions of KLs and FKs are prepared in a non-alcoholic solvent, like acetonitrile. The kava working standard solution and sample solution should be freshly prepared for LC analysis, to prevent or minimize the isomerization of yangonin and FKs.

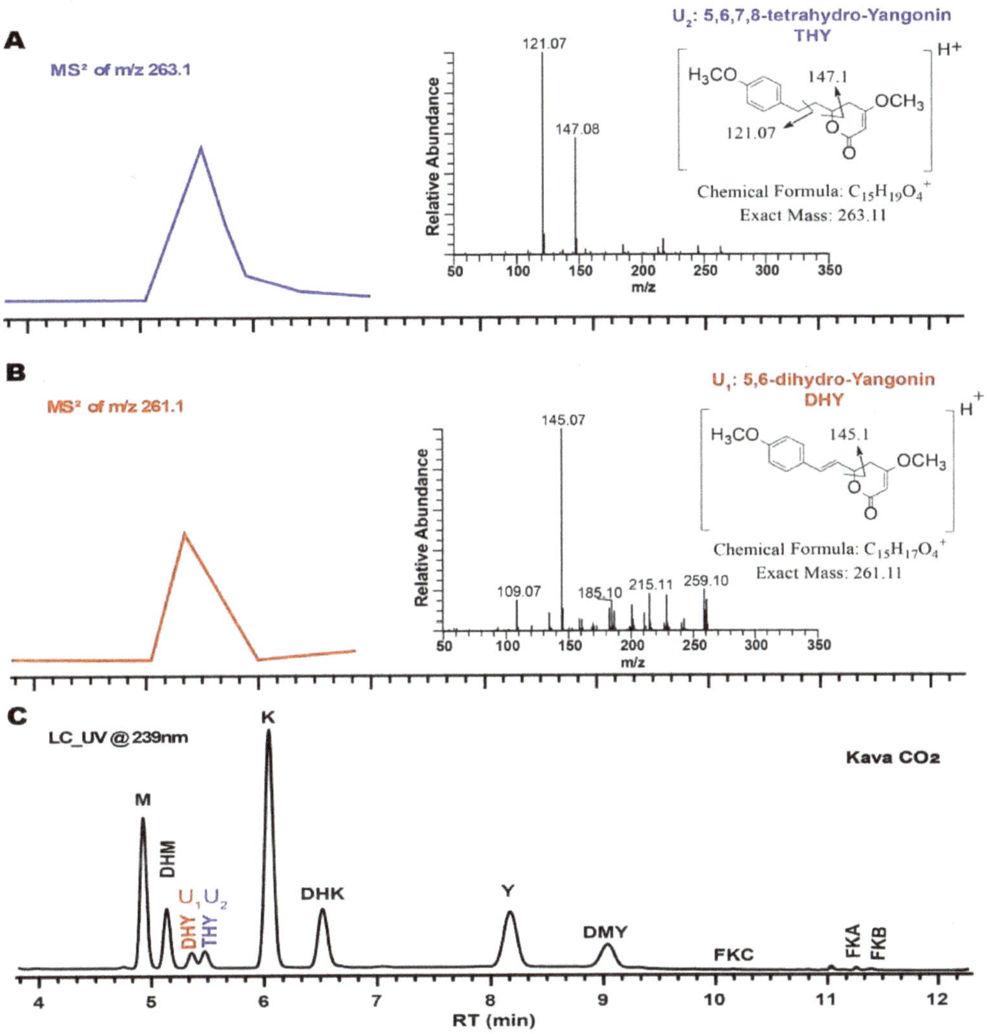

Figure 2. UHPLC-UV-MS data of kava CO_2 extract as follows: (**A**) MS/MS^2 trace of m/z 263.1 and MS^2 spectrum of peak U_2 (5,6,7,8-Tetrahydroyangonin, THY); (**B**) MS/MS^2 trace of m/z 261.1 and MS^2 spectrum of peak U_1 (5,6-Dihydroyangonin, DHY); (**C**) UV trace at the wavelength of 239 nm; M, methysticin; DHM, 7,8-Dihydromethysticin; K, kavain; DHK, 7,8-Dihydrokavain; Y, yangonin; DMY, desmethoxyyangonin; FKA (B or C) flavokavains A (B or C).

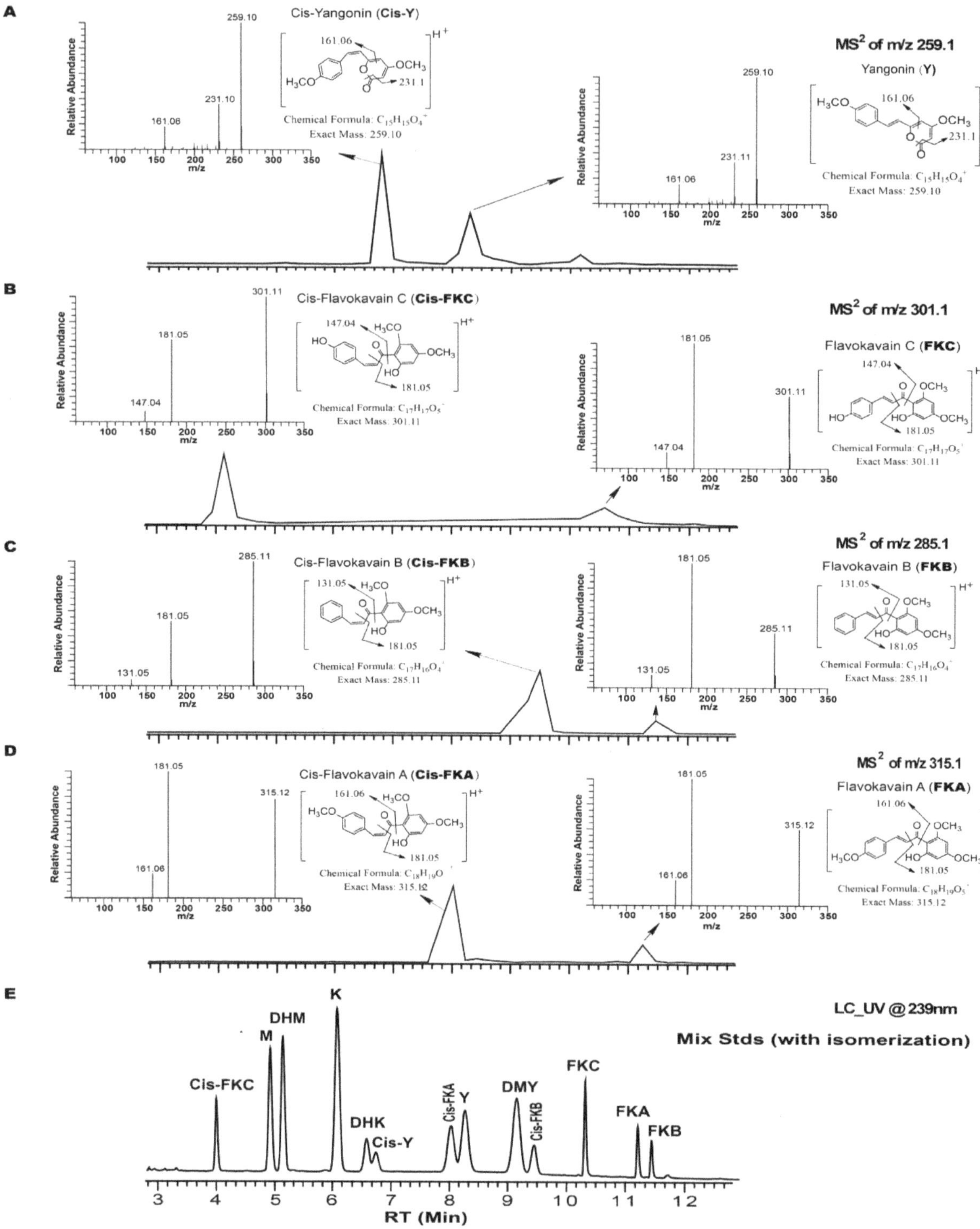

Figure 3. UHPLC-UV-MS data of degraded kava standard mixture, as follows: (**A**) MS/MS^2 trace of m/z 259.1, and MS^2 spectra of peaks of Y and cis-Y; (**B**) MS/MS^2 trace of m/z 301.1, and MS^2 spectra of peaks of FKC and cis-FKC; (**C**) MS/MS^2 trace of m/z 285.1, and MS^2 spectra of peaks of FKB and cis-FKB; (**D**) MS/MS^2 trace of m/z 315.1 and MS^2 spectra of peaks of FKA and cis-FKA; (**E**) UV trace at the wavelength of 239 nm; M, methysticin; DHM, 7,8-Dihydromethysticin; K, kavain; DHK, 7,8-Dihydrokavain; Y, Yangonin; Cis-Y, cis-yangonin; DMY, Desmethoxyyangonin; FKA (B or C), flavokavains A (B or C); cis-FKA (B, or C), flavokavains A (B or C).

2.2. Method Validation

2.2.1. Specificity/Resolution

Each individual reference standard was injected into the HPLC-UV to compare with the standard mixture (Support Figure S2). Identification of KLs in the test materials was determined by comparing peak retention times and UV spectra to the reference standards. Representative chromatograms of the standard mixture and kava products are displayed in Figure 4. Under the chromatographic conditions used in the present study, all six major KLs and three FKs were eluted separately following this following order: Methysticin, DHM, kavain, DHK, yangonin, DMY, FKC, FKA, and FKB. Values for the relative retention times (α), retention factors (k'), and chromatographic resolutions (R_s), calculated from the LC analysis of the kava standard mixtures (Figure 4), are summarized in Table 2. Retention factors (k') were calculated as $k' = (t_r - t_0)/t_0$, where t_r is the retention time of the analyte and t_0 is the retention time of unretained compounds (solvent front). The k' values were within the optimum range ($k' > 2$) for satisfactory chromatographic elution. An excellent chromatographic specificity was observed with the good resolution of the peaks ($R_s > 1.5$) and with no significant interfering peaks for all compounds in the mixed standard sample. The total chromatography run time was 15.5 min.

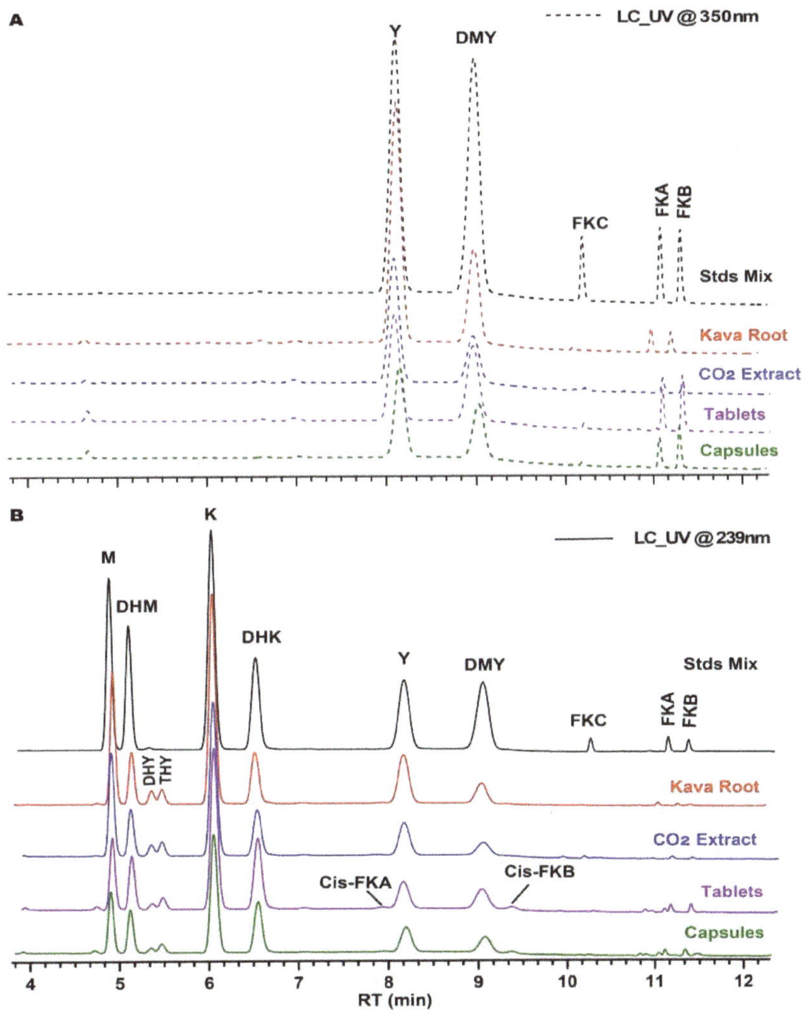

Figure 4. UHPLC chromatography of kava standard mixture and kava products. (**A**) UV trace at the wavelength of 350 nm; (**B**) UV trace at the wavelength of 239 nm; M, methysticin; DHM, 7,8-Dihydromethysticin; DHY, 5,6-Dihydroyangonin; THY, 5,6,7,8-Tetrahydroyangonin; K, kavain; DHK, 7,8-Dihydrokavain; DMY, Desmethoxyyangonin; Y, Yangonin; FKA (B or C), flavokavains A (B or C); cis-FKA (B), cis-flavokavains A (B).

Table 2. Parameters of kavalactones and flavokavains.

Compound	Retention Factor (k')	Relative RT (α)	Chromatographic Resolution (R_s)
Methysticin (1)	8.01	–	–
Dihydromethysticin (2)	8.41	$\alpha_{1/2} = 1.045$	$R_{1/2} = 1.78$
Kavain (3)	10.15	$\alpha_{2/3} = 1.184$	$R_{2/3} = 6.61$
Dihdyrokavain (4)	11.07	$\alpha_{3/4} = 1.082$	$R_{3/4} = 3.02$
Yangonin (5)	14.12	$\alpha_{4/5} = 1.253$	$R_{4/5} = 8.20$
Desmethoxyyangonin (6)	15.76	$\alpha_{5/6} = 1.109$	$R_{5/6} = 3.62$
Flavokavain C (7)	18.18	$\alpha_{6/7} = 1.144$	$R_{6/7} = 7.67$
Flavokavain A (8)	19.86	$\alpha_{7/8} = 1.088$	$R_{7/8} = 11.92$
Flavokavain B (9)	20.29	$\alpha_{8/9} = 1.020$	$R_{8/9} = 3.04$

2.2.2. Standard Linearity

To evaluate the linearity of the calibration curves, calibration standard mixed solutions, at the target concentrations, were prepared as described above. Each peak area of the chromatograms was recorded as the UV responses at 239 nm for methysticin, DHM, kavain, and DHK, and at 350nm for yangonin, DMY, and FKs. Calibration curves were plotted by the peak area vs. the concentration of the standard compounds (Support Figure S3). Regression analyses were processed by Lab Solution software. Calibration curves were linear over the concentration range used, with an $R^2 > 0.999$. The normalized intercept/slope of the regression line and the correlation coefficient were calculated for the whole data set. The method was evaluated by determining the coefficient of linearity and the intercept values, as summarized in Table 3.

Table 3. Calibration parameters for kavalactones and flavokavains from three different calibration curves.

	Calibration Range (μg/mL)	Slope (±SD) [a]	Y-Intercept (±SD) [a]	r^2 (±SD) [a]	LOD (μg/mL)	LOQ (μg/mL)
Methysitcin	0.50~75.0	106708.0 ± 700.0	−3443.1 ± 4846.8	0.99996 ± 0.00005	0.150	0.454
DHM		80132.1 ± 502.3	6666.6 ± 3847.5	0.99997 ± 0.00002	0.158	0.480
Kavain		176782.0 ± 1450.0	−5987.0 ± 4889.8	0.99996 ± 0.00004	0.091	0.277
DHK		80884.5 ± 381.5	11433.5 ± 5551.5	0.99995 ± 0.00002	0.226	0.686
Yangonin		167271.0 ± 1964.9	3729.7 ± 3167.9	0.99994 ± 0.00003	0.062	0.189
DMY		164787.0 ± 2352.4	−12618.3 ± 6952.1	0.99995 ± 0.00003	0.139	0.422
FKC	0.05~7.50	135720.0 ± 3576.4	−3194.0 ± 3663.5	0.99970 ± 0.00017	0.089	0.270
FKA		143817.0 ± 2873.9	768.8 ± 889	0.99980 ± 0.00018	0.020	0.062
FKB		150461.0 ± 3339.0	−1374.6 ± 4555.5	0.99988 ± 0.00017	0.100	0.303

Note: Calibration curves were performed three times on different days and established by measuring the concentration vs. the corresponding peak area. The given values are the mean of three replicates ± standard deviation [a].

2.2.3. Limit of Detection and Limit of Quantification

The limit of detection (LOD) and the limit of quantification (LOQ) of the kava standard assay were determined by the standard deviation of the y-intercepts over the slope of the regression lines from three replicated calibration curves on different days. The LODs and LOQs for six major KLs and three FKs were calculated and summarized in Table 3.

2.2.4. Recovery

A spike recovery study based on spiking six kavalactones and three flavokavains into a kava test material (kava root) at high, medium, and low levels, as well as non-spiked, were completed and shown in Table 4. The amounts of kavalactones and flavokavains were compared before and after spiking. As the AOAC guidelines suggest for single-laboratory validation, the recoveries were calculated in two ways, as follows: (1) Total recovery based on recovery of the native plus added analyte, and (2) marginal recovery based only on the added analyte (the native analyte is subtracted from both the numerator and denominator). The total recovery is used for the native analytes presented in amounts greater than about 10% of the amount added, otherwise, the marginal recovery is applied. The average recoveries for each analyte at each level were 99.0–102.3% for KLs and 98.1–102.9% for FKs, within the ranges of recovery limits from 95% to 102% at 10% concentration, 92% to 105% at 1% concentration, 90% to 108% at 0.1% concentration, and 85% to 110% for 0.01 at 0.01% concentration.

Table 4. Spike recovery results of HPLC-UV method for determination of kavalactones and flavokavains.

	Methysticin	DHM	Kavain	DHK	DMY	Yangonin	FKC	FKA	FKB
Spiked Level			3.0%					~0.3%	
Native (μg/mL) [a]	40.16 ± 0.62	22.14 ± 0.34	47.80 ± 0.74	29.75 ± 0.46	36.94 ± 0.57	14.52 ± 0.22	0.296 ± 0.005	1.27 ± 0.02	1.03 ± 0.02
Spiked (μg/mL) [b]	60.02 ± 0.80	58.73 ± 0.78	59.28 ± 0.79	59.23 ± 0.79	59.36 ± 0.79	57.73 ± 0.77	5.67 ± 0.08	7.99 ± 0.11	5.76 ± 0.08
After Spiked (μg/mL) [b]	100.2 ± 0.2	80.87 ± 0.44	107.1 ± 0.1	88.99 ± 0.33	96.30 ± 0.22	72.24 ± 0.54	5.97 ± 0.07	9.25 ± 0.09	6.79 ± 0.06
Detected (μg/mL) [a]	101.5 ± 0.3	81.43 ± 0.51	109.6 ± 0.6	89.63 ± 0.44	98.21 ± 0.41	73.42 ± 0.36	5.86 ± 0.16	9.48 ± 0.14	6.92 ± 0.11
Marginal recovery (%) [a]	102.1 ± 0.1	101.0 ± 0.1	104.2 ± 1.0	101.1 ± 0.4	103.2 ± 0.6	102.0 ± 0.4	**98.1 ± 1.6** [c]	**102.9 ± 0.6**	**102.2 ± 0.8**
Total Recovery (%) [a]	**101.3 ± 0.1**	**100.7 ± 0.1**	**102.3 ± 0.5**	**100.7 ± 0.3**	**102.0 ± 0.4**	**101.6 ± 0.3**	98.2 ± 1.5	102.5 ± 0.5	101.3 ± 0.1
Spiked Level			1.5%					~0.15%	
Native (μg/mL) [a]	41.04 ± 0.81	22.63 ± 0.45	48.86 ± 0.97	30.41 ± 0.60	37.76 ± 0.75	14.84 ± 0.29	0.302 ± 0.006	1.29 ± 0.03	1.05 ± 0.02
Spiked (μg/mL) [b]	30.21 ± 0.15	29.55 ± 0.15	29.83 ± 0.15	29.81 ± 0.15	29.87 ± 0.15	29.05 ± 0.14	2.86 ± 0.014	4.02 ± 0.02	2.90 ± 0.01
After Spiked (μg/mL) [b]	71.25 ± 0.66	52.18 ± 0.30	78.69 ± 0.82	60.22 ± 0.46	67.63 ± 0.60	43.89 ± 0.15	3.15 ± 0.01	5.31 ± 0.01	3.95 ± 0.01
Detected (μg/mL) [a]	70.50 ± 0.74	51.73 ± 0.42	78.02 ± 1.17	59.63 ± 0.51	67.21 ± 0.83	44.12 ± 0.36	3.10 ± 0.02	5.36 ± 0.02	3.97 ± 0.02
Marginal recovery (%) [a]	97.4 ± 2.5	99.0 ± 1.6	93.8 ± 2.5	98.6 ± 1.1	97.8 ± 2.9	100.8 ± 0.9	98.4 ± 0.5	101.0 ± 0.5	100.3 ± 0.9
Total Recovery (%) [a]	**99.0 ± 1.1**	**99.1 ± 0.9**	**99.2 ± 1.0**	**99.0 ± 0.5**	**99.4 ± 1.3**	**100.5 ± 0.6**	**98.1 ± 0.5**	**100.8 ± 0.4**	**100.3 ± 0.7**
Spiked Level			0.25%					~0.025%	
Native (μg/mL) [a]	40.83 ± 0.91	22.51 ± 0.50	48.60 ± 1.09	30.25 ± 0.68	37.56 ± 0.84	14.76 ± 0.33	0.300 ± 0.007	1.29 ± 0.03	1.05 ± 0.02
Spiked (μg/mL) [b]	5.01 ± 0.07	4.90 ± 0.07	4.95 ± 0.07	4.94 ± 0.07	4.95 ± 0.07	4.82 ± 0.07	0.473 ± 0.007	0.67 ± 0.01	0.48 ± 0.01
After Spiked (μg/mL) [b]	45.84 ± 0.85	27.41 ± 0.44	53.55 ± 1.02	35.20 ± 0.62	42.51 ± 0.78	19.58 ± 0.27	0.774 ± 0.004	1.95 ± 0.02	1.53 ± 0.02
Detected (μg/mL) [a]	45.51 ± 1.06	27.45 ± 0.51	53.47 ± 1.30	35.25 ± 0.78	42.21 ± 1.06	19.66 ± 0.32	0.779 ± 0.002	1.96 ± 0.03	1.53 ± 0.03
Marginal recovery (%) [a]	93.6 ± 4.8	100.7 ± 1.4	98.5 ± 5.7	101.1 ± 3.6	93.9 ± 5.6	101.7 ± 1.1	101.0 ± 0.7	101.7 ± 1.5	99.9 ± 3.2
Total Recovery (%) [a]	**99.3 ± 0.5**	**100.1 ± 0.2**	**99.9 ± 0.5**	**100.1 ± 0.5**	**99.3 ± 0.7**	**100.4 ± 0.3**	**100.6 ± 0.4**	**100.6 ± 0.5**	**100.0 ± 1.0**

Note: The given values are the mean of three replicated measurements ± standard deviation [a]. The values for all the spiked and after-spiked concentrations were calculated as the mean of the three replicated plus the standard deviation [b]. The bold recoveries were applied for the final evaluation [c].

2.2.5. Precision

The precision and accuracy of the method was assessed by determining the intraday precisions (n = 5) from repeating the analysis of the kava samples on the same day; and the interday precisions (n = 3 × 5, overall 15) from analyzing the same kava samples over the different days. Both intraday and interday precisions were calculated as RSD_r (%) = (standard deviation)/(mean) × 100. As shown in Table 5, all nine analytes had adequate precision in each of the four different solid matrices at different concentrations. The level of KLs were about 10–100 mg/g for CO_2 extract and root. The level of FKs in the kava CO_2 extract and the root were about 0.13–0.30 mg/g for FKC, 0.60–1.5 mg/g for FKA, and 0.50–0.90 mg/g for FKB. The intraday repeatability relative standard deviations (RSD_r) of the CO_2 extract and the root were 0.31% to 1.61% for KLs and 0.98% to 3.83% for FKs, which are <2% for KLs and <4% for FKs, as AOAC guidelines suggested. About 2–10 mg/g for KLs, 0.08–0.10 mg/g for FKC, 0.46–0.75 mg/g for FKA, and 0.60–1.00 mg/g for FKB were detected for kava products in tablets and capsules. The intraday RSD_r of kava products in tablets and capsules were 0.28% to 1.96% for KLs and 0.23% to 4.24% for FKs, which are also <3% for KLs and <6% for FKs, as AOAC guidelines suggested. Overall, the interday repeatability relative standard deviations (RSD_r) ranged from 0.50% to 2.56% for kavalactones and 2.44% to 5.52% for flavokavains. The Horwitz ratio (HorRat) values are used to evaluate method performance based on the ratios of actual precision to predicted precision. AOAC guidelines for single-laboratory validation accept a HorRat range from 0.5 to 2. In our method, the HorRat value for kavalactones ranged from 0.24 to 1.05 and for flavokavains ranged from 0.77 to 1.89. A HorRat value lower than 0.5 was considered acceptable, considering the analysis was performed under tightly controlled conditions.

Table 5. Precision summary of the HPLC-UV method for detecting kavalactones and flavokavains in kava products.

		Methysticin	DHM	Kavain	DHK	Yangonin	DMY	FKC	FKA	FKB
Tablets										
LOQ (mg/g)		0.076	0.080	0.046	0.114	0.032	0.070	0.045	0.010	0.051
D1	Mean ± SD (mg/g)	3.79 ± 0.03	3.94 ± 0.03	6.71 ± 0.07	7.17 ± 0.06	3.55 ± 0.03	2.76 ± 0.03	0.082 ± 0.001	0.477 ± 0.007	0.627 ± 0.008
	RSD_r (%) Intraday	0.86	0.80	1.06	0.82	0.98	1.08	1.28	1.39	1.32
D2	Mean ± SD (mg/g)	3.83 ± 0.01	3.96 ± 0.01	6.67 ± 0.03	7.05 ± 0.04	3.58 ± 0.04	2.78 ± 0.04	0.076 ± 0.001	0.452 ± 0.002	0.587 ± 0.002
	RSD_r (%) Intraday	0.34	0.34	0.45	0.50	1.22	1.30	1.64	0.49	0.28
D3	Mean ± SD (mg/g)	3.83 ± 0.02	3.96 ± 0.01	6.68 ± 0.02	7.06 ± 0.04	3.64 ± 0.03	2.75 ± 0.02	0.078 ± 0.001	0.459 ± 0.001	0.592 ± 0.002
	RSD_r (%) Intraday	0.39	0.28	0.32	0.60	0.81	0.70	0.77	0.23	0.49
Mean ± SD (mg/g)		3.81 ± 0.03	3.95 ± 0.02	6.69 ± 0.05	7.09 ± 0.07	3.59 ± 0.05	2.77 ± 0.03	0.079 ± 0.003	0.463 ± 0.011	0.602 ± 0.019
RSD (%) Interday		0.73	0.54	0.68	1.04	1.40	1.05	3.75	2.44	3.20
HorRat		0.32	0.24	0.32	0.50	0.60	0.43	0.91	0.77	1.05
Capsules										
LOQ (mg/g)		0.151	0.160	0.092	0.229	0.063	0.141	0.090	0.021	0.101
D1	Mean± SD (mg/g)	5.87 ± 0.07	5.80 ± 0.05	9.09 ± 0.09	9.48 ± 0.11	5.85 ± 0.09	3.74 ± 0.05	0.103 ± 0.002	0.702 ± 0.008	0.920 ± 0.010
	RSD_r (%) Intraday	1.14	0.82	0.97	1.20	1.50	1.33	2.18	1.16	1.08
D2	Mean± SD (mg/g)	5.98 ± 0.07	5.93 ± 0.08	9.33 ± 0.12	9.73 ± 0.14	6.06 ± 0.10	3.83 ± 0.07	0.094 ± 0.004	0.773 ± 0.011	0.999 ± 0.019
	RSD_r (%) Intraday	1.10	1.31	1.29	1.45	1.66	1.74	4.24	1.39	1.93
D3	Mean± SD (mg/g)	5.99 ± 0.12	5.98 ± 0.11	9.37 ± 0.12	9.85 ± 0.13	6.07 ± 0.10	3.87 ± 0.06	0.101 ± 0.002	0.796 ± 0.011	1.012 ± 0.012
	RSD_r (%) Intraday	1.96	1.82	1.32	1.32	1.59	1.58	2.04	1.43	1.21
Mean± SD (mg/g)		5.95 ± 0.10	5.90 ± 0.11	9.26 ± 0.17	9.69 ± 0.20	6.00 ± 0.14	3.81 ± 0.08	0.099 ± 0.004	0.757 ± 0.042	0.977 ± 0.044
RSD_r (%) Interday		1.67	1.84	1.80	2.06	2.26	2.06	4.51	5.56	4.49
HorRat		0.77	0.85	0.89	1.03	1.05	0.89	1.13	1.89	1.59
CO_2 Extract										
LOQ (mg/g)		0.454	0.480	0.277	0.686	0.189	0.422	0.270	0.062	0.303
D1	Mean± SD (mg/g)	42.39 ± 0.41	26.75 ± 0.29	50.5 ± 0.7	36.2 ± 0.5	32.5 ± 0.3	14.0 ± 0.1	0.29 ± 0.01	1.34 ± 0.02	0.84 ± 0.02
	RSD_r (%) Intraday	0.97	1.08	1.28	1.39	1.06	0.99	2.13	1.84	2.11
D2	Mean± SD (mg/g)	42.81 ± 0.21	27.0 ± 0.2	50.6 ± 0.3	36.2 ± 0.4	33.1 ± 0.2	14.2 ± 0.1	0.29 ± 0.01	1.43 ± 0.02	0.85 ± 0.01
	RSD_r (%) Intraday	0.50	0.54	0.55	0.97	0.56	0.38	2.09	1.34	1.22

Table 5. *Cont.*

		Methysticin	DHM	Kavain	DHK	Yangonin	DMY	FKC	FKA	FKB
D3	Mean± SD (mg/g)	42.75 ± 0.53	27.1 ± 0.3	50.8 ± 0.5	36.9 ± 0.4	33.3 ± 0.2	14.1 ± 0.1	0.270 ± 0.004	1.47 ± 0.02	0.89 ± 0.03
	RSD_r (%) Intraday	1.25	1.08	0.88	1.18	0.74	0.68	1.59	1.15	3.83
Mean± SD (mg/g)		42.65 ± 0.42	27.0 ± 0.3	50.6 ± 0.5	36.4 ± 0.5	33.0 ± 0.4	14.1 ± 0.1	0.28 ± 0.01	1.41 ± 0.06	0.86 ± 0.03
RSD_r (%) Interday		0.99	1.04	0.92	1.49	1.29	0.85	3.92	4.27	3.37
HorRat		0.62	0.60	0.59	0.91	0.78	0.45	1.15	1.60	1.17
						Root				
LOQ (mg/g)		0.151	0.160	0.092	0.229	0.063	0.141	0.090	0.021	0.101
D1	Mean± SD (mg/g)	20.2 ± 0.1	11.05 ± 0.05	25.9 ± 0.4	14.7 ± 0.1	18.8 ± 0.1	7.31 ± 0.05	0.128 ± 0.001	0.64 ± 0.01	0.52 ± 0.01
	RSD_r (%) Intraday	0.58	0.49	1.61	0.95	0.63	0.75	1.11	0.98	1.10
D2	Mean± SD (mg/g)	20.3 ± 0.1	11.11 ± 0.04	25.3 ± 0.2	14.8 ± 0.2	18.6 ± 0.2	7.33 ± 0.07	0.139 ± 0.003	0.70 ± 0.01	0.57 ± 0.01
	RSD_r (%) Intraday	0.31	0.38	0.87	1.14	1.01	0.97	2.14	1.51	1.39
D3	Mean± SD (mg/g)	20.00 ± 0.12	10.89 ± 0.06	24.68 ± 0.18	14.65 ± 0.08	18.55 ± 0.12	7.23 ± 0.06	0.137 ± 0.001	0.63 ± 0.01	0.514 ± 0.003
	RSD_r (%) Intraday	0.60	0.54	0.73	0.57	0.63	0.88	0.95	0.81	0.66
Mean± SD (mg/g)		20.19 ± 0.17	11.02 ± 0.11	25.30 ± 0.59	14.74 ± 0.15	18.62 ± 0.16	7.29 ± 0.07	0.135 ± 0.006	0.66 ± 0.03	0.53 ± 0.03
RSD_r (%) Interday		0.86	0.96	2.32	1.00	0.88	1.02	4.13	4.64	5.20
HorRat		0.48	0.49	1.33	0.53	0.48	0.49	1.09	1.55	1.69

3. Discussion

As a more applicable method, the RP-HPLC method was reported for quantitation analysis of kava products [18–22,26]. Over many studies, Alexander H. Schmidt and Imre Molnar [26] applied computer-assisted optimization in the development to achieve a great resolution for the separations of all the major KLs. The study also included the two minor KLs (DHY and THY), but not FKs. As the above description, the kava root (~ 10% KLs), the kava CO_2 extract product (~ 20% KLs), and commercial tablet and capsule kava products were analyzed for the quantitation of each kava compound, based on the standard responses of M, DHM, K, and DHK (at 239 nm) and Y, DMY, and FKs (at 350 nm). The calculated concentrations are shown in Table 5. In this study, the percentage of each KL and FK over the total KLs and FKs, calculated as follows:

$$\frac{x}{M + DHM + K + DHK + Y + DMY + FKs} \times 100 \quad (1)$$

The kavalactone and flavokavain profile were normalized to the percentage of total kavalactone content and presented in Figure 5. Although there is a large difference in the total content of kava lactones, the relative intensities of the individual compounds differ only slightly. AFS kava root and the CO_2 extract contain a very similar profile of KLs at different levels, about 100 mg KLs per gram in the root and at about 200 mg KLs per gram in the CO_2 extract. The capsule and tablet products contain about 40 and 30 mg KLs per gram.

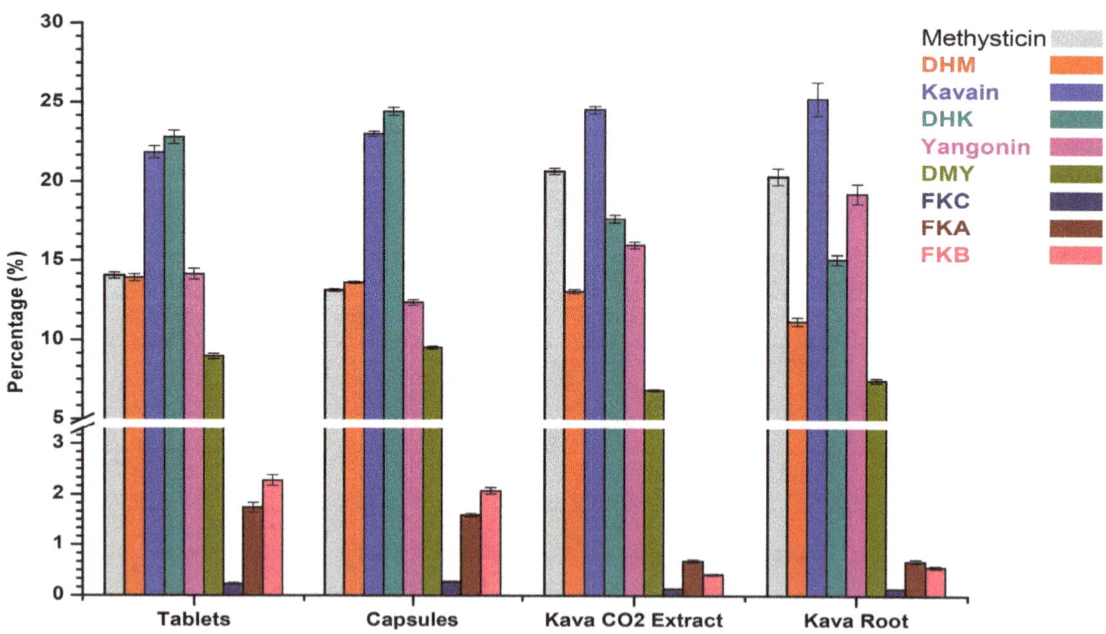

Figure 5. The profiles of kavalactones and flavokavains for kava products.

The quality of kava products was considered as a safety issue for consumers. The key criteria for the quality of kava product are the contents and chemotypes of six major KLs. Different quantity and ratio of KLs can impact their physiological action and safety [29]. The chemotypes of kava products were identified as noble or non-noble varieties following the simple system described by Lebot and Lévesque [30,31]. The six major KLs are used to define the chemotype (1 = DMY; 2 = DHK; 3 = yangonin; 4 = kavain; 5 = DHM; and 6 = methysticin). The different chemotypes of kava products are coded by listing, in decreasing order of proportion, the KLs. The noble cultivars have chemotypes rich in kavain, like 423561 or 423651. The chemotypes of 521634, 526341, or 254631 represent non-noble cultivars with very high proportions of DHM or DHK.

In this study, the chemotypes of kava products were determined by the percentage of each KL. The results show that the chemotypes of kava root (462351) were very similar to the CO_2 extract (463251), while the chemotypes of the capsule (245631) were very similar to the tablet (246531).

Due to the possible hepatotoxicities of FKs, another key criterion for the quality of kava products is to limit the amount of FKs, as FKs/KLs $\leq$ 0.29 and FKB% < 0.15% [16,17]. The ratios of FKs/KLs for kava products in this study were detected as 0.013 for the CO_2 extract, 0.014 for the root, 0.041 for the capsule, and 0.044 for the tablet. FKB was 0.086%, 0.055%, 0.06% and 0.096% for the kava CO_2 extract, the root, the tablets, and the capsules, respectively. In this study, it was noticed that cis-isomers of FKA and FKB were present at different levels in the powdered samples (Figure 4). The level of cis-isomers of FKA and FKB were found to be very low in both the kava CO_2 extract and the kava root, while a small amount of cis-isomers of FKA and FKB were observed in the capsule and tablet kava products.

4. Materials and Methods

4.1. Chemicals and Materials

The 2-propanol (HPLC grade), acetonitrile (ACN, HPLC grade), water (H2O, LC-MS grade), and methanol (MeOH, LC-MS grade) were purchased from Fisher Scientific (Hampton, NH, USA). The reference standard compounds of D, L-kavain (purity: 95% HPLC), yangonin (purity: 95% HPLC), flavokavain A (purity: 95% HPLC), flavokavain B (purity: 95% HPLC), and flavokavain C (purity: 95% HPLC) were purchased from Extrasynthese (Genay, France). The standard compounds of methysticin (purity: 99.63% HPLC), dihydromethysticin (purity 98.68% HPLC), desmethoxyyangonin (purity: 97.84% HPLC), and dihydrokavain (purity: 99.04% HPLC) products were from PhytoLab (Vestenbergsgreuth, Germany) and purchased from Cerilliant (Round Rock, TX, USA). Ultrapure (18 MΩ) water was produced using a Barnstead™ GenPure™ Pro Water Purification System from Thermo Fisher Scientific (Waltham, MA, USA).

4.2. Instrumentation

Method development and validation studies were performed on a Shimadzu Nexera-X2 UHPLC system (Shimadzu Scientific Instruments, Columbia, MD, USA), equipped with a LC-30AD pump, a SIL-30AC autosampler with a thermostated unit, a thermostated column compartment, and an SPD-M30A PDA detector. The UHPLC system was also interfaced with tandem Q-Exactive Orbitrap mass spectrometer (Thermo Fisher Scientific Inc., San Jose, CA, USA). High-resolution MS and MS^2 spectra were obtained on the Q-Exactive Orbitrap mass spectrometer equipped with a heated electrospray ionization, operated in both positive and negative ion mode. The optimized parameters were set as follows: Capillary voltage, 3.0 kV; sheath gas flow rate, 35 arbitrary unit; auxiliary gas flow rate, 5 arbitrary unit; sweep gas flow rate, 5 arbitrary unit; capillary temperature, 325 °C; and sheath gas heater temperature, 200 °C. MS scans were recorded in a mass range of m/z 100–1500 at a resolution of 70,000 with an AGC target of 3×10^6. After each MS scan, up to 5 of the most abundant multiply charged ions were selected for fragmentation. MS^2 scans were recorded in a mass range of m/z 50 to the parent ion at a resolution of 17,500, with an AGC target of 1×10^5 and a maximum fill time of 50 ms, using the stepped NCE of 25 and 35 for fragmentation in the HCD cell. Data were acquired from 50 to 1500 Da with dd-MS^2 or MS^2 in centroid mode. Raw data were acquired and processed using the Xcalibur software (Version 2.3.1, Thermo Electron Corporation, San Jose, CA, United States).

4.3. Chromatographic Condition

The chromatographic separation was extended from the previous HPLC study [24] and achieved under a gradient separation at 60 °C. Optimum separation of KLs and FKs was achieved using an UHPLC column (Acquity HSS T3, 100 mm × 2.1 mm, 1.8 μm). Gradient elution was performed using water (solvent A) and isopropanol (solvent B) with the gradient program listed in Table 1.

4.4. Standard Preparation

Individual standard stock solutions of 1000 µg/mL of each kava standard compound were accurately prepared by weighing about 10 mg of each compound and dissolving them into a 10 mL volumetric flask using acetonitrile. Volumetric flasks were sonicated for 10 min and wrapped with aluminum foil to protect them from light. Stock solutions were kept refrigerated. Working standard solutions were prepared fresh on a daily basis by pipetting aliquots of stock solutions and serial dilutions with 50% ACN were made at concentrations ranging from 0.5 µg/mL to 75 µg/mL for KLs and from 0.05 µg/mL to 7.5 µg/mL for FKs.

4.5. Test Materials and Sample Preparation

The kava CO_2 extract and root were obtained from Applied Food Sciences Inc. (AFS, Austin, TX, USA). Kava commercial products (dry-filled capsules and formulated tablets) were purchased from Foods market. All the samples were analyzed more than triplicate, unless stated otherwise.

4.5.1. Kava CO_2 Extract and Root Powders

A total of 100 mg of kava CO_2 extract or 300 mg of the root powder were first extracted with 15 mL ACN and sonicated for 30 min at 40 °C in a Fisher sonication bath. Following a 10-min centrifugation at 12,000× *g*, the supernatant was transferred to a 50 mL volumetric flask. The remaining residue was re-extracted twice with 15 mL ACN, following the same procedure. The volumetric flask was filled to the mark with ACN in the end. The samples were freshly diluted 1× with 18MW water and the extracts were filtered through a 3 mm syringe fitted with a 0.22 µm nylon filter (VWR) into an amber glass HPLC vial and readied for LC analysis.

4.5.2. Capsules and Phytocaps

The content of the 20 capsules or phytocaps were combined and mixed thoroughly. Two hundred milligrams of the capsules or phytocaps content were extracted with ACN and acetone following the same procedure as above.

4.6. Method Validation Parameters

This method was validated following the AOAC and ICH (Q2) guidelines [24,25] for conducing single-laboratory validation. For all the standards, 1000 µg/mL stock solutions were prepared by dissolving individual reference materials in ACN in volumetric flasks. The stock solutions of the reference materials were stored at −20 °C for long-term storage. The stock solution for each standard was mixed to the kava working standard solution at the concentration 100 µg/mL for KLs and 10 µg/mL for FKs, then diluted to the appropriate concentration to establish the retention time and combined at different concentration levels for external calibration.

4.6.1. Specificity/Resolution

The mixed reference standard was injected into the HPLC-UV to establish the selectivity of the method. The resolution for each reference standard was calculated. The value of $R_s > 1.5$ between closely eluting components was considered acceptable for FKs and major KLs.

4.6.2. Linearity

The linearity for the reference standard was determined by seven-point standard calibration curves. The standard curve for the six KLs ranged from 0.5 µg/mL to 75 µg/mL (0.5, 1.0, 5.0, 10.0, 25.0, 50.0, and 75.0 µg/mL). The standard curve for the three FKs ranged from 0.05 µg/mL to 7.5 µg/mL (0.05, 0.1, 0.5, 1.0, 2.5, 5.0, and 7.5 µg/mL). A simple linear regression was used to calculate R^2 value, the slope, and the y-intercept of each curve for each analyte. An $R^2 \geq 99.9\%$ value was considered acceptable. The calibration standards of seven KLs and three FKs were triplicated at the seven concentrations and analyzed over three days.

4.6.3. LOD and LOQ

The limit of detection (LOD) and limit of quantification (LOQ) of the kava standard assay were determined from the calibration curve method, as ICH Q2 (R1) recommendations [25], analyzing at least three replicates of the calibration standards. The LOD and LOQ of the proposed method were calculated using the following equations:

$$LOD = \frac{3.3 \times \text{Stdev y–intercept of Calibration Curve}}{\text{Slope of calibration curve } (A_{ave})} \quad (2)$$

$$LOQ = \frac{10 \times \text{Stdev y–intercept of Calibration Curve}}{\text{Slope of calibration curve } (A_{ave})} \quad (3)$$

4.6.4. Recovery

Spike recovery experiments were performed at three levels (high, 30 μg/mg; medium, 15 μg/mg; and low, 2.5 μg/mg) for KLs and three levels (high, 3 μg/mg; medium, 1.5 μg/mg; and low, 0.25 μg/mg) for FKs. Powdered kava root material was analyzed for KLs and FKs prior to the standards being spiked. The appropriate amount of reference standards was used to spike the powdered kava root material, followed by the extraction process. Considering the cost of the reference standards, for high-level spike recovery experiments a 5 mg sample was extracted with 5 mL extraction solvent (ACN). For medium-and low-level spike recovery experiments, a 10 mg sample was extracted with 10 mL extraction solvent (ACN). Three replicates were performed at each level and the mean recovery was calculated.

4.6.5. Precision

Four independent replicates of the same sample were prepared and analyzed on three separate days ($n = 5 \times 3$). The within-day, between-day, overall precision for all nine target compounds were calculated for single-laboratory validation.

5. Conclusions

In conclusion, the UHPLC-UV method described herein for the determination of six major KLs and three FKs in kava raw materials and finished products was validated based on AOAC Guidelines for Single-Laboratory Validation of Chemical Methods for Dietary Supplements and Botanicals. This method maximized efficiency and chromatographic resolution under short analysis times. Both DHY and THY were found in all the kava products. The two minor KLs were well-separated from the major KLs under the excellent LC resolution. The isomerizations of yangonin and FKs were prevented or limited by the usage of non-alcoholic solvents, like acetonitrile, for sample preparation in this method. This suggested method of kava analysis is free of interference from minor KLs, like THY & DHY, and at a very low level of interferences from cis-isomers of yangonin and FKs. The results of the study demonstrate that this UHPLC-UV analytical method is a successful approach to determine methysticin, DHM, kavain, DHK, yangonin, DMY, FKA, FKB, and FKC in kava raw materials (kava root powder and kava CO_2 extract) and finished products (dry-filled capsules or formulated tablets) under a quick analysis time of 15 min and it therefore expands the scope to analyze a broad variety of market samples.

Author Contributions: Y.T.—Study concept and design, acquisition of data, analysis and interpretation of data, and drafting of the manuscript. C.F.—Study concept and design, analysis and interpretation of data, and critical review of the manuscript.

References

1. Singh, Y.N. Kava-an Overview. *J. Ethnopharmacol.* **1992**, *37*, 13–45. [CrossRef]
2. Pluskal, T.; Torrens-Spence, M.P.; Fallon, T.R.; de Abreu, A.; Shi, C.H.; Weng, J. The biosynthetic origin of psychoactive kavalactones in kava. *Febs. Open Bio.* **2018**, *8*, 87.
3. Zhang, N.; Wu, L.; Liu, X.; Shen, X. Plant-Derived Kavalactones and Their Bioactivities. *Med. Res.* **2018**, 2. [CrossRef]
4. Naumov, P.; Dragull, K.; Yoshioka, M.; Tang, C.S.; Ng, S.W. Structural characterization of genuine (−)-pipermethystine, (−)-epoxypipermethystine, (+)-dihydromethysticin and yangonin from the kava plant (Piper methysticum). *Nat. Product Commun.* **2008**, *3*, 1333–1336. [CrossRef]
5. Dragull, K.; Yoshida, W.Y.; Tang, C.S. Piperidine alkaloids from Piper methysticum. *Phytochemistry* **2003**, *63*, 193–198. [CrossRef]
6. Wu, D.; Nair, M.G.; DeWitt, D.L. Novel compounds from Piper methysticum Forst (*Kava Kava*) roots and their effect on cyclooxygenase enzyme. *J. Agric. Food Chem.* **2002**, *50*, 701–705. [CrossRef] [PubMed]
7. Dharmaratne, H.R.W.; Nanayakkara, N.P.D.; Khan, I.A. Kavalactones from Piper methysticum, and their C-13 NMR spectroscopic analyses. *Phytochemistry* **2002**, *59*, 429–433. [CrossRef]
8. World Health Organization. *Assessment of the Risk of Hepatotoxicity with Kava Products*; World Health Organization: Geneva, Switzerland, 2007.
9. Furbee, R.B.; Barlotta, K.S.; Allen, M.K.; Holstege, C.P. Hepatotoxicity associated with herbal products. *Clin. Lab. Med.* **2006**, *26*, 227. [CrossRef]
10. Ulbricht, C.; Basch, E.; Boon, H.; Ernst, E.; Hammerness, P.; Sollars, D.; Tsourounis, C.; Woods, J.; Bent, S. Safety review of kava (*Piper methysticum*) by the Natural Standard Research Collaboration. *Expert. Opin. Drug Saf.* **2005**, *4*, 779–794. [CrossRef]
11. Schmidt, M. Are kavalactones the hepatotoxic principle of kava extracts? The pitfalls of the glutathione theory. *J. Altern. Complement. Med.* **2003**, *9*, 183–187. [CrossRef]
12. Wu, D.; Yu, L.; Nair, M.G.; DeWitt, D.L.; Ramsewak, R.S. Cyclooxygenase enzyme inhibitory compounds with antioxidant activities from Piper methysticum (*kava kava*) roots. *Phytomedicine* **2002**, *9*, 41–47. [CrossRef] [PubMed]
13. Mathews, J.M.; Etheridge, A.S.; Black, S.R. Inhibition of human cytochrome P450 activities by kava extract and kavalactones. *Drug Metab. Dispos.* **2002**, *30*, 1153–1157. [CrossRef] [PubMed]
14. Zhou, P.; Gross, S.; Liu, J.H.; Yu, B.Y.; Feng, L.L.; Nolta, J.; Sharma, V.; Piwnica-Worms, D.; Qiu, S.X. Flavokawain B, the hepatotoxic constituent from kava root, induces GSH-sensitive oxidative stress through modulation of IKK/NF-kappaB and MAPK signaling pathways. *FASEB J.* **2010**, *24*, 4722–4732. [CrossRef] [PubMed]
15. Jhoo, J.W.; Freeman, J.P.; Heinze, T.M.; Moody, J.D.; Schnackenberg, L.K.; Beger, R.D.; Dragull, K.; Tang, C.S.; Ang, C.Y.W. In vitro cytotoxicity of nonpolar constituents from different parts of kava plant (*Piper methysticum*). *J. Agr. Food Chem.* **2006**, *54*, 3157–3162. [CrossRef]
16. Fratini Vergano Law Firm. Scientific and Legal assistance for the Development of a Quality and Safety Standard for Kava Production and Trade in the Pacific Region. Available online: http://www.acp-eu-tbt.org/pagenewsarc.cfm?id=279CC3E89126DCC6C78B90A8D1ED66FFD4E480FBF584FA6D1FCEE8E906FEFEC8DD (accessed on 30 November 2015).
17. Lebot, V.; Do, T.K.T.; Legendre, L. Detection of flavokavins (A, B, C) in cultivars of kava (*Piper methysticum*) using high performance thin layer chromatography (HPTLC). *Food Chem.* **2014**, *151*, 554–560. [CrossRef]
18. Meissner, O.; Haberlein, H. HPLC analysis of flavokavins and kavapyrones from Piper methysticum Forst. *J. Chromatogr. B Analyt. Technol. Biomed. Life Sci.* **2005**, *826*, 46–49. [CrossRef] [PubMed]
19. Murauer, A.; Ganzera, M. Quantitative Determination of Lactones in Piper methysticum (*Kava-Kava*) by Supercritical Fluid Chromatography. *Planta Med.* **2017**, *83*, 1053–1057. [CrossRef]
20. Wang, J.; Qu, W.Y.; Jun, S.J.; Bittenbender, H.C.; Li, Q.X. Rapid determination of six kavalactones in kava root and rhizome samples using Fourier transform infrared spectroscopy and multivariate analysis in comparison with gas chromatography. *Anal. Methods UK* **2010**, *2*, 492–498. [CrossRef]
21. Bilia, A.R.; Scalise, L.; Bergonzi, M.C.; Vincieri, F.F. Analysis of kavalactones from Piper methysticum (kava-kava). *J. Chromatogr. B Analyt. Technol. Biomed. Life Sci.* **2004**, *812*, 203–214. [CrossRef]

22. Liu, Y.; Lund, J.A.; Murch, S.J.; Brown, P.N. Single-Lab Validation for Determination of Kavalactones and Flavokavains in Piper methysticum (Kava). *Planta Med.* **2018**, *84*, 1213–1218. [CrossRef]
23. Wang, Y.; Eans, S.O.; Stacy, H.M.; Narayanapillai, S.C.; Sharma, A.; Fujioka, N.; Haddad, L.; McLaughlin, J.; Avery, B.A.; Xing, C. A stable isotope dilution tandem mass spectrometry method of major kavalactones and its applications. *PLoS ONE* **2018**, *13*, e0197940. [CrossRef]
24. Horwitz, W. *AOAC Guidelines for Single Laboratory Validation of Chemical Methods for Dietary Supplements and Botanicals*; AOAC International: Gaithersburg, MD, USA, 2002; p. 38.
25. Guideline, I.H.T. Validation of Analytical Procedures: Text and Methodology. In Proceedings of the International Conference on Harmonization, Geneva, Switzerland, November 2005.
26. Schmidt, A.H.; Molnar, I. Computer-assisted optimization in the development of a high-performance liquid chromatographic method for the analysis of kava pyrones in Piper methysticum preparations (vol 948, pg 51, 2002). *J. Chromatogr. A* **2006**, *1110*, 272. [CrossRef]
27. Smith, R.M.; Thakrar, H.; Arowolo, T.A.; Shafi, A.A. High-Performance Liquid-Chromatography of Kava Lactones from Piper-Methysticum. *J. Chromatogr.* **1984**, *283*, 303–308. [CrossRef]
28. Bobeldijk, I.; Boonzaaijer, G.; Spies-Faber, E.J.; Vaes, W.H.J. Determination of kava lactones in food supplements by liquid chromatography-atmospheric pressure chemical ionisation tandem mass spectrometry. *J. Chromatogr. A* **2005**, *1067*, 107–114. [CrossRef] [PubMed]
29. Lebot, V.; Merlin, M.; Lindstrom, L. *Kava: The Pacific Drug*; Yale University Press: New Haven, CT, USA, 1992; p. 255.
30. Lebot, V.; Le´vesque, J. The origin and distribution of kava (Piper methysticum Forst. f. Piperaceae): A phytochemical approach. *Allertonia* **1989**, *5*, 223–380.
31. Lebot, V.; Le´vesque, J. Genetic control of kavalactone chemo-types in Piper methysticum cultivars. *Phytochemistry* **1996**, *43*, 397–403. [CrossRef]

12

Simultaneous Characterization and Quantification of Varied Ingredients from *Sojae semen praeparatum* in Fermentation using UFLC–TripleTOF MS

Chuan Chai, Xiaobing Cui, Chenxiao Shan, Sheng Yu, Xinzhi Wang and Hongmei Wen *

School of Pharmacy, Nanjing University of Chinese Medicine, Nanjing 210029, Jiangsu, China; echo_0523@hotmail.com (C.C.); xiaobingcui@163.com (X.C.); thomastiger@163.com (C.S.); yusheng1219@163.com (S.Y.); wxzatnj@sina.com (X.W.)

* Correspondence: njwenhm@126.com

Academic Editors: Alessandra Gentili and Chiara Fanali

Abstract: Systematic comparison of active ingredients in *Sojae semen praeparatum* (SSP) during fermentation was performed using ultra-fast liquid chromatography (UFLC)–TripleTOF MS and principal component analysis (PCA). By using this strategy, a total of 25 varied compounds from various biosynthetic groups were assigned and relatively quantified in the positive or negative ion mode, including two oligosaccharides, twelve isoflavones, eight fatty acids, *N*–(3–Indolylacetyl)–DL–aspartic acid, methylarginine, and sorbitol. Additionally, as the representative constituents, six targeted isoflavones were sought in a targeted manner and accurately quantified using extracted ion chromatograms (XIC) manager (AB SCIEX, Los Angeles, CA, USA) combined with MultiQuant software (AB SCIEX, Los Angeles, CA, USA). During the fermentation process, the relative contents of oligoses decreased gradually, while the fatty acids increased. Furthermore, the accurate contents of isoflavone glycosides decreased, while aglycones increased and reached a maximum in eight days, which indicated that the ingredients converted obviously and regularly throughout the SSP fermentation. In combination with the morphological changes, which meet the requirements of China Pharmacopoeia, this work suggested that eight days is the optimal time for fermentation of SSP from the aspects of morphology and content.

Keywords: *Sojae semen praeparatum* (SSP); fermentation; conversion; ultra-fast liquid chromatography (UFLC)–TripleTOF MS; principal component analysis (PCA)

1. Introduction

Fermentation is one of the major processes used in the production of food from soybeans and has played an important role in human life for centuries [1,2]. Many studies have reported the components that are converted and how bioactivities increased in soybean products during the fermentation process [3–7].

Sojae semen praeparatum (SSP), whose Chinese herbal name is dandouchi, which is a product of Chinese fermented preparation obtained from the ripe seed of soybean (*Glycine max* (L.) Merr.), has been used as an important component in traditional diets and as an effective traditional Chinese medicine (TCM) among the Chinese community worldwide. More people are expected to consume SSP if the fermentation process includes quality assessment and quality control. Other studies have focused on the active ingredients in SSP, such as isoflavone [8–11], peptides [12], biogenic amines [13,14], and volatile components [15], the physiological properties of SSP such as anti-oxidative activity [16], anti-proliferative activity [17], anti-α-glucosidase activity [18] and anti-hypertensive effects [19], and species and quantities of fermenting bacteria in SSP spontaneous fermentation, such as bacterial fermentation and fungus fermentation [20,21]. However, no systematic comparison has been conducted of the active ingredients among the raw materials and the SSP products collected at different fermentation stages.

Isoflavones were reported to be representative constituents affecting soybean due to their significant estrogen-like bioactivity [22,23], which increased after fermentation [24,25]. The main isoflavones found in soybean are daidzein, genistein and glycitein, which are present either in glycosidic or aglycone form, mainly with β–glycosides and some 6″–*O*–malonyl or 6″–*O*–acetylglucose [26,27]. Aglyca were reported to have more bioactivity compared to the corresponding glycosides [28]. Some studies suggested that bacterial or fungal β–glycosidases are attractive candidates for use in converting β–glycosides isoflavone to their aglycones, thus enhancing the nutritional value of soy products [29,30]. However, the composition and contents change trends of isoflavones contained in SSP during fermentation have not yet been reported.

Therefore, we aimed to characterize the conversion of ingredients associated with the SSP fermentation process using ultra–fast liquid chromatography-triple time of flight mass spectrometry (UFLC–TripleTOF MS) [31], and accurately quantify the major components that vary using extracted ion chromatograms (XIC) manager (AB SCIEX, Los Angeles, CA, USA) with standard injections, thereby providing some technological supports for the optimization and quality control of SSP fermentation.

2. Results and Discussion

2.1. Morphologic Changes

Morphological changes in soybean and SSP products during fermentation were shown in Figure 1. With the increase in fermentation time, black soybean was overgrown with white hyphae, which then changed to yellow in SSP fermented for six days (S6), turned yellow completely in SSP fermented for eight days (S8), and finally hardened. In accordance with the 2015 edition of China Pharmacopoeia [32], moisturized soybeans should be fermented with boiled Artemisiae annuae herba and Mori folium until "yellow cladding". The morphological changes in S8 were consistent with these requirements, so we speculated that eight days is the optimal time for the fermentation of SSP.

2.2. Qualitative Analysis and Principle Component Analysis (PCA)

Using UFLC–TripleTOF MS analysis, information on intact precursors and fragment ions were obtained from a single injection.

The base peak chromatograms (BPCs) of soybean and S8 using both positive and negative ion modes are shown in Figure 2. The BPCs of S8 were significantly different from those of soybeans. Compared to the BPC of soybean in positive ion mode (Figure 2A), the BPC of S8 showed much higher peak intensities at t_R = 10–20 min (Figure 2C). Much higher peak intensities occurred at t_R = 0–10 min of the BPC for S8 (Figure 2D) in negative ion mode compared to soybean (Figure 2B), which shows that the ingredients converted during the SSP fermentation.

Figure 3A shows that the score plot of soybean and SSP products in positive ion mode is separated into three significant clusters ($P < 0.05$) for the first and the second principal components (PCs). Here, the green cluster (S2 (SSP fermented for two days), S4 (SSP fermented for four days) and S6) and the red cluster (S8, S10 (SSP fermented for ten days) and S15 (SSP fermented for fifteen days)) are separated by the first PC, whereas the blue cluster (soybean and S0) and the green cluster (S2, S4 and S6) are separated by the second PC. The first and second PCs' values are both 14.5%.

Similarly, Figure 3C illustrates the score plot of soybean and SSP products in negative ion mode, which is separated into three significant clusters for the first and the second PCs. The blue cluster (soybean) and the red cluster (S6, S8, S10 and S15) are separated by the first PC, whereas the blue cluster (soybean) and the green cluster (S0, S2 and S4) are separated by the second PC. The first and second PCs' values are both 14.5%.

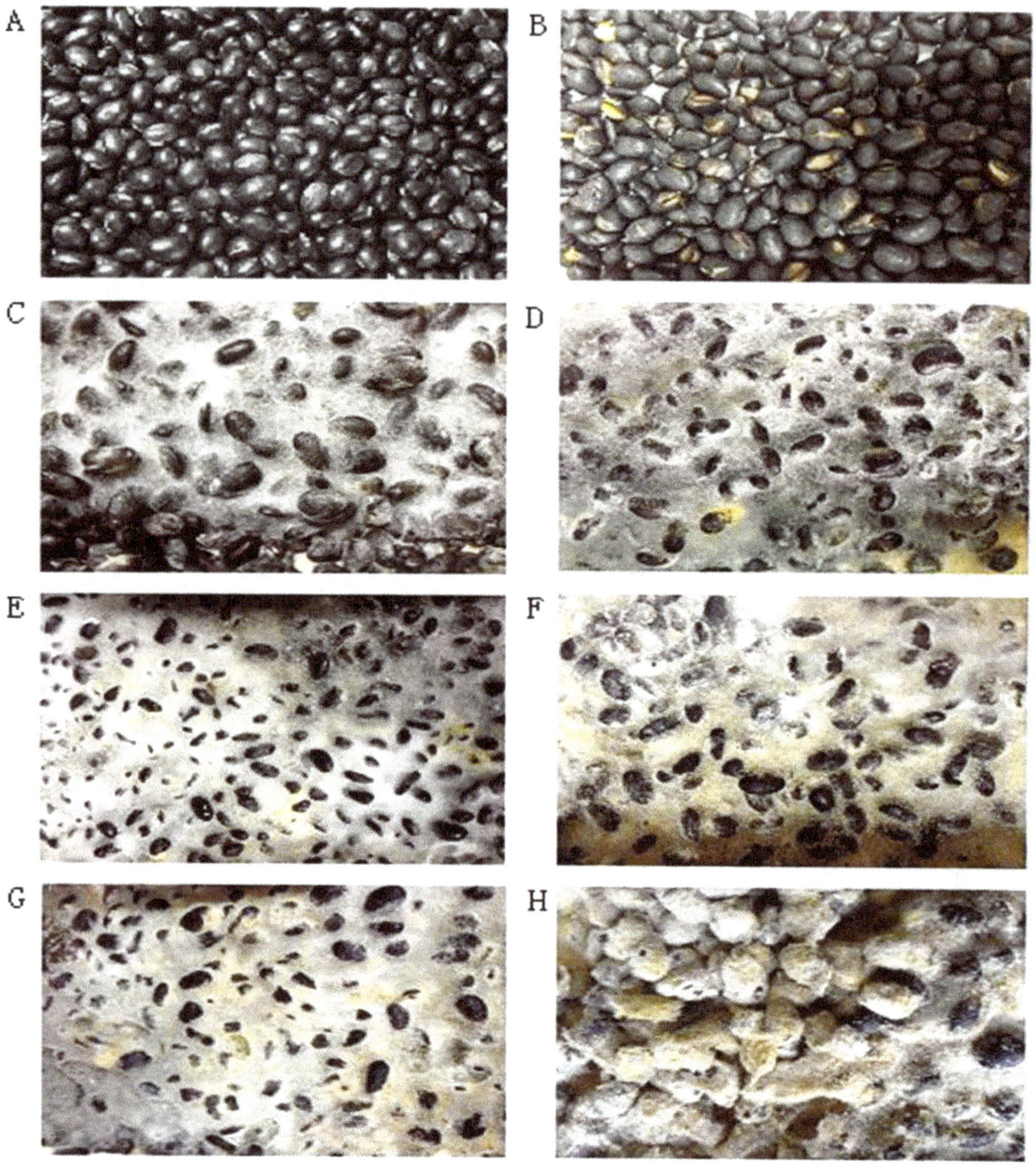

Figure 1. The morphology of soybean (**A**), S0 (**B**), S2 (**C**), S4 (**D**), S6 (**E**), S8 (**F**), S10 (**G**), S15 (**H**) days.

From the corresponding loadings plots (Figure 3B,D), a significant number of variables are located around the observations of the samples, indicating that SSP converted significantly throughout the fermentation process.

The ion species, retention times, molecular formulas, mean measured mass, mass accuracies and assigned identities of the significantly variables are shown in Table 1. $[M + H]^+$, $[M + Na]^+$, $[M + K]^+$, $[M + NH_4]^+$, $[M + H + CH_3OH]^+$, and $[M + H - H_2O]^+$ ion species were found in positive ion mode and $[M - H]^-$, $[M - H - H_2O]^-$, $[M + CH_3COO]^-$, and $[M + HCOO]^-$ were found in negative ion mode.

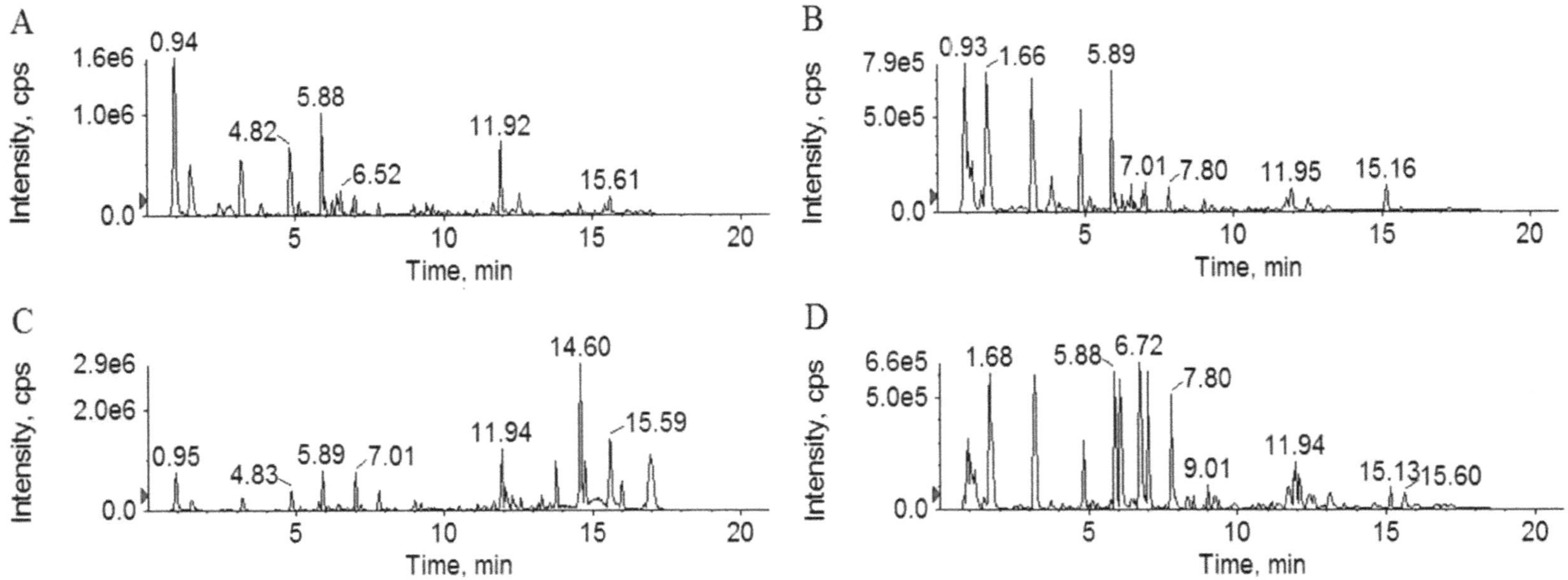

Figure 2. UFLC–TripleTOF MS base peak chromatograms (BPCs) of soybean in positive (**A**) and negative (**B**) ion modes and S8 in positive (**C**) and negative (**D**) ion modes.

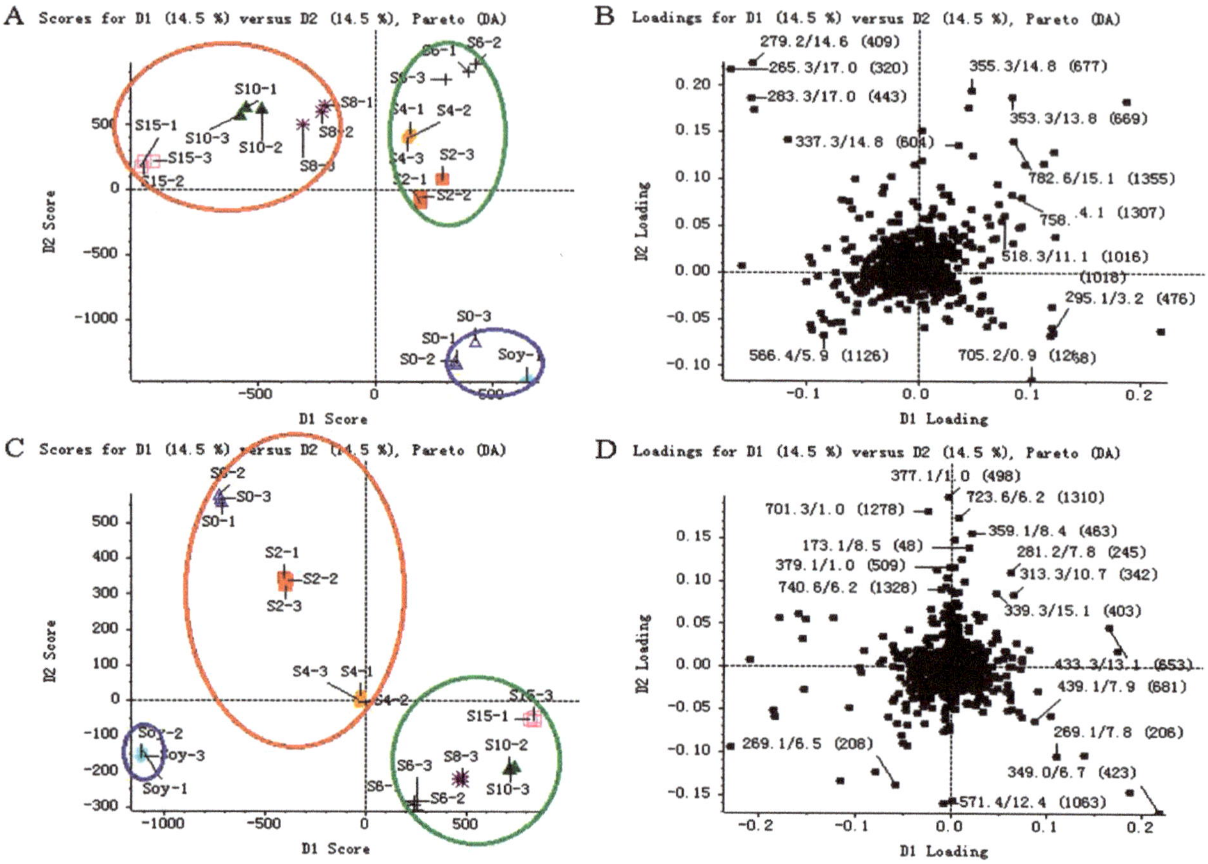

Figure 3. Score plots (**A**,**C**) and loading plots (**B**,**D**) of metabolites determined in soybean (Soy) and *Sojae semen praeparatum* products (S0, S2, S4, S6, S8, S10 and S15) by UFLC–TripleTOF MS in both positive (A and B) and negative (C and D) ion modes. The three clusters (green, blue and red) were used for the color coding of different groups separated by the first and second principal components.

We found 29 components of varied classes in positive ion mode, and 22 of them were inferred to be raffinose (**2**), stachyose (**3**), *N*–(3–Indolylacetyl)–DL–aspartic acid (**5**), daidzin (**6**), glycitin (**7**), genistin (**8**), 6″–*O*–malonyldaidzin (**9**), 6″–*O*–malonylglycitin (**10**), 6″–*O*–acetyldaidzin (**11**), 6″–*O*–acetylglycitin (**12**), 6″–*O*–malonylgenistin (**13**), daidzein (**14**), 6″–*O*–acetylgenistin (**15**), glycitein (**16**), methylarginine (**17**), genistein (**18**), dimorphecolic acid (**20**), α–linolenic acid (**22**), linoleic acid (**24**), oleic acid (**25**), palmitic acid (**27**) and stearic acid (**28**) with the help of Peakview® software (AB SCIEX, Los Angeles, CA, USA). We identified 23 of varied classes in negative ion mode, of which 10 were putatively identified as stachyose (**3**), sorbitol (**32**), *N*–(3–Indolylacetyl)–DL–aspartic acid (**5**), daidzin (**6**), genistin (**8**), 6″–*O*–acetyldaidzin (**11**), 6″–*O*–acetylgenistin (**15**), glycitein (**16**), gheddic acid (**42**) and nonadecanoic acid (**43**) by linking the masses of ions to structures. We found 7 assigned and 1 unassigned variable in both positive and negative ion modes.

All chemical structures, selected ion intensity trend plots, mass spectra, and mass spectral interpretation of putatively assigned identities were listed in Figure 4, Figures S1 and S2, and Table 2.

Table 1. Varied components putatively identified from soybean and *Sojae semen praeparatum* products in both positive and negative ion modes.

Ionization Mode	Compound No.	t_R [a] (min)	Molecular Formula	Mass (Da)	Ion Species	Mean Measured Mass (Da)	Mass Accuracy (ppm)	Assigned Identity	References
Positive	**1**	0.93	$C_{12}H_{18}N_6O_6$	342.1288	$[M + K]^+$	381.0907	−0.4	/[b]	–[c]
	2	0.93	$C_{18}H_{32}O_{16}$	504.1690	$[M + K]^+$	543.1474	−1.9	Raffinose	[33]
	3	0.95	$C_{24}H_{42}O_{21}$	666.2213	$[M + K]^+$	705.2045	−1.2	Stachyose	[33]
	4	1.44	$C_{13}H_{15}N_3O_5$	293.1012	$[M + NH_4]^+$	311.1328	−2.2	/[b]	–[c]
	5	3.86	$C_{14}H_{14}N_2O_5$	290.0903	$[M + H]^+$	291.0955	−4.6	*N*-(3-Indolylacetyl)-DL-aspartic acid	–[c]
	6	4.81	$C_{21}H_{20}O_9$	416.1107	$[M + H]^+$	417.1307	0.8	Daidzin	[37]
	7	5.14	$C_{22}H_{22}O_{10}$	446.1213	$[M + H]^+$	447.1429	−0.3	Glycitin	[37]
	8	5.89	$C_{21}H_{20}O_{10}$	432.1057	$[M + H]^+$	433.1266	0.9	Genistin	[37]
	9	6.03	$C_{24}H_{22}O_{12}$	502.1111	$[M + H]^+$	503.1335	−0.6	6″–*O*–malonyldaidzin	[37]
	10	6.04	$C_{25}H_{24}O_{13}$	532.1217	$[M + H]^+$	533.1451	−1.3	6″–*O*–malonylglycitin	[37]
	11	6.44	$C_{23}H_{22}O_{10}$	458.1213	$[M + H]^+$	459.1435	−0.2	6″–*O*–acetyldaidzin	[37]
	12	6.51	$C_{24}H_{24}O_{11}$	488.1319	$[M + H]^+$	489.1535	−1.0	6″–*O*–acetylglycitin	[37]
	13	6.57	$C_{24}H_{22}O_{13}$	518.1060	$[M + H]^+$	519.1285	−0.8	6″–*O*–malonylgenistin	[37]
	14	7.00	$C_{15}H_{10}O_4$	254.0579	$[M + H]^+$	255.0726	0.9	Daidzein	[37]
	15	7.05	$C_{23}H_{22}O_{11}$	474.1162	$[M + H]^+$	475.1379	−1.0	6″–*O*–acetylgenistin	[37]
	16	7.17	$C_{16}H_{12}O_5$	284.0685	$[M + H]^+$	285.0840	0.3	Glycitein	[37]
	17	7.69	$C_7H_{16}N_4O_2$	188.1268	$[M + H + H_2O]^+$	207.1452	−3.9	Methylarginine	–[c]
	18	7.80	$C_{15}H_{10}O_5$	270.0528	$[M + H]^+$	271.0680	1.0	Genistein	[37]
	19	11.94	$C_{21}H_{45}N_9O_6$	519.3493	$[M + H]^+$	520.3534	1.0	/[b]	–[c]
	20	13.29	$C_{18}H_{32}O_3$	296.2710	$[M + H]^+$	297.2517	−3.9	Dimorphecolic acid	–[c]
	21	13.77	$C_{23}H_{44}O_2$	352.3336	$[M + H]^+$	353.2797	0.1	/[b]	–[c]
	22	14.61	$C_{18}H_{30}O_2$	278.2246	$[M + H]^+$	279.2408	−0.4	α–Linolenic acid	[34]
	23	14.76	$C_{23}H_{46}O_2$	354.3492	$[M + H]^+$	355.2950	0.8	/[b]	–[c]
	24	15.62	$C_{18}H_{32}O_2$	280.2402	$[M + H]^+$	281.2567	0.7	Linoleic acid	[34]
	25	15.70	$C_{18}H_{34}O_2$	282.2559	$[M + H + CH_3OH]^+$	315.2915	0.6	Oleic acid	[34]
	26	16.01	$C_{21}H_{40}O_2$	324.3023	$[M + H + CH_3OH]^+$	357.3101	2.9	/[b]	–[c]
	27	16.67	$C_{16}H_{32}O_2$	256.2402	$[M + H]^+$	257.2565	−0.3	Palmitic acid	[34]
	28	17.04	$C_{18}H_{36}O_2$	284.2715	$[M + H]^+$	285.3050	−0.8	Stearic acid	[34]
Negative	**29**	0.92	$C_{12}H_{18}N_6O_6$	342.1288	$[M + HCOO]^-$	387.1623	1.7	/[b]	–[c]
	3	0.95	$C_{24}H_{42}O_{21}$	666.2213	$[M + HCOO]^-$	711.3077	1.3	Stachyose	[33]
	30	1.05	$C_{30}H_{32}N_6O_9$	620.2220	$[M - H - H_2O]^-$	601.2133	−1.1	/[b]	–[c]
	31	3.12	$C_{16}H_{18}N_6$	294.1582	$[M - H]^-$	293.1552	−1.4	/[b]	–[c]
	32	3.21	$C_6H_{14}O_6$	182.0790	$[M + CH_3COO]^-$	241.0915	0.8	Sorbitol	[35]
	5	3.85	$C_{14}H_{14}N_2O_5$	290.0903	$[M - H]^-$	289.1078	−1.3	*N*-(3-Indolylacetyl)-DL-aspartic acid	–[c]

Table 1. *Cont.*

Ionization Mode	Compound No.	t_R [a] (min)	Molecular Formula	Mass (Da)	Ion Species	Mean Measured Mass (Da)	Mass Accuracy (ppm)	Assigned Identity	References
	6	4.81	$C_{21}H_{20}O_9$	416.1107	$[M + HCOO]^-$	461.1650	0.2	Daidzin	[37]
	33	5.64	$C_{18}H_{10}N_2O_6$	350.0528	$[M - H]^-$	349.0451	−2	/ [b]	– [c]
	8	5.89	$C_{21}H_{20}O_{10}$	432.1057	$[M + HCOO]^-$	477.1629	−2.5	Genistin	[37]
	34	6.15	$C_{18}H_4N_6$	304.0497	$[M + CH_3COO]^-$	363.0635	−0.5	/ [b]	– [c]
	11	6.42	$C_{23}H_{22}O_{10}$	458.1213	$[M + HCOO]^-$	503.1805	−2.3	6″-*O*-acetyldaidzin	[37]
	35	6.47	$C_{13}H_{10}O_6$	262.0466	$[M - H]^-$	261.0400	2.4	/ [b]	– [c]
	36	6.68	$C_{19}H_6N_6O_3$	366.0490	$[M - H]^-$	365.0430	−1.3	/ [b]	– [c]
	15	7.04	$C_{23}H_{22}O_{11}$	474.1162	$[M + HCOO]^-$	519.1784	2.3	6″-*O*-acetylgenistin	[37]
Negative	**16**	7.15	$C_{16}H_{12}O_5$	284.0685	$[M - H]^-$	283.0969	2.1	Glycitein	[37]
	37	7.91	$C_{18}H_{18}O_{10}$	394.0895	$[M + HCOO]^-$	439.0878	−2.0	/ [b]	– [c]
	38	10.83	$C_{31}H_{50}N_2O_9$	594.3505	$[M - H]^-$	593.3490	−0.6	/ [b]	– [c]
	39	11.29	$C_{19}H_{36}O_3$	312.2654	$[M - H - H_2O]^-$	293.2483	1.7	/ [b]	– [c]
	40	11.66	$C_{32}H_{48}N_6O_5$	596.3675	$[M - H]^-$	595.3613	1.9	/ [b]	– [c]
	41	12.06	$C_{17}H_{36}N_8O_5$	432.2798	$[M - H]^-$	431.2728	0.1	/ [b]	– [c]
	42	12.54	$C_{34}H_{68}O_2$	508.3389	$[M - H]^-$	507.3340	0.1	Gheddic acid	[36]
	43	12.58	$C_{19}H_{38}O_2$	298.2861	$[M - H]^-$	297.2814	4.0	Nonadecanoic acid	– [c]
	44	13.57	$C_{18}H_{42}N_{10}O$	414.3543	$[M + HCOO]^-$	459.3520	−1.3	/ [b]	– [c]
	45	15.55	$C_{12}H_{28}N_8O_6$	380.2121	$[M - H]^-$	379.2049	−2.6	/ [b]	– [c]

[a] t_R, retention time; [b] Not assigned; [c] No reference.

Table 2. Mass spectral interpretation of assigned compounds in soybean and *Sojae semen praeparatum* products.

Compound No.	Assigned Identity (Ion Mode)	MS/MS Fragments Ions
2	Raffinose (+)	$543.1474[M + K]^+$, $381.0905[M + K - glu + H_2O]^+$,
3	Stachyose (+)	$705.2045[M + K]^+$, $543.1471[M{+}K - glu]^+$,
	(−)	$711.3077[M + HCOO]^-$, $665.3014[M - H]^-$, $485.2130[M - H - glu]^-$, $341.1514[M - H - 2glu + 2H_2O]^-$,
5	*N*-(3-Indolylacetyl)-DL-aspartic acid (+)	$291.0955[M + H]^+$, $161.0645[C_5H_6O_2N + H]^+$, $139.0428[M + H - C_8H_6N - 2H_2O]^+$,
	(−)	$289.1078[M - H]^-$, $271.0968[M - H - H_2O]^-$, $245.1127[M - H - CO_2]^-$, $227.1008[M - H - CO_2 - H_2O]^-$,
6	Daidzin (+)	$439.1128[M + Na]^+$, $417.1307[M + H]^+$, $277.0551[M + Na - glu + H_2O]^+$, $255.0728[M + H - glu + H_2O]^+$,
	(−)	$461.1650[M + HCOO]^-$, $415.1556[M - H]^-$, $253.0825[M - H - glu + H_2O]^-$,
7	Glycitin (+)	$469.1235[M + Na]^+$, $447.1429[M + H]^+$, $307.0662[M + Na - glu + H_2O]^+$, $285.0839[M + H - glu + H_2O]^+$,
8	Genistin (+)	$433.1266[M + H]^+$, $271.0674[M + H - glu + H_2O]^+$, $243.0712[M + H - glu + H_2O - CO]^+$, $215.0752[M + H - glu + H_2O - 2CO]^+$, $153.0218[M + H - C_{13}H_{12}O_7]^+$,
	(−)	$477.1629[M + HCOO]^-$, $431.1529[M - H]^-$, $269.0795[M - H - glu + H_2O]^-$,
9	6″-*O*-malonyldaidzin (+)	$525.1146[M + Na]^+$, $503.1335[M + H]^+$, $481.1244[M + Na - CO_2]^+$, $439.1133[M + Na - malonyl - H_2O]^+$, $277.0549[M + Na - malonyl - glu]^+$, $255.0728[M + H - malonyl - glu]^+$,

Table 2. *Cont.*

Compound No.	Assigned Identity (Ion Mode)	MS/MS Fragments Ions
10	6″–*O*–malonylglycitin (+)	533.1451[M + H]$^+$, 285.0845[M + H – malonyl – glu]$^+$,
11	6″–*O*–acetyldaidzin (+)	459.1435[M + H]$^+$, 255.0726[M + H – acetyl – glu]$^+$,
	(−)	503.1805[M + HCOO]$^-$, 457.1720[M – H]$^-$, 253.0822[M – H – acetyl – glu]$^-$,
12	6″–*O*–acetylglycitin (+)	489.1535[M + H]$^+$, 285.0839[M + H – acetyl – glu]$^+$,
13	6″–*O*–malonylgenistin (+)	541.1093[M + Na]$^+$, 523.0995[M + Na – H_2O]$^+$, 519.1285[M + H]$^+$, 497.1167[M + Na – CO_2]$^+$, 455.1096[M + Na – malonyl – H_2O]$^+$, 293.0505[M + Na – malonyl – glu]$^+$, 271.0678[M + H – malonyl – glu]$^+$,
14	Daidzein (+)	277.0550[M + Na]$^+$, 255.0726[M + H]$^+$, 237.0611[M + H – H_2O]$^+$, 227.0762[M + H – CO]$^+$, 199.0806[M + H – 2CO]$^+$, 181.0695[M + H – 2CO – H_2O]$^+$, 137.0273[M + H – C_8H_6O]$^+$, 91.0582[M + H – H_2O – $C_9H_6O_2$]$^+$,
15	6″–*O*–acetylgenistin (+)	475.1379[M + H] $^+$, 271.0682[M + H – acetyl – glu]$^+$,
	(−)	519.1784[M + HCOO]$^-$, 473.1689[M – H]$^-$, 269.0795[M – H – acetyl – glu]$^-$,
16	Glycitein (+)	307.0672[M + Na]$^+$, 285.0840[M + H]$^+$, 270.0599[M + H – CH_3]$^+$, 242.0642[M + H – CH_3 – CO]$^+$, 169.0614[M + H – C_8H_4O]$^+$, 141.0740[M + H – $C_9H_4O_2$]$^+$,
	(−)	283.0969[M – H]$^-$, 268.0714[M – H – CH_3]$^-$, 240.0729[M – H – CH_3 – CO]$^-$, 196.0776[M – C_2H_3O – OH – OH]$^-$,
17	Methylarginine (+)	207.1452[M + H + H_2O]$^+$, 189.1321[M + H]$^+$, 161.1377[M + H – H_2O]$^+$,
18	Genistein (+)	271.0680[M + H]$^+$, 253.0565[M + H – H_2O]$^+$, 243.0716[M + H – CO]$^+$, 215.0757[M + H – 2CO]$^+$, 153.0223[M + H – C_8H_6O]$^+$,
20	Dimorphecolic acid (+)	297.2517[M + H]$^+$, 279.2398[M + H – H_2O]$^+$, 261.2287[M + H – $2H_2O$]$^+$, 233.2328[M + H – $2H_2O$ – CO]$^+$, 109.1051[M + H – $C_9H_{17}COOH$ – CH_4]$^+$, 97.1051[M + H – $C_9H_{17}COOH$ – C_2H_4]$^+$, 81.0743[M + H – $C_9H_{17}COOH$ – C_3H_8]$^+$. 67.0590[M + H – $C_9H_{17}COOH$ – C_4H_{10}]$^+$
22	α–Linolenic acid (+)	297.2517[M + H + H_2O]$^+$, 279.2408[M + H]$^+$, 149.0276[M + H – C_6H_3COOH]$^+$, 135.1206[M + H – $C_7H_{15}COOH$]$^+$, 125.1000[M + H – $C_8H_{13}COOH$]$^+$, 123.1200[M + H – $C_8H_{15}COOH$]$^+$, 109.1045[M + H – $C_9H_{17}COOH$]$^+$, 95.0892[M + H – $C_{10}H_{19}COOH$]$^+$, 81.0738[M + H – $C_{11}H_{21}COOH$]$^+$, 67.0588[M + H – $C_{12}H_{23}COOH$]$^+$,
24	Linoleic acid (+)	313.2827[M + H + CH_3OH]$^+$, 281.2567[M + H]$^+$, 263.2443[M + H – H_2O]$^+$, 239.2433[M + H + CH_3OH – H_2O – C_4H_8]$^+$, 221.2318[M – H – CH_3COOH]$^+$, 147.1210[M + H + CH_3OH – $C_{12}H_{22}$]$^+$ 133.1051[M + H + CH_3OH – $C_{13}H_{24}$]$^+$, 109.1049[M + H – $C_9H_{19}COOH$]$^+$, 95.0895[M + H – $C_{10}H_{21}COOH$]$^+$, 71.0902[M + H – $C_{12}H_{21}COOH$]$^+$, 57.0755[M + H – $C_{13}H_{23}COOH$]$^+$,
25	Oleic acid (+)	315.2915[M + H + CH_3OH]$^+$, 283.2461[M + H]$^+$, 271.5449[M + H + CH_3OH – CO_2]$^+$, 267.0220[M + H – CH_4]$^+$, 265.2757[M + H – H_2O]$^+$, 187.1151[M + H + CH_3OH – C_9H_{20}]$^+$, 171.0300[M + H – C_8H_{16}]$^+$, 114.9656[M + H – $C_9H_{15}COOH$]$^+$, 96.9540[M + H – $C_{10}H_{21}COOH$]$^+$, 83.0875[M + H – $C_{11}H_{23}COOH$]$^+$, 57.0750[M + H – $C_{13}H_{25}COOH$]$^+$
27	Palmitic acid (+)	257.2565[M + H]$^+$, 201.1894[M + H – C_4H_8]$^+$, 97.1025[M + H – H_2O – $C_{10}H_{22}$]$^+$, 71.0902[M + H – $C_{10}H_{21}COOH$]$^+$, 57.0753[M + H – $C_{11}H_{23}COOH$]$^+$,
28	Stearic acid (+)	285.3050[M + H]$^+$, 267.2656[M + H – H_2O]$^+$, 126.9036[M + H – $C_8H_{17}COOH$]$^+$, 83.0895[M + H – $C_{10}H_{21}COOH$ – CH_4]$^+$, 69.0739[M + H – $C_{11}H_{23}COOH$ – CH_4]$^+$, 57.0742[M + H – $C_{13}H_{27}COOH$]$^+$,
32	Sorbitol (−)	241.0915[M + CH_3COO]$^-$, 223.0801[M + CH_3COO – H_2O]$^-$, 181.0622[M – H]$^-$, 149.0803[M – H – CH_3OH]$^-$,
42	Gheddic acid (−)	507.3340[M – H]$^-$, 279.2671[M – H – $C_{13}H_{27}COOH$]$^-$, 153.0148[M – H – $C_{22}H_{45}COOH$]$^-$,
43	Nonadecanoic acid (−)	297.2814[M – H]$^-$, 279.2672[M – H – H_2O]$^-$, 183.1614[M – H – C_5H_9COOH]$^-$,

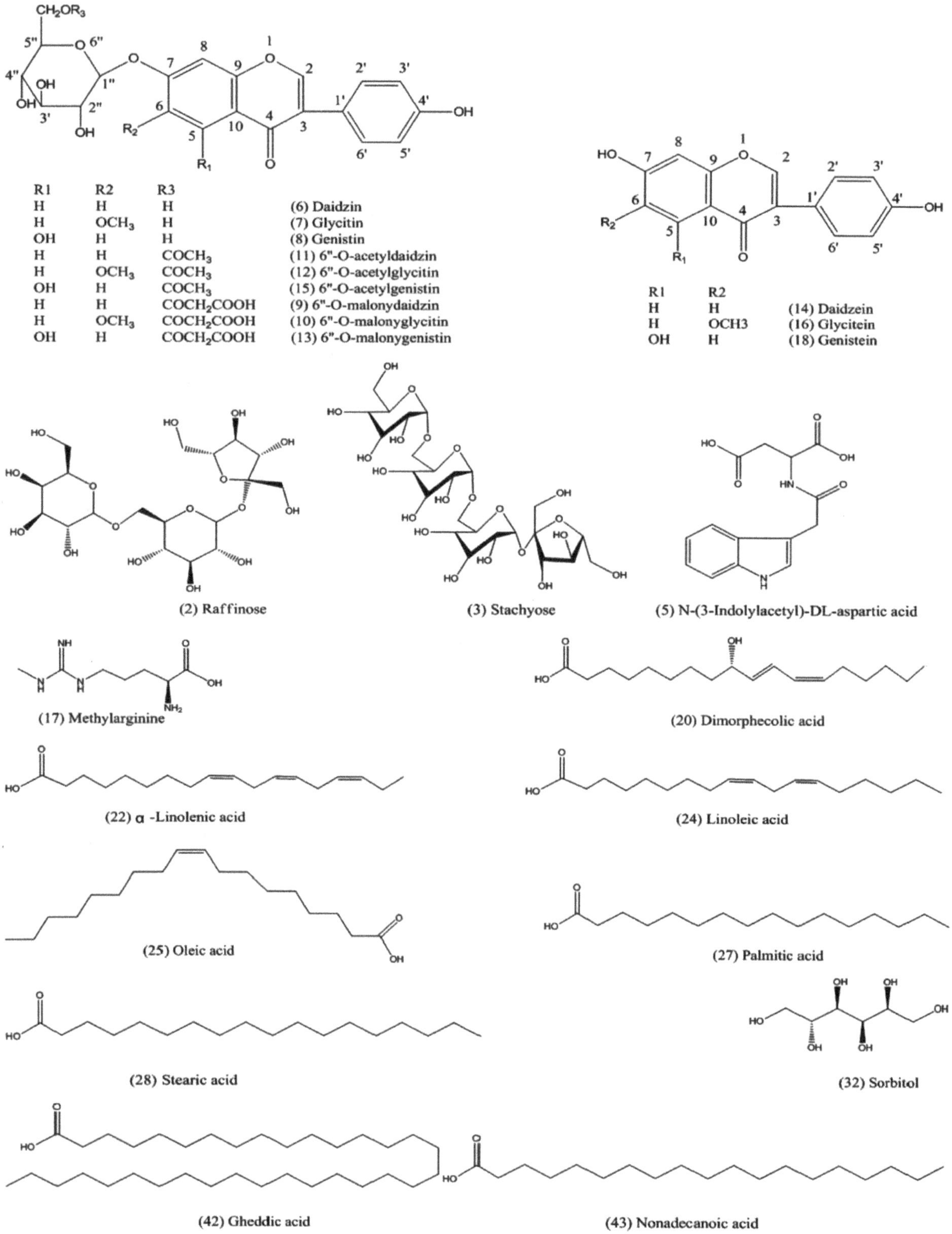

Figure 4. Chemical structures of assigned compounds.

Compounds **2** and **3** were inferred as raffinose and stachyose, respectively, for the loss of aglyca. Compounds **14**, **16** and **18** were putatively identified as isoflavone aglycones for their fragment ions at *m*/*z* 137, 153 and 169, respectively, after retro–Diels–Alder reaction. Compounds **6**–**8** were assigned as isoflavone glycosides for their glycones $[M + H - glu + H_2O]^+$ at *m*/*z* 255, 271 and 285 and $[M - H - glu + H_2O]^-$ at *m*/*z* 253, 269 and 283 after deglycosylation. Further dissociation of the glycones yielded a serial of fragments in agreement with the aglycones. The six isoflavone compounds were also confirmed by injecting a mix of standard solutions (Figure 6). Compounds **9**, **10** and **13** were confirmed as isoflavone glycoside malonates: compounds **11**, **12** and **15** were identified as isoflavone acetyl glycosides for their common glycones in comparison with glycosides. Compounds **20**, **22**, **24**, **25**, **27** and **28** were assumed to be a series of fatty acids for the homologous fragment ions at *m*/*z* (67, 81, 95, 109 and 123); (83, 97 and 111); and (57, 71 and 85), the difference between every pair of the fragment ions was 14 (–CH_2–). Twenty of the assigned compounds were previously reported in soybean [33–37]. However, *N*–(3–Indolylacetyl)–DL–aspartic acid, methylarginine, dimorphecolic acid, gheddic acid, and nonadecanoic acid have never been reported in soybean; sorbitol was only detected in germinating soybean seeds [35], and gheddic acid was identified in *Mori folium* [36]. Six constituents were presumed to be introduced from processing adjuvants or produced during the SSP fermentation process.

2.3. Relative Quantitative Analysis

In agreement with previous results [38–40], the relative contents of raffinose and stachyose decreased gradually during the entire fermentation process due to degradation by bacteria. As stachyose and raffinose cause indigestion and flatulence in animals after ingestion, the reduction of oligosaccharides is an indication that fermentation can promote the absorption of soybean nutrients. Isoflavone was inferred to be the principle difference among the products obtained from the SSP fermentation process due to its high proportions of varied components. With increasing fermentation time, the relative contents of isoflavone glycosides decreased while the isoflavone aglycones increased, reaching a maximum in S8. The isoflavone glycoside malonates decreased while the isoflavone acetyl glycosides increased to a maximum in S2 and then dropped. We assumed that the isoflavone glycoside malonates were transformed to acetyl glycosides in the early fermentation period due to their heat instability, and that all the isoflavone glycosides were converted to isoflavone aglycones. The increase in fatty acids showed that lipids could be degraded during fermentation. All the results provide some technological supports for the optimization and quality control of SSP fermentation. Given of the regular component conversion in SSP during fermentation, further identification of the unassigned varied ingredients present in SSP is needed.

2.4. Accurate Quantitative Analysis

XIC manager combined with MultiQuant software was used to automatically highlight all findings above a defined thresholds at an exactive mass of ± 0.02 Da and to quantitatively compare samples with a series of standard injections.

The validation values are summarized in Table 3. The calibration curves show satisfactory linearity. The correlation coefficient (*r*) ranged from 0.9840–0.9981 for all the isoflavones. Limit of detection (LOD) and limit of quantitation (LOQ) values were 0.1–50.0 and 2.0–250.0 ng/mL, respectively. The intra and inter-day precisions were less than 0.48% and 2.87%, respectively. The repeatability was within 2.53–4.82%. The recoveries were between 97.61 ± 3.73% and 104.84 ± 2.58% at different spiking concentration levels. The short-term stability analyzed at various periods was less than 4.10%. The above results demonstrate that the established method is accurate and reproducible for determining the six isoflavones in SSP.

Controlled by mix standard solution, the accurate quantitative results of six isoflavones in SSP were summarized in Figure 5. All six isoflavones were identified in soybean and SSP products. With increasing fermentation time, daidzin, glycitin and genistin decreased while daidzein, glycitein and genistein increased and raised to the top in S8 at 74.50, 13.52 and 47.42 mg/100 g dry weight respectively. Total glycoside and aglycone were also calculated as the sum of each individual isoflavone and presented in Figure 6. As the fermentation time increased, total glycoside contents decreased, while total aglycone contents increased significantly and rose to the top in the S8. The total glycoside content in S8 was less than half a percent of soybean's while the total aglycone content in S8 was 4.8 times higher than that in soybean, indicating that the ingredients converted regularly during the fermentation process.

Table 3. Validation data of targeted analytes.

Isoflavone	RT [a] (min)	Regression Equation [b]	Linear Range (μg/mL)	r	LOD [c] (ng/mL)	LOQ [d] (ng/mL)	Precision (RSD % [e]) Intra Day	Precision (RSD % [e]) Inter Day	Repeatability (RSD, %)	Recovery (Mean [f] ± RSD %)	Stability (RSD % [a])
Daidzin	4.81	$Y = 58544X + 384206$	0.010–100.0	0.9979	5.0	10.0	1.14	4.24	3.98	103.83 ± 3.15	4.10
Glycitin	5.14	$Y = 49463X + 421236$	0.010–100.0	0.9981	5.0	10.0	0.57	2.87	3.27	104.84 ± 2.58	1.84
Genistin	5.89	$Y = 60953X + 286019$	0.100–600.0	0.9980	10.0	50.0	0.48	4.48	4.41	97.76 ± 4.70	3.04
Daidzein	7.02	$Y = 76972X + 705619$	0.100–100.0	0.9840	0.1	3.0	1.61	3.50	2.53	97.61 ± 3.73	3.34
Glycitein	7.17	$Y = 52697X + 643182$	0.010–200.0	0.9973	0.1	2.0	1.42	2.61	4.19	103.71 ± 2.69	3.78
Genistein	7.80	$Y = 78899X + 342189$	0.50–200.0	0.9973	50.0	250.0	2.35	3.99	4.82	99.67 ± 3.16	2.39

[a] RT, retention time; [b] Y, peak area; X, concentration (μg/mL); [c] LOD, Limit of detection (S/N = 3); [d] LOQ, Limit of quantification (S/N = 10). [e] Relative standard deviation (%) = (standard deviation / mean) × 100. ($n = 3$); [f] Mean extraction yield (%) = (detected amount − original amount)/spiked amount × 100. ($n = 3$).

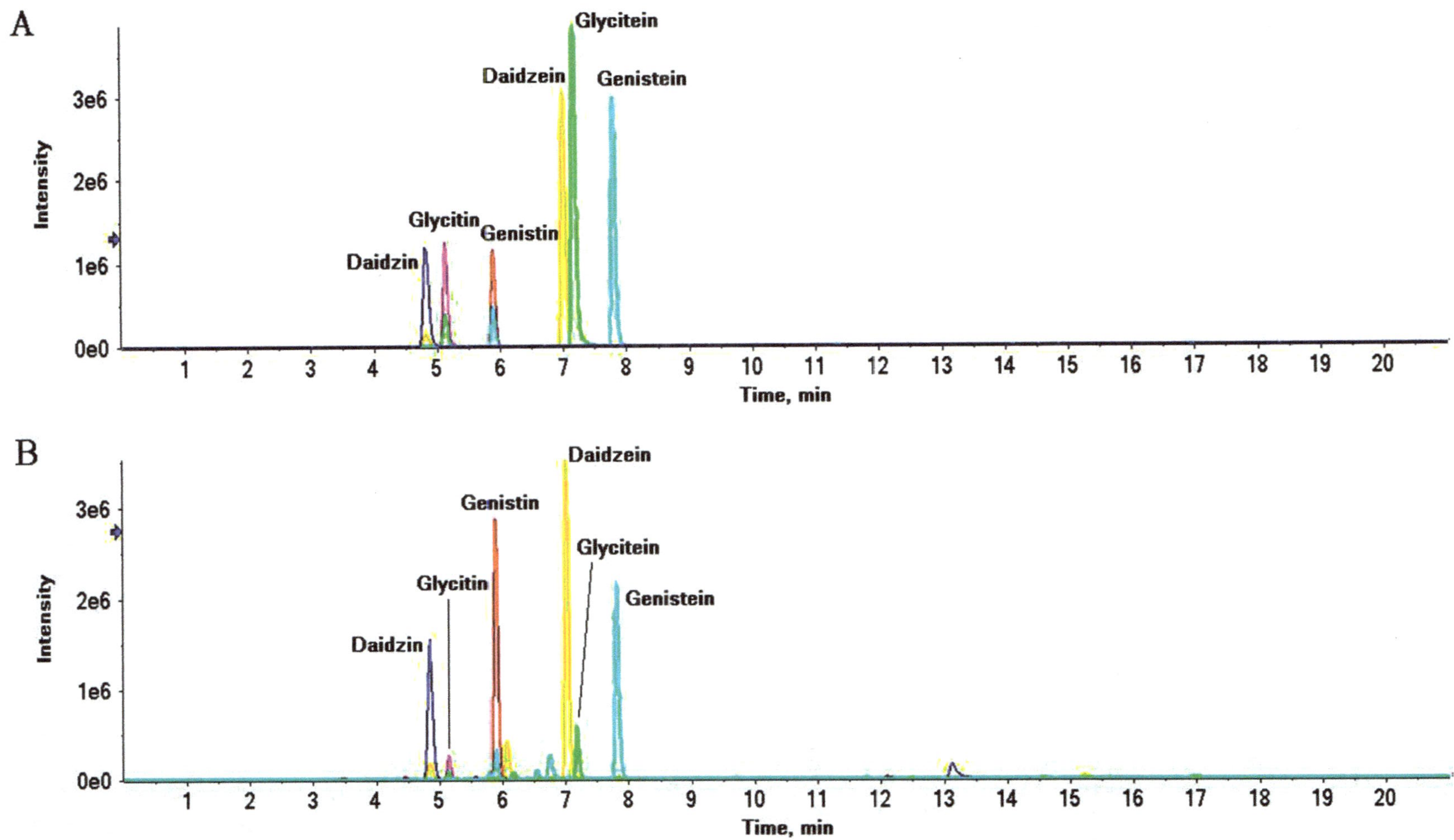

Figure 5. Representative extract ions chromatograms (XIC) of mix standard solution (**A**) and S8 (**B**) at *m/z* 417.118 ± 0.02 (daidzin), 447.129 ± 0.02 (glycitin), 433.113 ± 0.02 (genistin), 255.065 ± 0.02 (daidzein), 285.076 ± 0.02 (glycitein) and 271.060 ± 0.02 (genistein).

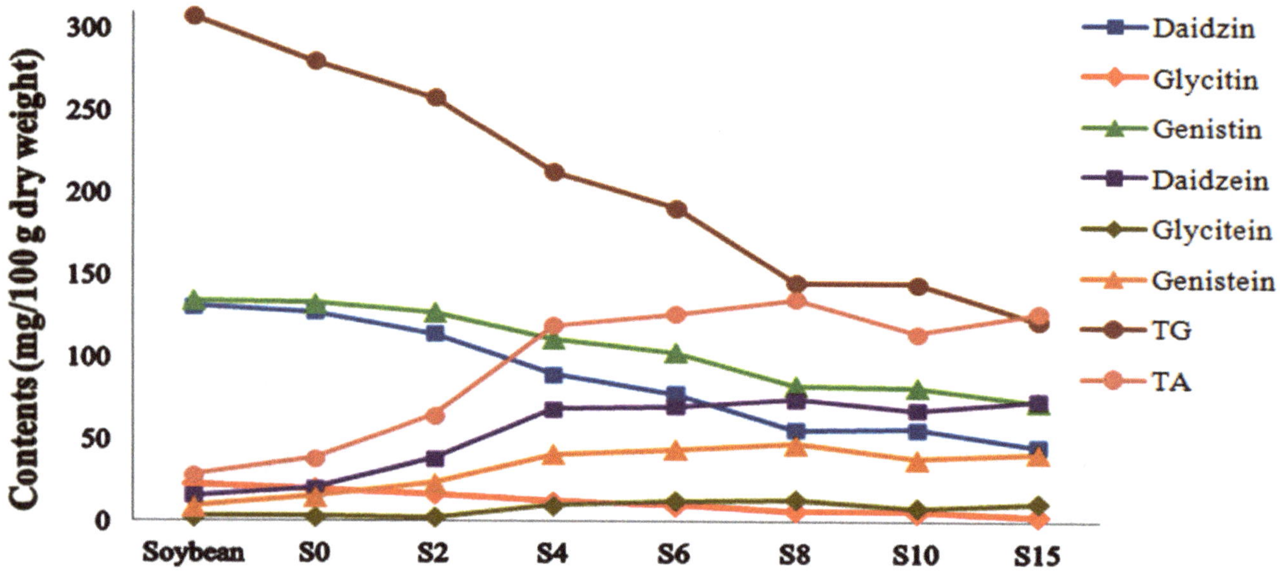

Figure 6. Content profile plots for six isoflavones, total glycoside (TG), and aglycone (TA) in SSP products collected at different fermentation stages.

Previous research has suggested that the increase in aglycone content and β−glucosidase activity during the fermentation of soybean show a similar trend [41]. As an attractive candidate to convert isoflavone glycosides to their aglycones, β−glucosidase reached its maximum activity on the eighth days of fermentation.

As aglycone possesses higher bioactivity and bioavailability compared to the β−glycosides isoflavone, the quantitative results of the representative constituents illustrated that eight days is the optimal time for the fermentation of SSP, which agrees with the morphologic changes (Figure 1). We reasoned that the bioactivities may be related to the variations in isoflavone content and β−glucosidase activity during SSP fermentation. Moreover, we created a perfect setup for SSP fermentation quality assessment and quality control. To determine the impact of bacterial or fungi on SSP fermentation, −further study is required.

3. Materials and Methods

3.1. Chemicals and Reagents

The reference standards of daidzin, glycitin, genistin, daidzein, glycitein and genistein were all purchased from Sigma−Aldrich (St. Louis, MO, USA)

Liquid chromatography (LC)/MS−grade acetonitrile, formic acid, methanol, and water were purchased from Merck Co. (Darmstadt, Germany).

Soybean, Artemisiae annuae herba and Mori folium used in the fermentation were purchased from YiFeng TCM shop (Nanjing, China) and authenticated by Associate Professor Jianwei Chen (Department of Pharmacy, Nanjing University of Chinese Medicine, Nanjing, Jiangsu, China).

3.2. SSP Fermentation

SSP was fermented in the laboratory and the preparation was performed as described in detail by the 2015 edition of China Pharmacopoeia as illustrated in Figure 7. The steps and parameters were as follows:

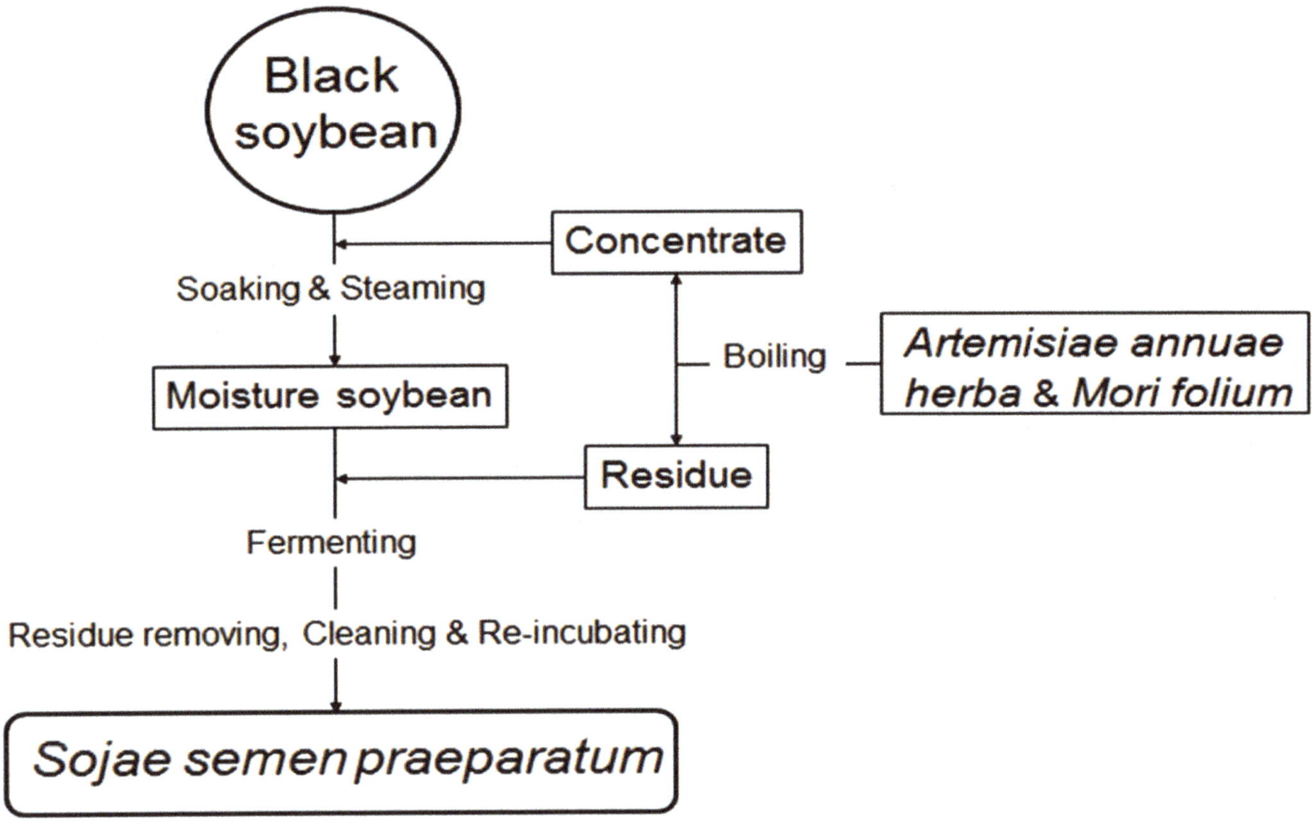

Figure 7. Flow diagram for fermentation of soybean to *Sojae semen praeparatum* (SSP).

Artemisiae annuae herba (100 g dry weight) and Mori folium (90 g dry weight) were washed and boiled with water (3600 mL) for 1 h in triplicate, and the decoction was concentrated to a relative density of 1.10–1.12 g/cm^3 concentrate. Black soybean (1000 g dry weight) was soaked in the concentrate overnight and steamed for 1.5 h while covered with wet cheesecloth. Aliquots (100 g wet weight) of the moisture soybean were placed on enamel trays covered with wet cheesecloth, which were covered with the residue of boiled Artemisiae annuae herba and Mori folium and incubated at 37 °C with 60–80% humidity. After fermenting for 0, 2, 4, 6, 8, 10, 15 days and removal of residues, wash cleaning, and re-incubating until a sweet smell drifting out, the SSP samples (S0, S2, S4, S6, S8, S10, S12, and S15) were then dried and pulverized into powder using an electric mill and sieved through 80 mesh sieves.

3.3. Sample Extraction

One gram of the powdered samples was accurately weighed and extracted with 25 mL of 75% methanol at 80 °C for 30 min using a Soxhlet extractor. This was followed by centrifugation for 15 min at 12000 rpm. The extraction supernatants were then diluted 10 times and filtered through a 0.45 μm filter unit.

3.4. Standard Solutions Preparation

Mix standard solutions were prepared by accurately weighing the standard substances and mixing them in 75% methanol. This standard mixture was filtered through a 0.45 μm filter unit.

3.5. LC-MS Spectrometric Conditions

An ultra-fast liquid chromatography system (Shimadzu Corporation UFLC XR; Kyoto, Japan) was connected to a triple time–of–flight mass spectrometer (TripleTOF 5600 system, AB SCIEX, Los Angeles, CA, USA) with an electrospray ionization source.

All samples were separated by an ACQUITY UPLC BEH C_{18} column (2.1 mm × 100 mm, 1.7 μm, Waters Corp., Milford, MA, USA). A binary solvent gradient consisting of solvent A (water with 0.2% formic acid) and solvent B (acetonitrile with 0.2% formic acid) was used. The flow rate was 300 μL/min. The total run time was 21 min with a gradient as follows: 0–3 min, 10–16% B; 3–7 min, 16–50% B; 7–12 min, 50–80% B; 12–15 min, 80–90% B; 15–17 min, 90–10% B; and 17–21 min, 10% B for column equilibration before the next run. An injection volume of 1 μL was used. The column temperature was 40 °C.

The samples were analyzed by acquiring full scan MS data in both positive and negative ion modes. The automatic data-dependent information product ion spectra (IDA−MS/MS) without any predefinition of the ions was checked. A calibrated delivery system was used to ensure the accuracy error of masses less than 1 ppm. The settings were nitrogen gas for nebulization at 55 psi, heater gas pressure at 55 psi, curtain gas at 35 psi, temperature of 500 °C, and ion spray voltage at 5500 V in positive ion mode, and −4500 V in negative ion mode. The acquisition of a survey tripleTOF MS spectrum was operated under high-resolution settings. The optimized declustering potential (DP) and collision energy (CE) were set at 80 eV and 15 eV in positive ion mode, and to −80 eV and −15 eV in negative ion mode, respectively. A sweeping collision energy setting at 35/−35 eV ± 15 eV was applied for collision-induced dissociation (CID).

3.6. *Method Validation*

The method was fully validated in accordance with guidelines on linearity, precision, recovery, detection limit, quantification limit, and stability. Calibration curves were generated by plotting peak area against the concentration of standard solutions. The intra-day precision was examined for six replicates of injections with the mixed standard solutions in one day, and the inter-precision was determined by injection in duplicates over three consecutive days. All the results are expressed using the relative standard deviation (RSD). The LOD and LOQ were calculated based on the peak−to−noise ratios of 3:1 and 10:1, respectively. The repeat, recovery, and stability were tested on the analytes in S8, and the repeatability was analyzed on six sample solutions from the same sample in parallel. The recovery was used to evaluate the accuracy at different spiking concentration levels (80%, 100%, and 120% as compared to the nominal concentration) of standard solutions. and the sample stability was tested by periodic analysis at room temperature for various periods (0, 2, 4, 8, 12, 16, 20, and 24 h).

3.7. *Data Processing*

TOF−MS data were collected using Analyst® version 1.6 software (AB SCIEX, Los Angeles, CA, USA) and processed using PeakView® version 1.2 software (AB SCIEX, Los Angeles, CA, USA) with the XIC Manager (AB SCIEX, Los Angeles, CA, USA) add-in and MultiQuant™ version 2.1 software (AB SCIEX, Los Angeles, CA, USA).

The PeakView® software contained a simple fragment ion predictor to help link the MS/MS spectrum to structures (saved as .mol files) and to provide insights into fragmentation mechanisms.

XIC Manager was used for targeted and non-targeted data processing, which consisted of a table for defining a list of masses or formulae to generate extracted ion chromatograms (XIC), and to review the identification of detected compounds. Our high confidence in results is based on retention times, accurate mass, isotopic pattern and MS/MS library searching.

The PCA was performed using MarkerView® software, where three repeated spectra for each sample were imported and analyzed with Pareto scaling. The T−value and corresponding P−value were calculated between each group and all the other eight groups. The program was linked back to the raw data so that differences could be directly visualized in spectra or chromatograms. The converted components were putatively identified by PeakView® software.

4. Conclusions

This was the first systematic comparison of active ingredients in the raw materials and processed products obtained during *Sojae semen praeparatum* (SSP) fermentation. Simultaneous characterization and quantification were performed using ultra-fast liquid chromatography (UFLC)−TripleTOF MS combined with XIC manager. The quantitative results verified that the components converted during the SSP fermentation, and we identified 45 components in positive ion mode and negative ion mode, in which 25 were putatively identified and a high proportion was isoflavone. *N*-(3-Indolylacetyl)-DL-aspartic acid, methylarginine, dimorphecolic acid, sorbitol, gheddic acid, and nonadecanoic acid were presumed to be introduced from processing adjuvants or produced during the fermentation process. The relative contents of raffinose and stachyose decreased gradually, while the fatty acids and isoflavone aglycones increased, which indicated that fermentation promotes the absorption of soybean nutrition and lipids degradation. The accurate quantitation of isoflavone, the representative constituents in soybean, revealed that fermentation for eight days produced a marked increase in the content of aglycone, the bioactive isoflavone, and a significant reduction in the content of β−glycosides isoflavone compared with unfermented soybean. This illustrated that eight days is the optimal time for the fermentation of SSP from the aspects of content, in agreement with the morphologic changes. We reasoned that the bioactivities of SSP might be related to isoflavone. Our study has provided some technological support for the optimization and quality control of SSP fermentation.

Author Contributions: Conceptualization, C.C., X.C. and H.W.; data curation, C.C. and C.S.; formal analysis, C.C. and C.S.; funding acquisition, H.W.; methodology, C.C.; resources, S.Y. and H.W.; software, C.C. and X.W.; supervision, X.C. and S.Y.; validation, C.C.; writing—original draft, C.C.; writing—review & editing, H.W.

References

1. Choi, M.-Y.; Chai, C.; Park, J.H.; Lim, J.; Lee, J.; Kwon, S.W. Effects of storage period and heat treatment on phenolic compound composition in dried citrus peels (chenpi) and discrimination of chenpi with different storage periods through targeted metabolomic study using hplc-dad analysis. *J. Pharm. Biomed. Anal.* **2011**, *54*, 638–645. [CrossRef]
2. Mortera, P.; Zuljan, F.A.; Magni, C.; Bortolato, S.A.; Alarcon, S.H. Multivariate analysis of organic acids in fermented food from reversed-phase high-performance liquid chromatography data. *Talanta* **2018**, *178*, 15–23. [CrossRef] [PubMed]
3. Chai, C.; Ju, H.K.; Kim, S.C.; Park, J.H.; Lim, J.; Kwon, S.W.; Lee, J. Determination of bioactive compounds in fermented soybean products using gc/ms and further investigation of correlation of their bioactivities. *J. Chromatogr. B* **2012**, *880*, 42–49. [CrossRef] [PubMed]
4. Zhu, H.; Bi, K.; Han, F.; Guan, J.; Zhang, X.; Mao, X.; Zhao, L.; Li, Q.; Hou, X.; Yin, R. Identification of the absorbed components and metabolites of zhi-zi-da-huang decoction in rat plasma by ultra-high performance liquid chromatography coupled with quadrupole-time-of-flight mass spectrometry. *J. Pharm. Biomed. Anal.* **2015**, *111*, 277–287. [CrossRef]
5. Zhang, P.; Zhang, P.; Xie, M.; An, F.; Qiu, B.; Wu, R. Metaproteomics of microbiota in naturally fermented soybean paste, da-jiang. *J. Food Sci.* **2018**, *83*, 1342–1349. [CrossRef] [PubMed]
6. Kim, M.S.; Kim, B.; Park, H.; Ji, Y.; Holzapfel, W.; Kim, D.Y.; Hyun, C.K. Long-term fermented soybean paste improves metabolic parameters associated with non-alcoholic fatty liver disease and insulin resistance in high-fat diet-induced obese mice. *Biochem. Biophys. Res. Commun.* **2018**, *495*, 1744–1751. [CrossRef] [PubMed]
7. Ali, M.W.; Shahzad, R.; Bilal, S.; Adhikari, B.; Kim, I.D.; Lee, J.D.; Lee, I.J.; Kim, B.O.; Shin, D.H. Comparison of antioxidants potential, metabolites, and nutritional profiles of korean fermented soybean (cheonggukjang) with bacillus subtilis kctc 13241. *J. Food Sci. Technol.* **2018**, *55*, 2871–2880. [CrossRef]
8. Qu, L.; Fan, G.; Peng, J.; Mi, H. Isolation of six isoflavones from semen sojae praeparatum by preparative HPLC. *Fitoterapia* **2007**, *78*, 200–204. [CrossRef] [PubMed]
9. Chai, C.; Bai, Y.T.; Wen, H.M.; Song, L.H.; Tu, J.Y.; Shan, C.X. Determination of 3 kinds of isoflavone aglycones in semen sojae praeparatum by uplc. *Chin. J. Ethnomed. Ethnopharm.* **2012**, *12*, 2.

10. Guo, H.; Zhang, Z.; Yao, Y.; Liu, J.; Chang, R.; Liu, Z.; Hao, H.; Huang, T.; Wen, J.; Zhou, T. A new strategy for statistical analysis-based fingerprint establishment: Application to quality assessment of semen sojae praeparatum. *Food Chem.* **2018**, *258*, 189–198. [CrossRef] [PubMed]
11. Qiu, F.; Shi, L.; Wang, S.; Wu, S.; Wang, M. Simultaneous high-performance liquid chromatography with diode array detection and time-of-flight mass spectrometric confirmation of the ten bioactive compounds in semen sojae preparatum. *J. Sep. Sci.* **2018**, *41*, 3360–3371. [CrossRef]
12. Zhang, J.-H.; Tatsumi, E.; Ding, C.-H.; Li, L.-T. Angiotensin i-converting enzyme inhibitory peptides in douchi, a chinese traditional fermented soybean product. *Food Chem.* **2006**, *98*, 551–557. [CrossRef]
13. Tsai, Y.-H.; Kung, H.-F.; Chang, S.-C.; Lee, T.-M.; Wei, C.-I. Histamine formation by histamine-forming bacteria in douchi, a chinese traditional fermented soybean product. *Food Chem.* **2007**, *103*, 1305–1311. [CrossRef]
14. Chai, C.; Cui, X.; Shan, C.; Yu, S.; Wen, H. Contents variation analysis of free amino acids, nucleosides and nucleobases in semen sojae praeparatum fermentation using uflc-qtrap ms. *Biomed. Chromatogr.* **2017**, *31*, 1–11. [CrossRef]
15. Chai, C.; Yu, S.; Cui, X.B.; Zhang, A.H.; Zhu, D.; Shan, C.X.; Wen, H.M. Analysis of volatile components in semen sojae praepatum with automatic static headspace and gas chromatography-mass spectrometry. *Food Res. Dev.* **2013**, *34*, 4.
16. Suo, H.; Feng, X.; Zhu, K.; Wang, C.; Zhao, X.; Kan, J. Shuidouchi (fermented soybean) fermented in different vessels attenuates hcl/ethanol-induced gastric mucosal injury. *Molecules* **2015**, *20*, 19748–19763. [CrossRef]
17. Qu, K.; Zhao, L.; Luo, X.; Zhang, C.; Hou, P.; Bi, K.; Chen, X. An LC-MS method for simultaneous determination of five iridoids from zhi-zi-chi decoction in rat brain microdialysates and tissue homogenates: Towards an in depth study for its antidepressive activity. *J. Chromatogr. B Anal. Technol. Biomed. Life Sci.* **2014**, *965*, 206–215. [CrossRef]
18. Chen, J.; Cheng, Y.-Q.; Yamaki, K.; Li, L.-T. Anti-α-glucosidase activity of chinese traditionally fermented soybean (douchi). *Food Chem.* **2007**, *103*, 1091–1096. [CrossRef]
19. McClean, S.; Beggs, L.B.; Welch, R.W. Antimicrobial activity of antihypertensive food-derived peptides and selected alanine analogues. *Food Chem.* **2014**, *146*, 443–447. [CrossRef]
20. Chen, L.; Liu, Q.; Sun, Y.; Wang, P.; Zhang, L.; Wang, W.J.C.P. Screening and enzymatic activity analysis of dominant fermentive bacteria of sojae semen praeparatum from different production places. *China Pharma.* **2017**, *28*, 4359–4361.
21. Li, G.; Long, K.; Su, M.S.; Liang, Y.H.; Yang, A.J.; Xie, X.M. Preliminary study on dynamic change of microbial flora in fermentation process to 'yellow cladding' of sojoe semen praeparatum. *Chin. J. Exp. Tradit. Med. Formulae* **2014**, *20*, 139–142.
22. Blay, M.; Espinel, A.E.; Delgado, M.A.; Baiges, I.; Bladé, C.; Arola, L.; Salvadó, J. Isoflavone effect on gene expression profile and biomarkers of inflammation. *J. Pharm. Biomed. Anal.* **2010**, *51*, 382–390. [CrossRef]
23. Kao, T.H.; Chen, B.H. Effects of different carriers on the production of isoflavone powder from soybean cake. *Molecules* **2007**, *12*, 917–931. [CrossRef]
24. Saha, J.; Biswas, A.; Chhetri, A.; Sarkar, P.K. Response surface optimisation of antioxidant extraction from kinema, a bacillus-fermented soybean food. *Food Chem.* **2011**, *129*, 507–513. [CrossRef]
25. Wang, L.; Yin, L.; Li, D.; Zou, L.; Saito, M.; Tatsumi, E.; Li, L. Influences of processing and nacl supplementation on isoflavone contents and composition during douchi manufacturing. *Food Chem.* **2007**, *101*, 1247–1253. [CrossRef]
26. Hurtado-Fernández, E.; Gómez-Romero, M.; Carrasco-Pancorbo, A.; Fernández-Gutiérrez, A. Application and potential of capillary electroseparation methods to determine antioxidant phenolic compounds from plant food material. *J. Pharm. Biomed. Anal.* **2010**, *53*, 1130–1160. [CrossRef]
27. Gaya, P.; Medina, M.; Sanchez-Jimenez, A.; Landete, J.M. Phytoestrogen metabolism by adult human gut microbiota. *Molecules* **2016**, *21*, 1034. [CrossRef]
28. Chung, I.-M.; Seo, S.-H.; Ahn, J.-K.; Kim, S.-H. Effect of processing, fermentation, and aging treatment to content and profile of phenolic compounds in soybean seed, soy curd and soy paste. *Food Chem.* **2011**, *127*, 960–967. [CrossRef]
29. Yang, S.; Wang, L.; Yan, Q.; Jiang, Z.; Li, L. Hydrolysis of soybean isoflavone glycosides by a thermostable β-glucosidase from paecilomyces thermophila. *Food Chem.* **2009**, *115*, 1247–1252. [CrossRef]
30. Kaya, M.; Ito, J.; Kotaka, A.; Matsumura, K.; Bando, H.; Sahara, H.; Ogino, C.; Shibasaki, S.; Kuroda, K.;

Ueda, M.; et al. Isoflavone aglycones production from isoflavone glycosides by display of beta-glucosidase from aspergillus oryzae on yeast cell surface. *Appl. Microbiol. Biotechnol.* **2008**, *79*, 51–60. [CrossRef]

31. Zeng, X.; Su, W.; Zheng, Y.; Liu, H.; Li, P.; Zhang, W.; Liang, Y.; Bai, Y.; Peng, W.; Yao, H. UFLC-Q-TOF-MS/MS-based screening and identification of flavonoids and derived metabolites in human urine after oral administration of exocarpium citri grandis extract. *Molecules* **2018**, *23*, 895. [CrossRef] [PubMed]
32. The Pharmacopoeia Commission of the Ministry of Health of the People's Republic of China. *Pharmacopoeia of People's Republic of China; Part I*; Medical Science and Technology Press: Beijing, China, 2015; p. 328.
33. Huang, Y.; Zhang, H.; Ben, P.; Duan, Y.; Lu, M.; Li, Z.; Cui, Z. Characterization of a novel gh36 alpha-galactosidase from bacillus megaterium and its application in degradation of raffinose family oligosaccharides. *Int. J. Biol. Macromol.* **2018**, *108*, 98–104. [CrossRef]
34. Yu, F.X.; Chen, X.; Chen, Z.W.; Wei, X.J. Fatty acid analysis of edible oils. *Adv. Mater. Res.* **2014**, *962–965*, 1222–1225. [CrossRef]
35. Kuo, T.M.; Doehlert, D.C.; Crawford, C.G. Sugar metabolism in germinating soybean seeds. Evidence for the sorbitol pathway in soyabean axes. *Plant Physiol.* **1990**, *93*, 1514–1520. [CrossRef] [PubMed]
36. Wang, X.J.; Wang, Y.S.; Qiu, J. Study on chemical composition of folium mori. *Food Drug* **2007**, *30*, 1–3.
37. Li, S.; Li, S.; Liu, C.; Liu, C.; Zhang, Y. Extraction and isolation of potential anti-stroke compounds from flowers of pueraria lobata guided by in vitro pc12 cell model. *J. chromatogr. B Anal. Technol. Biomed. Life Sci.* **2017**, *1048*, 111–120. [CrossRef] [PubMed]
38. Yang, Y.J.; Yao, Y.S.; Qin, Y.C.; Qiu, J.; Li, J.G.; Li, J.; Gu, X. Investigation and analysis of main afn in soybean meal and fermented soybean meal. *Sci. Agric. Sin.* **2016**, *49*, 573–580.
39. Liu, Y.C.; Cao, Y.; Zhang, J.X.; Zhang, N.; Lin-Jie, B.I.; Liu, T.; Chen, J.J. The condition optimization of bacterial combinations for fermentation of bean and determination of the content of stachyose and raffinose. *Hubei Agric. Sci.* **2015**, *54*, 4007–4011.
40. Fu, T.; Qi, W.; Li, A.; Liu, J.; Liang, X.; Yun, T.; Wang, Y. Effects of different bacteria on fermentative degradation of stachyose and raffinose in soybean meal and detection technology. *J. Chin. Cereals Oils Assoc.* **2014**, *29*, 111–118.
41. Lee, I.H.; Chou, C.C. Distribution profiles of isoflavone isomers in black bean kojis prepared with various filamentous fungi. *J. Agric. Food Chem.* **2006**, *54*, 1309–1314. [CrossRef]

13

Morphology and Molecular Identification of Twelve Commercial Varieties of Kiwifruit

Qiaoli Xie [1], Hongbo Zhang [1], Fei Yan [2], Chunxia Yan [1], Shuguang Wei [1], Jianghua Lai [1], Yunpeng Wang [1,*] and Bao Zhang [1,*]

[1] School of Forensic Medicine, Xi'an Jiaotong University, 76 Yanta West Road, Xi'an 710061, China; sunshineqiaoer@xjtu.edu.cn (Q.X.); zhanghb@mail.xjtu.edu.cn (H.Z.); yanchunxia@mail.xjtu.edu.cn (C.Y.); weisg@mail.xjtu.edu.cn (S.W.); laijh1011@mail.xjtu.edu.cn (J.L.)

[2] School of Energy and Power Engineering, Chongqing University, 174 Shapingba Main Street, Chongqing 400030, China; yanfei0506@cqu.edu.cn

* Correspondence: wyp033@xjtu.edu.cn (Y.W.); zhangbao_814@mail.xjtu.edu.cn (B.Z.)

Abstract: The quality and safety of food are important guarantees for the health and legal rights of consumers. As an important special fruitcrop, there are frequently shoddy practices in the kiwifruit (*Actinidia chinensis*) market, which harms the interests of consumers. However, there is lack of rapid and accurate identification methods for commercial kiwifruit varieties. Here, twelve common commercial varieties of kiwifruit were morphologically discriminated. DNA barcodes of chloroplast regions *psbA-trnH*, *rbcL*, *matK*, *rpoB*, *rpoC1*, *ycf1b*, *trnL* and *rpl32_trnL*(*UAG*), the nuclear region *At103* and intergenic region *ITS2* were amplified. Divergences and phylogenetic trees were used to analyze the phylogenetic relationship of these twelve commercial kiwifruit varieties. The results showed that *matK*, *ITS2* and *rpl32_trnL*(*UAG*) can be utilized as molecular markers to identify CuiYu, JinYan, HuangJinGuo, ChuanHuangJin, HuaYou, YaTe, XuXiang and HongYang. This provides experimental and practical basis to scientifically resolve kiwifruit-related judicial disputes and legal trials.

Keywords: food safety; kiwifruit (*Actinidia chinensis*); molecular identification; phylogeny; DNA barcode

---

1. Introduction

The quality and safety of food are important guarantees for the health and legal rights of consumers. Kiwifruit (*Actinidia chinensis*), also called 'the king of fruits', is an important economical crop because of its exceedingly high content of ascorbic acid (vitamin C), dietary fiber, nutritional minerals compositions and other health beneficial metabolites [1]. China, the origin of kiwifruit, possesses the largest planted area of kiwifruit in the world. According to statistics, in 2016, the national kiwifruit cultivation area reached 0.365 million acres, and the output was 2.15 million tons (China Industry Report Network) [2] Since the commercial establishment of kiwifruit, its classification has been controversial. These kiwifruits, named with geographical indications, cannot represent the species of kiwifruit. Besides, many commercial varieties vary greatly in market demand and price due to differences in taste and nutritional value. Therefore, there are frequent problems in the kiwifruit industry, such as false labelling, and lack of origin confirmation and identification. In order to protect consumer rights, we are looking for ways to quickly and accurately identify the commercial varieties of kiwifruit in the market and scientifically resolve judicial disputes and legal trials. However, morphological-based identification methods have great difficulties for non-professionals, and the methods of omics or chemistry are complex, time consuming and susceptible to environmental factors [3,4].

The DNA barcoding technique is a quick and effective molecular marker technology for classification and identification of organisms by using a standard gene region [5]. Since Hebert et al. recommended mitochondrial *cox1* gene as a DNA barcode for animal species identification in 2003 [5], it has been widely and effectively applied in the classification, identification, and phylogenetic analysis of thousands of species [6,7]. DNA barcode technology is an ideal identification method because of its accurate identification and simple operation. To date, the Consortium for the Barcode of Life (CBOL) formally proposed chloroplast markers *rbcL* and *matK* as the core barcodes for plant species identification [8]. Noncoding intergenic spacer *psbA-trnH* barcode was used to identify species of medicinal pteridophytes and members of Dendrobium of Orchid [9,10]. *ITS2* was selected as a standard barcode for identifying medical plants [11]. Two chloroplast genome markers, coding *rpoB* and *rpoC1* were utilized to discriminate 92 species in 32 diverse genera of land plants [12]. *ycf1b* was reported as the most variable plastid genome region and can serve as a core barcode of land plants [13]. *trnL* was chosen as the barcoding gene for reference library constructing and high-throughput sequencing for wetland plants [14]. *At103* (Mgprotoporphyrin IX monomethyl ester cyclase) developed by Li et al. [15] as universally amplifiable marker for phylogenetic reconstructions and that together with *matK* to be used to distinguish toxic hybrids form parental species [16]. Fu et al. [17] found that *rpl32_trnL* (*UAG*) had a greater degree of variation and could be used as the core barcode sequence of cherry plants. These genes have the potential to be as powerful as mitochondrial CO1 gene in identifying species [10]. However, as far as we have been concerned, there have been no reports on DNA barcode for common kiwifruit commercial varieties in the market. Previous studies have provided us with reference to the feasibility of using DNA barcodes to identify kiwifruit commercial varieties. Lee et al. made use of *ITS2* to identify the varieties and provenances of Taiwan's domestic and imported made teas [18]. Enan and Ahmed using chloroplast DNA barcode *psbK-psbI* spacers for identification of Emirati date palm (*Phoenix dactylifera* L.) varieties (cultivar-level) [19]. Jaakola et al. successfully identified the blueberry varieties "Northcountry" and "Northblue" using DNA barcode technology combined with high-resolution dissolution profiles [20]. Through DNA barcodes, He et al. authenticated cultivars of *Angelica anomala* Ave'-Lall [21].

Single nucleotide polymorphism (SNP) is a single nucleotide variation with a specific and determined genetic location in at least 1% of the population [22]. SNP is one of the richest and most stable genetic polymorphisms in the genome, which is suitable for solving the differences between closely related species [23]. SNP typing has been successfully used to conveniently and accurately identify plant origin [24], medicinal plants [25], and bacteria [26].

In this study, we attempted to use the morphological method combined with molecular biological methods for rapid cultivar identification of twelve kiwifruit commercial varieties in China. SNP typing method was evaluated by using ten candidate DNA barcoding markers (chloroplast genome *psbA-trnH, rbcL, matK, rpoB, rpoC1, ycf1b, trnL* and *rpl32_trnL*(*UAG*), the nuclear region *At103* and intergenic region *ITS2*), which were used for molecular identification. This study will not only lay a foundation for phylogenetic analysis but also provides experimental and practical basis for the rapid identification of kiwifruit.

2. Results

2.1. Morphological Identification

In order to enable consumers to visually identify the commercial varieties of kiwifruit in the first place, by referring to the classification criteria of "Flora of China", the morphology of 12 commercial kiwifruit were analyzed. We mainly made statistics on fruit type, fruit shape, fruit size, peel color, peel spots, hair presence, hair length, hair softness and hardness, hair shedding, pulp color, pulp taste, seed color, seed number and shape, beak prominence, beak diameter, fruit picking time and so on. The results are shown in Figure 1 and Supplementary Table S1. The 72 samples of the 12 commercial varieties of kiwifruit were harvested from 150–160 days after pollination. Their fruits are all bacca.

The shape of the fruit is mostly cylindrical and spherical. HuangJinGuo and ChuanHuangJin are long oval. JinYan is cylindrical. HongYang is short cylindrical and CuiYu is oblate cone. Statistical analysis of fruit size showed that the fruit size of FengXianLou is the largest, followed by HuangJinGuo, CuiXiang, and HongYang is the smallest. For the peel, only CuiYu has no spots on the peel surface, other commercial varieties all have spots. The color of mostly commercial varieties' peel is brown. The fruit peel of HuangJinGuo, ChuanHuangJin and JinYan is yellower. The peel of HongYang is greenish. Hair analysis found that the skin of HuangJinGuo, ChuanHuangJin, CuiYu and HongYang have no hair, other commercial varieties all have hair. QinMei and HaiWoDe have very dense hair on the surface. HuaYou has few hair on the surface. The surface of FengXianLou has the longest hair, soft and easy to fall off. HuaYou's surface hair is the shortest and easy to fall off.

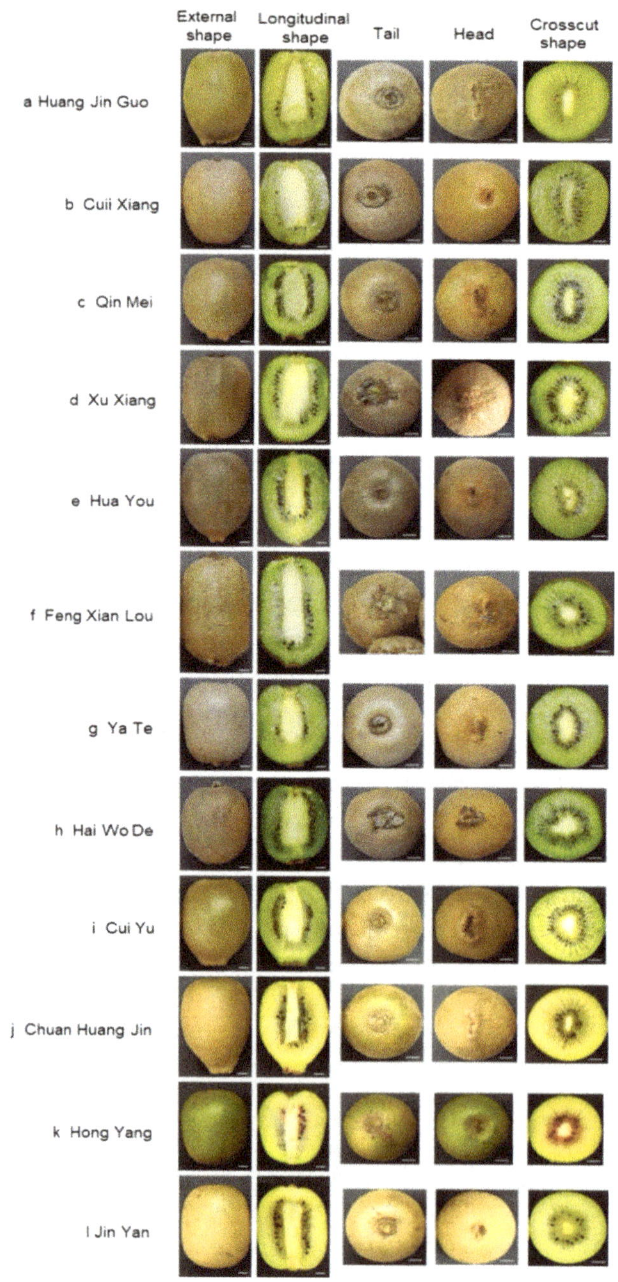

Figure 1. External and internal forms of twelve commercial kiwifruit. (**a**) HuangJinGuo, (**b**) CuiXiang, (**c**) QinMei, (**d**) XuXiang, (**e**) HuaYou, (**f**) FengXianLou, (**g**) YaTe, (**h**) HaiWoDe, (**i**) CuiYu, (**j**) ChuanHuangJin, (**k**) HongYang, (**l**) JinYan. Each variety contains external shape, longitudinal shape, head, tail, crosscut shape. Scale bar represents 6 mm.

All statistical results are in Supplementary Table S1. For the color of the flesh, the HongYang is the easiest to distinguish, which has a radial red color in the pulp center. The flesh color of HuangJinGuo, ChuanHuangJin and JinYan are yellow. The flesh color of HuaYou is yellow-green and other commercial varieties are all green. The fruit of HuangJinGuo, ChuanHuangJin, JinYan, CuiXiang, HuaYou, CuiYu are sweet, no sour taste, and other commercial varieties are sweet and sour. Most of the seeds are dark brown and flat oval. HaiWoDe has the largest number of seeds, and HuangJinGuo has the fewest seeds. Statistical results are in Supplementary Table S1. Kiwifruit has a beak at the top. Except CuiXiang and HongYang, the beaks of other commercial varieties are prominent. The longest beak exists on HuangJinGuo, followed by QinMei, and HongYang has the smallest beak.

The above analysis shows that HongYang is the most discernible variety by the peel color and flesh color. JinYan is easy to identify because of its yellow peel and flesh. The shape of CuiYu is similar to that of HuangJinGuo, while there are differences between them. The surface of CuiYu has no spots and its flesh is green. The surface of HuangJinGuo is spotted and its flesh is yellow. Therefore, CuiYu is easier to be distinguished. The hair of Huayou is short and small and its flesh is yellow-green, which made it easier to be discriminated. FengXianLou is a long cylindrical shape with many hairs easy to fall off makes it easy to be identified. It is not easy to distinguish HuangJinGuo and ChuanHuangJin. CuiXiang, QinMei, XuXiang, YaTe and HaiWoDe all have green pulp, but the shape of CuiXiang is slightly flat, and its beak is the smallest and not prominent. CuiXiang is sweet, while QinMei, XuXiang, YaTe and HaiWoDe are sour and sweet. YaTe's skin is white and brown compared with QinMei, XuXiang and HaiWoDe. QinMei is greener than XuXiang and HaiWoDe. For ordinary consumers, XuXiang and HaiWoDe are relatively difficult to be differentiated.

2.2. Analysis of Variable Sites in Different Commercial Varieties

The sequencing results showed that the product length of *rbcL*, *matK*, *psbA-trnH*, *ITS2*, *rpoB*, *rpoC1*, *trnL* and *rpl32_trnL(UAG)* are 743 bp, 889 bp, 502 bp, 491 bp, 512 bp, 529 bp, 193bp and 1010 bp, respectively. Sequences of *rbcL*, *psbA-trnH*, *rpoB*, *rpoC1* and *trnL* have no difference in these twelve kiwi commercial varieties (Supplementary Figure S1). *ycf1b* and *At103* get no results of amplification. The amplification efficiency of 72 *ITS2* sequences is 100%. Sequencing results showed that the sequences of 6 repetitive samples from each commercial variety are consistent and stable. There are two haplotypes in *ITS2* sequence. Bases at 115 bp, 132 bp and 310 bp of HuangJinGuo, ChuanHuangJin and HongYang are "C", that of the other nine commercial varieties are "T". Bases at 206 bp and 215 bp of HuangJinGuo, ChuanHuangJin and HongYang are "G", that of the other nine commercial varieties are "A" (Figure 2a, Table 1 and Supplementary Table S2).

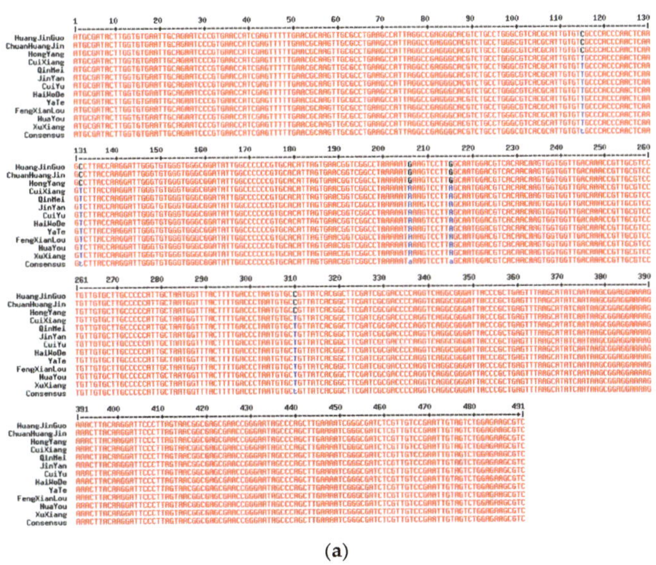

(a)

Figure 2. *Cont.*

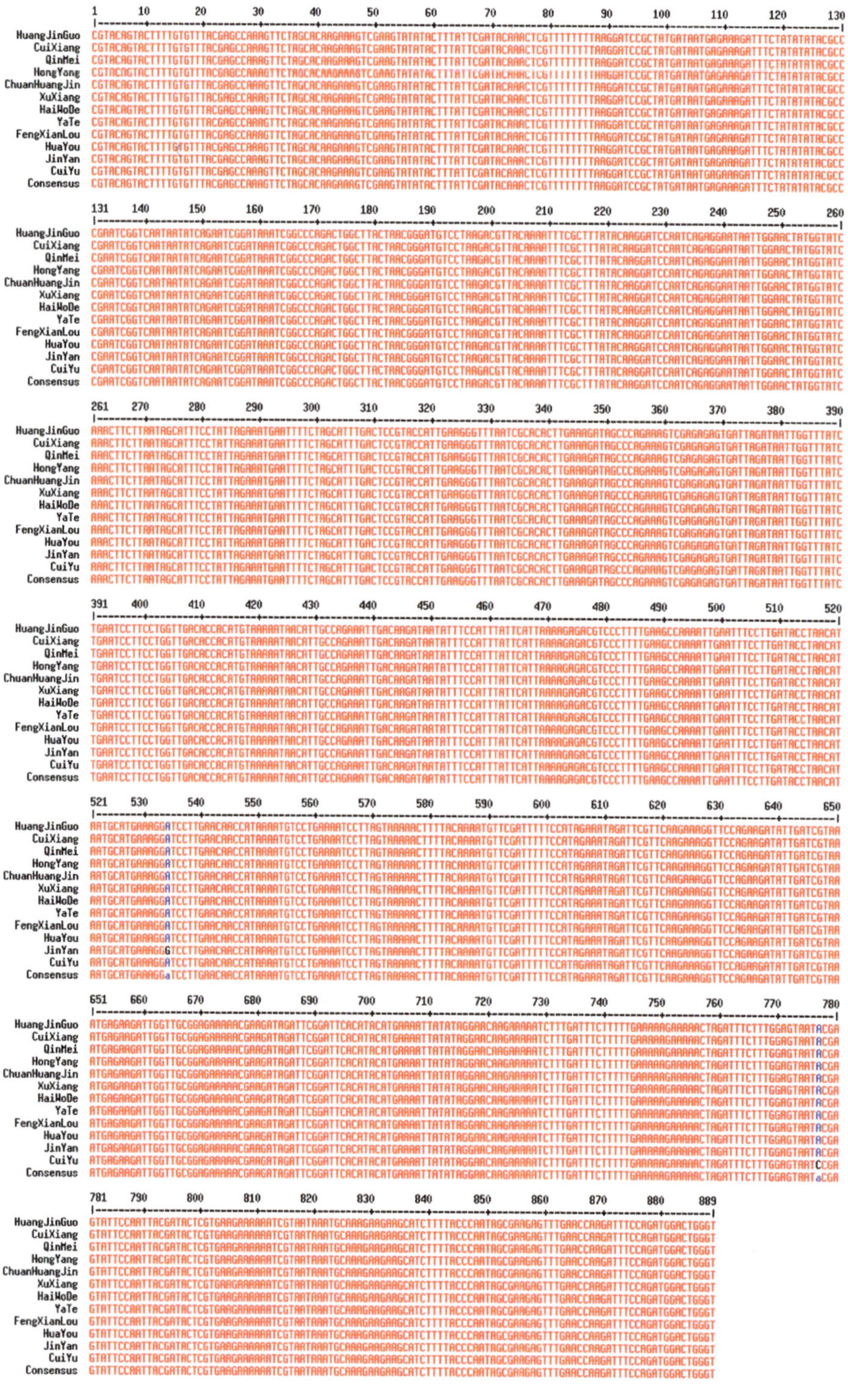

(b)

Figure 2. *Cont.*

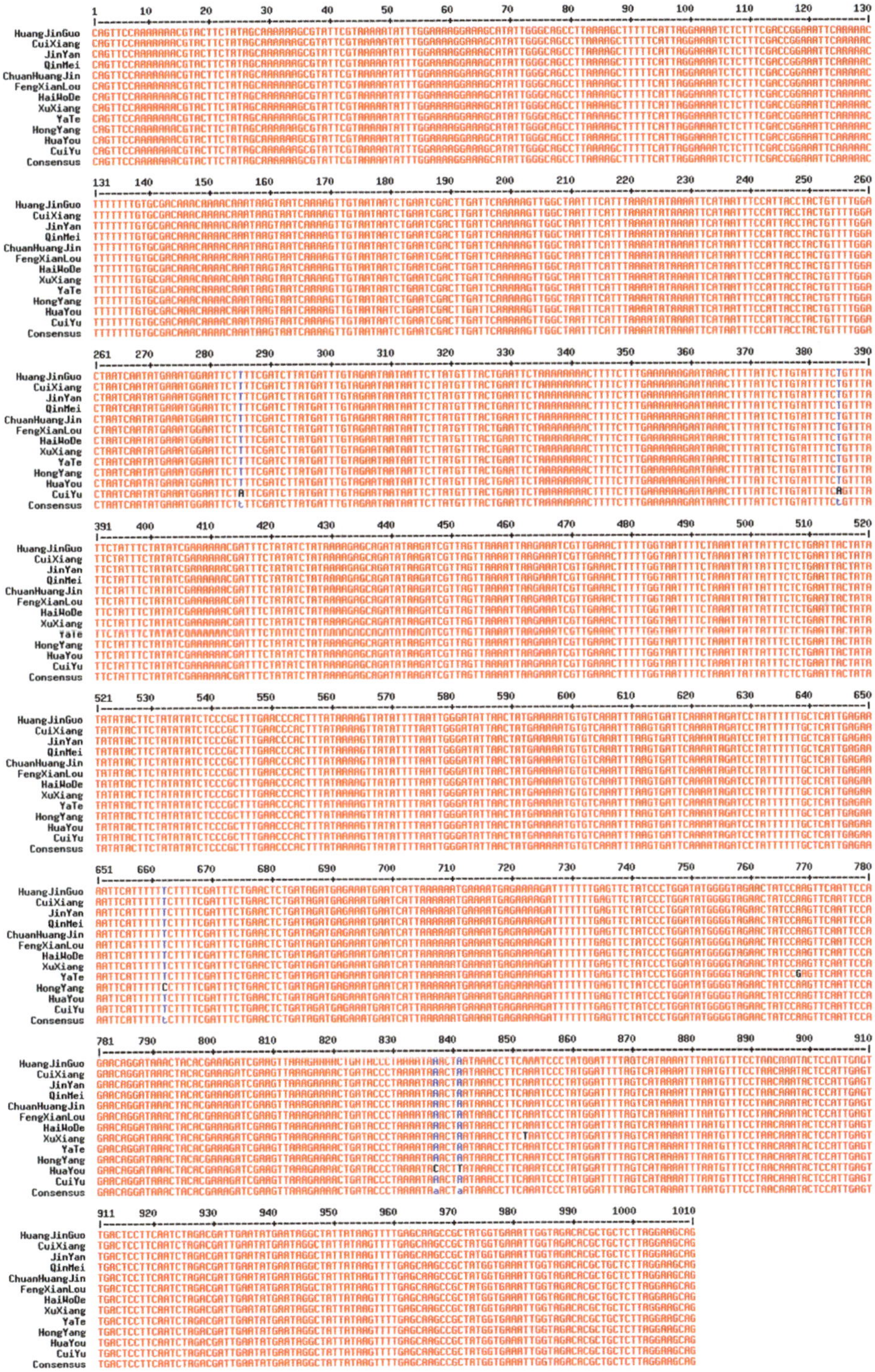

(c)

Figure 2. DNA barcode sequence alignment of 12 commercial kiwifruits. (**a**) *ITS2* sequence alignment 491 bp; (**b**) *matK* sequence alignment 889 bp; (**c**) *rpl32_trnL(UAG)* sequence alignment 1010 bp.

Table 1. SNP site information for two DNA barcodes.

Variety	SNP Site Information (bp)													
	ITS2					*matK*		*rpl32*						
	115	132	206	215	310	534	777	285	385	662	768	837	841	852
HuangJinGuo	C	C	G	G	C	A	A	T	T	T	A	A	A	A
ChuanHuangJin	C	C	G	G	C	A	A	T	T	T	A	A	A	A
HongYang	C	C	G	G	C	A	A	T	T	C	A	A	A	A
CuiiXiang	T	T	A	A	T	A	A	T	T	T	A	A	A	A
QinMei	T	T	A	A	T	A	A	T	T	T	A	A	A	A
JinYan	T	T	A	A	T	G	A	T	T	T	A	A	A	A
CuiYu	T	T	A	A	T	A	C	A	A	T	A	A	A	A
HaiWoDe	T	T	A	A	T	A	A	T	T	T	A	A	A	A
YaTe	T	T	A	A	T	A	A	T	T	T	G	A	A	A
FengXianLou	T	T	A	A	T	A	A	T	T	T	A	A	A	A
HuaYou	T	T	A	A	T	A	A	T	T	T	A	C	T	A
XuXiang	T	T	A	A	T	A	A	T	T	T	A	A	A	T

For *matK*, the amplification efficiency of 72 *matK* sequences is 100%. Sequencing results showed that the sequences of 6 repetitive samples of each commercial variety are consistent and stable. There are three haplotypes in *matK* sequences. Bases at 534 bp of JinYan are "G", which of the other eleven commercial varieties are "A", and bases at 777 bp of CuiYu are "C", the other eleven commercial varieties are "A" (Figure 2b, Table 1 and Supplementary Table S3). The amplification efficiency of 72 *rpl32_trnL*(*UAG*) sequences is 100%. There are six haplotypes in *rpl32_trnL*(*UAG*) sequence. Bases at 285 bp and 385 bp of CuiYu are "A", that of the other 11 commercial varieties are "T". Bases at 662 bp of HongYang are "C", that of the other 11 commercial varieties are "T". Bases at 768 bp of YaTe are "G", that of the other 11 commercial varieties are "A". Bases at 837 bp and 841 bp of HuaYou are "C" and "T", respectively, that of the other 10 commercial varieties are "A". Bases at 852 bp of XuXiang are "T", that of the other 11 commercial varieties are "A". (Figure 2c, Table 1 and Supplementary Table S4).

2.3. K2P Genetic Distance Analysis

Genetic distances of all samples were calculated by using the MEGA 7.0 [27] software (Pennsylvania State University, USA). The results of *ITS2*, *matK*, *rpl32_trnL*(*UAG*) and *ITS2*+*matK*+*rpl32_trnL*(*UAG*) are shown in Supplementary Table S6. For *ITS2*, there is no difference among HuangJinGuo, ChuanHuangJin and HongYang. The genetic distance between them and nine other commercial varieties is 0.01. There is no difference among the nine commercial varieties (Supplementary Table S5a). For *matK*, the genetic distance between JinYan and CuiYu is 0.002, and between JinYan and other 10 commercial varieties is 0.001. The genetic distance between CuiYu and other 10 cultivars is 0.001, and there is no difference among the other 10 cultivars (Supplementary Table S5b). For *rpl32_trnL*(*UAG*), the genetic distance between HuaYou and HongYang, YaTe, XuXiang is 0.003. The genetic distance between HuaYou and JinYan, CuiXiang, FengXianLou, ChuanHuangJin, HaiWoDe, QinMei and HuangJinGuo is 0.002. The genetic distance between HongYang and CuiYu is 0.003. The genetic distance between HongYang and YaTe, XuXiang is 0.002. The genetic distance between HongYang and JinYan, CuiXiang, FengXianLou, ChuanHuangJin, HaiWoDe, QinMei and HuangJinGuo is 0.001. The genetic distance between CuiYu and YaTe, XuXiang is 0.003. The genetic distance between CuiYu and JinYan, CuiXiang, FengXianLou, ChuanHuangJin, HaiWoDe, QinMei and HuangJinGuo is 0.001. The genetic distance between JinYan and YaTe, XuXiang is 0.001. There is no difference between JinYan and CuiXiang, FengXianLou, ChuanHuangJin, HaiWoDe, QinMei and HuangJinGuo. The genetic distance between CuiXiang and YaTe, XuXiang is 0.001. There is no difference between CuiXiang and FengXianLou, ChuanHuangJin, HaiWoDe, QinMei and

HuangJinGuo. The genetic distance between YaTe and XuXiang is 0.002. There is no difference between YaTe and FengXianLou, ChuanHuangJin, HaiWoDe, QinMei and HuangJinGuo. The genetic distance between FengXianLou and XuXiang is 0.001. There is no difference between FengXianLou and ChuanHuangJin, HaiWoDe, QinMei and HuangJinGuo. The genetic distance between ChuanHuangJin and XuXiang is 0.001. There is no difference between ChuanHuangJin and HaiWoDe, QinMei and HuangJinGuo. The genetic distance between HaiWoDe and XuXiang is 0.001. There is no difference between HaiWoDe and QinMei and HuangJinGuo. The genetic distance between QinMei and XuXiang is 0.001. There is no difference between QinMei and HuangJinGuo. The genetic distance between HuangJinGuo and XuXiang is 0.001 (Supplementary Table S5c). For *ITS2+matK+rpl32_trnL(UAG)*, the genetic distance between YaTe and JinYan is 0.001. The genetic distance between XuXiang and YaTe, JinYan is 0.001. There is no difference between QinMei and YaTe, JinYan, XuXiang. There is no difference between HaiWoDe and YaTe, JinYan, XuXiang, QinMei. There is no difference between CuiXiang and YaTe, JinYan, XuXiang, QinMei, HaiWoDe. The genetic distance between HuaYou and YaTe, JinYan, XuXiang, QinMei, HaiWoDe, CuiXiang is 0.001. There is no difference between FengXianLou and YaTe, JinYan, XuXiang, QinMei, HaiWoDe and CuiXiang. The genetic distance between FengXianLou and HuaYou is 0.001. The genetic distance between CuiYu and YaTe, JinYan, XuXiang, HuaYou is 0.002. The genetic distance between CuiYu and QinMei, HaiWoDe, CuiXiang, FengXianLou is 0.001. The genetic distance between HuangJinGuo and YaTe, JinYan, XuXiang, HuaYou, CuiYu is 0.003. The genetic distance between HuangJinGuo and QinMei, HaiWoDe and CuiXiang, FengXianLou is 0.002. The genetic distance between HongYang and YaTe, JinYan, XuXiang, QinMei, HaiWoDe, CuiXiang, HuaYou, FengXianLou is 0.003. The genetic distance between HongYang and CuiYu is 0.004. There is no difference between HongYang and CuiYu. The genetic distance between ChuanHuangJin and YaTe, JinYan, XuXiang, HuaYou, CuiYu is 0.003. The genetic distance between ChuanHuangJin and QinMei, HaiWoDe and CuiXiang, FengXianLou is 0.002. There is no difference between ChuanHuangJin and HuangJinGuo, HongYang (Supplementary Table S5d).

2.4. Phylogenetic Analysis

ITS2, *matK* and *rpl32_trnL(UAG)* sequences and *ITS2+matK+rpl32_trnL(UAG)* combination sequences were tested 1000 times by bootstrap method to build NJ phylogenetic tree (Figure 3 and Supplementary Figure S2). Phylogenetic analysis of *ITS2* gene sequences (Figure 3a and Supplementary Figure S2a) showed that HuangJinGuo, ChuanHuangJin and HongYang were clustered into one group, and the other nine commercial varieties were grouped into another category. For the *matK* gene sequences, JinYan and CuiYu were separately distinguished, and the other ten commercial varieties were clustered into one category (Figure 3b and Supplementary Figure S2b). Figure 3c and Supplementary Figure S2c indicated that, using *rpl32_trnL(UAG)*, HuaYou, YaTe, XuXiang, HongYang, CuiYu could be distinguished from the other seven commercial varieties, respectively.

For *ITS2+matK+rpl32_trnL(UAG)*, CuiYu, JinYan, YaTe, XuXiang, HuaYou, HongYang, HuangJinGuo and ChuanHuangJin can be distinguished from the other four commercial varieties, respectively(Figure 3d).

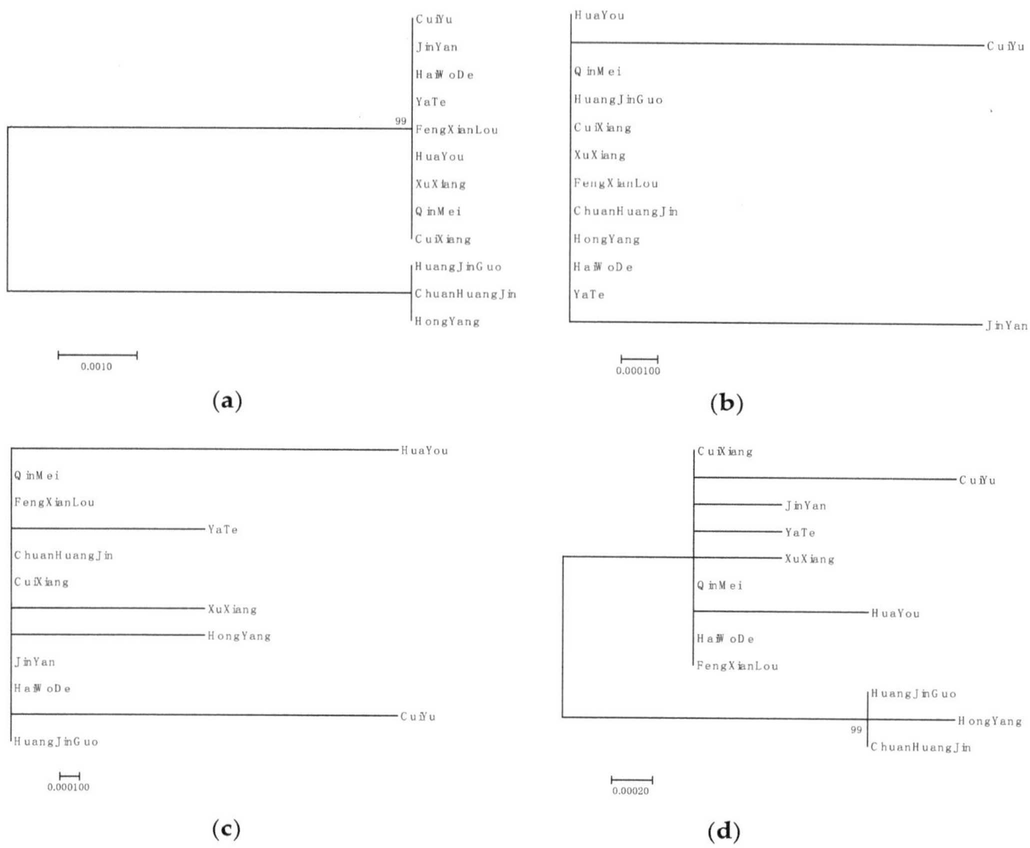

Figure 3. Phylogenetic analysis of 12 kiwifruit commercial varieties. (**a**) Analysis of *ITS2* sequence fragments; (**b**) Analysis of *matK* sequence fragments; (**c**) Analysis of *rpl32_trnL*(*UAG*) sequence fragments; (**d**) Analysis of *ITS2+matK+rpl32_trnL*(*UAG*) sequence fragments.

3. Discussion

The accurate classification of kiwifruit is one of the problems in the current kiwifruit plant resource utilization and market regulation. It is very difficult to classify kiwifruit only by morphology, because many shapes of kiwifruit are very close that requires the participation of professionals. If the suitable DNA barcode can be screened for kiwifruit, it is very meaningful to classification and identification, resource protection, variety selection and market regulation of kiwifruit.

In this study, 72 samples of 12 commercial varieties were collected in the market. We firstly conducted a morphological analysis of twelve kiwifruit commercial varieties that are easily confused in the market. Suggestions for morphological identification are proposed. By referring to the classification criteria of "Flora of China (http://foc.iplant.cn/)", the morphology of 12 commercial kiwifruit was analyzed in Supplementary Table S1. We mainly made statistics on fruit type, fruit shape, fruit size, peel color, peel spots, hair presence, hair length, hair softness and hardness, hair shedding, pulp color, pulp taste, seed color, seed number and shape, beak prominence, beak diameter, fruit picking time (Supplementary Table S1). The analysis results showed that HongYang is most discernible by the peel color and flesh color. Therefore, the expensive HongYang, whose price is 4 to 5 times than that of QinMei, HuaYou, YaTe and HaiWoDe in the market, can be identified morphologically. JinYan whose price is 3 to 4 times than that of QinMei, HuaYou, YaTe and HaiWoDe in the market, is easy to identify because of its yellow peel and flesh. The shape of CuiYu is similar to that of HuangJinGuo, while there are differences between them. The surface of HuangJinGuo is spotted and its flesh is yellow. Therefore, CuiYu is easier to be distinguished. The hair of Huayou is short and small and its flesh is yellow-green, which made it easier to be discriminated. FengXianLou is a long cylindrical shape with many hairs easy to fall off makes it easy to be identified. It is not easy to distinguish HuangJinGuo and ChuanHuangJin. CuiXiang, QinMei, XuXiang, YaTe and HaiWoDe all have green pulp, but the shape of CuiXiang is slightly flat, and its beak is the smallest and not prominent. CuiXiang is sweet, while QinMei, XuXiang, YaTe and HaiWoDe are sour and sweet. YaTe's skin is white and brown compared with QinMei, XuXiang and HaiWoDe. QinMei is greener than XuXiang and HaiWoDe. For ordinary consumers, XuXiang and HaiWoDe are relatively difficult to be differentiated. As the identification of morphology is very demanding on the professional, there are limitations in the identification of morphology.

Recently, DNA barcodes are often used as a standard for identifying plant species [28,29]. The DNA region, as an effective barcode, should contain sufficient variability for identification, contain conserved regions for the development of universal primers and be short enough to be sequenced in one reaction. In animals, a portion of mitochondrial cytochrome C oxidation enzyme I gene sequence is often used as a general barcode and is also used in forensic identification [16,28]. Currently, there is no common area available in plants. In the nuclear and plastid genomes, there are multiple loci were selected as DNA barcodes in plants [12,30–32].

Ten DNA barcoding *rbcL*, *matK*, *psbA-trnH*, *rpoB*, *rpoC1*, *ITS2*, *ycf1b*, *trnL*, *rpl32_trnL*(*UAG*) and *At103* were explored to quickly and accurately identify twelve kiwifruit commercial varieties in this study. The primer sequences used for amplification of these ten regions, the length of the products, and the amplification procedures are listed in Table 2. The *psbA-trnH* sequence of the intergenic region is one of the most mutated non-coding regions in the plant chloroplast genome and is often used as a barcode for material identification [32–34]. However, the results of this experiment found that the *psbA-trnH* sequences of the 12 commercial kiwifruits are identical, which means that the *psbA-trnH* is not suitable for identifying the 12 commercial kiwifruits. Intergenic region *ITS2* (internal transcribed spacer 2) and the cp gene *rbcL* may have potential as universal plant barcodes [34]. Three regions *matK*, *rpoB* and *rpoC1* were outlined as viable markers for land plant barcoding [33]. CBOL formally proposed chloroplast markers *rbcL* and *matK* as the core barcodes for plant species identification [8]. The results, in this study, showed that the *rbcL*, *rpoB* and *rpoC1* sequences of the 12 commercial kiwifruits are unanimous, which indicates the three barcodes sequences are improper for the identification of the 12 commercial kiwifruits. *ycf1b* was reported as the most variable plastid genome region and can serve as a core barcode of land plants [13]. *At103* together with *matK* to be used to distinguish toxic hybrids form parental species [16]. Our results showed that these two DNA barcodes could not be amplified in the 12 commercial varieties of kiwifruit, indicating they are not suitable to be DNA barcodes of kiwifruit. *trnL* was chosed as the barcoding gene for reference library constructing for wetland plants [14], whose amplification results were indistinguishable among the 12 commercial varieties of kiwifruit. Therefore, *trnL* could not be used as the barcode to discriminate these 12 commercial varieties of kiwifruit.

There are two SNP sites in the *matK* sequences of 12 commercial kiwifruits. For *matK*, bases at 534 bp of Jin Yan are "G", that of the other eleven commercial varieties are "A", and bases at 777 bp of Cui Yu are "C", that of the other eleven commercial varieties are "A" (Figure 2b, Table 1 and Supplementary Table S3). It can be applied to distinguish Jin Yan, CuiYu and other 10 commercial varieties. Genetic distance analysis discovered that the genetic distance between JinYan and CuiYu is 0.002, and between JinYan and 10 other commercial varieties is 0.001. The genetic distance between CuiYu and other 10 cultivars is 0.001, and there is no difference among the other 10 cultivars (Supplementary Table S5b). The phylogenetic analysis also found that Jin Yan, CuiYu and other ten commercial varieties is clustered into three branches. JinYan and CuiYu is separately distinguished from other ten commercial varieties, respectively (Figure 3b).

Table 2. Primers and amplification procedures used for PCR in this study.

Gene Amplification Region	Primer Name	Sequence (5′ →3′)	Length of Amplified Fragment	Amplification Procedure
*rbc*L	*rbc*L-F *rbc*L-R	ATGTCACCACAAACAGA TCGCATGTACCTGCAGTA	743 bp	94 °C 5 min; [94 °C 30S, 54.5 °C 30S, 72 °C 45S]* 35 cycles; 72 °C 10 min; 4 °C 10 min
*mat*K	*mat*K-F *mat*K-R	CGTACAGTACTTTTGTGTTTAC ACCCAGTCCATCTGGAAATC	889 bp	94 °C 5 min; [94 °C 30S, 56 °C 30S, 72 °C 55S]* 35 cycles; 72 °C 10 min; 4 °C 10 min
*psb*A-*trn*H	*psbA-trnH*-F *psbA-trnH*-R	TATGCATGAACGTAATGCT GCATGGTGGATTCACAAT	502 bp	94 °C 5 min; [94 °C 30S, 55 °C 30S, 72 °C 30S]* 35 cycles; 72 °C 10 min; 4 °C 10 min
ITS2	*ITS2*-F *ITS2*-R	ATGCGATACTTGGTGTG GACGCTTCTCCAGACTA	491 bp	94 °C 5 min; [94 °C 30S, 55 °C 30S, 72 °C 30S]* 35 cycles; 72 °C 10 min; 4 °C 10 min
rpoB	*rpoB*-F *rpoB*-R	AAGTGCATTGTTGGAACTGG CCCAGCATCACAATTCC	512 bp	94 °C 5 min; [94 °C 30S, 55 °C 30S, 72 °C 35S]* 35 cycles; 72 °C 10 min; 4 °C 10 min
rpoC1	*rpoC1*-F *rpoC1*-R	CAAAGAGGGAAGAT TAAGCATATCTTGAGT	529 bp	94 °C 5 min; [94 °C 30S, 54.5 °C 30S, 72 °C 40S]* 35 cycles; 72 °C 10 min; 4 °C 10 min
rpl32_trnL(UAG)	*rpl32* *trnL(UAG)*	CAGTTCCAAAAAAACGTACTTC CTGCTTCCTAAGAGCAGCGT	1010 bp	94 °C 5 min; [94 °C 30S, 57.8 °C 30S, 72 °C 45S]* 35 cycles; 72 °C 10 min; 4 °C 10 min
trnL	*trnL-F* *trnL-R*	CGAAATCGGTAGACGCTACG CCATTGAGTCTCTGCACCTATC	193 bp	94 °C 5 min; [94 °C 30S, 56.5 °C 30S, 72 °C 30S]* 35 cycles; 72 °C 10 min; 4 °C 10 min
ycf1b	*ycf1b-F* *ycf1b-R*	TCTCGACGAAAATCAGATTGTTGTGAAT ATACATGTCAAAGTGATGGAAAA	No result	94 °C 5 min; [94 °C 30S, 56.5 °C 30S, 72 °C 45S]* 35 cycles; 72 °C 10 min; 4 °C 10 min 56.5 °C
At103	*At103-F* *At103-R*	CTTCAAGCCMAAGTTCATCTTCTA ATCATTGAGGTACATNGTMACATA	No result	94 °C 5 min; [94 °C 30S, 54 °C 30S, 72 °C 45S]* 35 cycles; 72 °C 10 min; 4 °C 10 min 56.5 °C

In addition, *ITS2* sequence analysis of 12 commercial kiwifruit results showed that compared to *rbcL, matK, psbA-trnH, rpoB, rpoC1* and *trnL*, the *ITS2* sequence difference is better. It was found that there are five SNPs in the 491 bp of *ITS2* sequence, bases at 115 bp, 132 bp and 310 bp of HuangJinGuo, ChuanHuangJin and HongYang are "C", that of the other nine commercial varieties are "T". Bases at 206 bp and 215 bp of HuangJinGuo, ChuanHuangJin and HongYang are "G", that of the other nine commercial varieties are "A" (Figure 2a, Table 1 and Supplementary Table S2). It can be used to distinguish HuangJinGuo, ChuanHuangJin, HongYang and other 9 commercial varieties. Genetic distance analysis discovered that for *ITS2*, there is no difference among HuangJinGuo, ChuanHuangJin and HongYang. The genetic distance between them and other 9 commercial varieties was 0.01. There is no difference among the nine commercial varieties (Supplementary Table S5a). *ITS2* showed more variable sites than the plastid regions *matK* and *rbcL*. This result confirmed that *ITS2* showed greater discriminatory power than plastid regions [35]. Phylogenetic analysis also showed that HuangJinGuo, ChuanHuangJin and HongYang were clustered into one group with the *ITS2* sequence, and the other nine commercial varieties were grouped into another category (Figure 3a).

rpl32_trnL(*UAG*) had a greater degree of variation and could be used as the core barcode sequence of cherry plants [17]. *rpl32_trnL*(*UAG*) sequence analysis of 12 commercial kiwifruit results showed that compared to *rbcL, matK, psbA-trnH, rpoB, rpoC1, trnL* and *ITS2*, the *rpl32_trnL*(*UAG*) sequence difference is the maximum. It was found that there are six SNPs in the 1010bp of *rpl32_trnL*(*UAG*) sequence. Bases at 285 bp and 385 bp of CuiYu are "A", that of the other 11 commercial varieties are "T". Bases at 662 bp of HongYang are "C", that of the other 11 commercial varieties are "T". Bases at 768 bp of YaTe are "G", that of the other 11 commercial varieties are "A". Bases at 837 bp and 841 bp of HuaYou are "C" and "T", respectively, that of the other 10 commercial varieties are "A". Bases at 852 bp of XuXiang are "T", that of the other 11 commercial varieties are "A". (Figure 2c, Table 1 and Supplementary Table S4). It can be used to distinguish CuiYu, JinYan, YaTe, XuXiang, HuaYou, HongYang, HuangJinGuo and ChuanHuangJin from the other four commercial varieties, respectively. Genetic distance analysis discovered that, for *rpl32_trnL*(*UAG*), the genetic distance between HuaYou and JinYan, CuiXiang, FengXianLou, ChuanHuangJin, HaiWoDe, QinMei and HuangJinGuo is 0.002. The genetic distance between HongYang and CuiYu is 0.003. The genetic distance between HongYang and YaTe, XuXiang is 0.002. The genetic distance between HongYang and JinYan, CuiXiang, FengXianLou, ChuanHuangJin, HaiWoDe, QinMei and HuangJinGuo is 0.001. The genetic distance between CuiYu and YaTe, XuXiang is 0.003. The genetic distance between CuiYu and JinYan, CuiXiang, FengXianLou, ChuanHuangJin, HaiWoDe, QinMei and HuangJinGuo is 0.001. The genetic distance between JinYan and YaTe, XuXiang is 0.001. There is no difference between JinYan and CuiXiang, FengXianLou, ChuanHuangJin, HaiWoDe, QinMei and HuangJinGuo. The genetic distance between CuiXiang and YaTe, XuXiang is 0.001. There is no difference between CuiXiang and FengXianLou, ChuanHuangJin, HaiWoDe, QinMei and HuangJinGuo. The genetic distance between YaTe and XuXiang is 0.002. There is no difference between YaTe and FengXianLou, ChuanHuangJin, HaiWoDe, QinMei and HuangJinGuo. The genetic distance between FengXianLou and XuXiang is 0.001. There is no difference between FengXianLou and ChuanHuangJin, HaiWoDe, QinMei and HuangJinGuo. The genetic distance between ChuanHuangJin and XuXiang is 0.001. There is no difference between ChuanHuangJin and HaiWoDe, QinMei and HuangJinGuo. The genetic distance between HaiWoDe and XuXiang is 0.001. There is no difference between HaiWoDe and QinMei and HuangJinGuo. The genetic distance between QinMei and XuXiang is 0.001. There was no difference between QinMei and HuangJinGuo. The genetic distance between HuangJinGuo and XuXiang is 0.001 (Supplementary Table S5c). *rpl32_trnL(UAG)* showed more variable sites than the other plastid regions *matK* and *rbcL*. This result confirmed that *rpl32_trnL(UAG)* showed better discriminatory power than other plastid regions. Phylogenetic analysis further indicated that HuaYou, YaTe, XuXiang, HongYang, CuiYu can be distinguished from the other seven commercial varieties, respectively (Figure 3c).

Multi-site combination methods based on plastid (chloroplast) sequences are considered as effective strategies in the identification of plant species [31,33,34]. Therefore, we comprehensively analyzed the sequences of *ITS2*, *matK* combined with *rpl32_trnL(UAG)*, and then found that, for *ITS2+matK+rpl32_trnL(UAG)*, there is no difference between QinMei and YaTe, JinYan, XuXiang. There is no difference between HaiWoDe and YaTe, JinYan, XuXiang, QinMei. There is no difference between CuiXiang and YaTe, JinYan, XuXiang, QinMei, HaiWoDe. The genetic distance between YaTe and JinYan is 0.001. The genetic distance between XuXiang and YaTe, JinYan is 0.001. The genetic distance between HuaYou and YaTe, JinYan, XuXiang, QinMei, HaiWoDe, CuiXiang are 0.001. There is no difference between FengXianLou and YaTe, JinYan, XuXiang, QinMei, HaiWoDe and CuiXiang. The genetic distance between FengXianLou and HuaYou is 0.001. The genetic distance between CuiYu and YaTe, JinYan, XuXiang, HuaYou is 0.002. The genetic distance between CuiYu and QinMei, HaiWoDe, CuiXiang, FengXianLou is 0.001. The genetic distance between HuangJinGuo and YaTe, JinYan, XuXiang, HuaYou, CuiYu is 0.003. The genetic distance between HuangJinGuo and QinMei, HaiWoDe and CuiXiang, FengXianLou is 0.002. The genetic distance between HongYang and YaTe, JinYan, XuXiang, QinMei, HaiWoDe, CuiXiang, HuaYou, FengXianLou is 0.003. The genetic distance between HongYang and CuiYu is 0.004. There is no difference between HongYang and CuiYu. The genetic distance between ChuanHuangJin and YaTe, JinYan, XuXiang, HuaYou, CuiYu is 0.003. The genetic distance between ChuanHuangJin and QinMei, HaiWoDe and CuiXiang, FengXianLou is 0.002. There is no difference between ChuanHuangJin and HuangJinGuo, HongYang (Supplementary Table S5d). These results suggest that CuiYu, JinYan, YaTe, XuXiang, HuaYou, HongYang, HuangJinGuo and ChuanHuangJin can also be differentiated by the genetic distance of *ITS2+matK+rpl32_trnL(UAG)* combination. We found that the genetic distance between samples of HaiWoDe, FengXianLou, QinMei and CuiXiang in this study is 0.000, which means that most of them have a unique *ITS2*, *matK* and *rpl32_trnL(UAG)* sequences. This feature is thus useful for identifying these twelve commercial varieties and related species, which is similar to previous study [36].

The analysis of SNP loci is an important tool for analyzing small differences between populations [37–39]. In this study, by analyzing the sequence variable sites of 8 DNA barcodes including *ITS2*, *psbA-trnH*, *rbcL*, *matK*, *rpoB*, *rpoC1* and *rpl32_trnL(UAG)* in 12 commercial kiwifruit, stable SNP loci of DNA barcode *ITS2* are found between HuangJinGuo, ChuanHuangJin, HongYang and the other nine commercial varieties. Likewise, DNA barcode *matK* has stable SNP loci between JinYan, CuiYu and the other 10 commercial varieties. *rpl32_trnL(UAG)* has stable SNP loci between CuiYu, JinYan, YaTe, XuXiang, HuaYou, HongYang, HuangJinGuo and ChuanHuangJin and the other four commercial varieties. The presence of these SNP loci makes the identification of CuiYu, JinYan, YaTe, XuXiang, HuaYou, HongYang, HuangJinGuo and ChuanHuangJin kiwifruits possible. These results indicate that intergenic region *ITS2* could distinguish HuangJinGuo and HongYang from the nine other commercial varieties, and *matK* could be utilized to distinguish JinYan and CuiYu from 10 the other commercial varieties. *rpl32_trnL(UAG)* could be utilized to distinguish HuaYou, YaTe, XuXiang, HongYang, CuiYu from the other seven commercial varieties. SNPs of *ITS2*, *matK* and *rpl32_trnL(UAG)* combining with the current plant DNA barcode systems could be used to accurately identify kiwifruit, which comprises samples of closely-related or subspecies from different cultivation areas. These results are good agreements with previous reports [38,39].

In conclusion, morphologically, JinYan, CuiYu, HuangJinGuo and HongYang can be discriminated. At the molecular level, SNPs based on DNA barcoding intergenic region *ITS2* and cp gene *matK* and *rpl32_trnL(UAG)* combined together can distinguish CuiYu, JinYan, YaTe, XuXiang, HuaYou, HongYang, HuangJinGuo and ChuanHuangJin from the other four commercial varieties, respectively, that is consistent with the results of morphology and better than the results of morphology. These results will provide an experimental basis for the classification and identification of commercially available kiwifruits and will offer a theoretical support for scientific trials and forensic botany of such disputes.

4. Materials and Methods

4.1. Samples Collection and DNA Extraction

Seventy two samples from twelve kiwifruit commercial varieties were collected in this study. Their names are HuangJinGuo, CuiXiang, QinMei, XuXiang, HuaYou, FengXianLou, YaTe, HaiWoDe, CuiYu, ChuanHuangJin, HongYang and JinYan. Six replicates were randomly selected for each commercial variety in the local market. According to our survey, the price of HongYang is 4 to 5 times than that of QinMei, HuaYou, YaTe and HaiWoDe in the market. Similarly, ChuanHuangJin, HuangJinGuo and JinYan, whose price are 3 to 4 times than that of QinMei, HuaYou, YaTe and HaiWoDe. HuangJinGuo, CuiXiang, QinMei, XuXiang, HuaYou, FengXianLou, YaTe and HaiWoDe were collected from Shaanxi province of China. CuiYu, ChuanHuangJin, HongYang and JinYan were collected from Si Chuan province of China (Detailed information in Supplementary Table S6). Among the 12 commercial varieties, HongYang is diploid, HaiWoDe is hexaploid, HuangJinGuo, CuiXiang, XuXiang, HuaYou, CuiYu, ChuanHuangJin and JinYan are tetraploid. QinMei, FengXianLou and YaTe are unknown, which is needed further research. DNA of each kiwifruit was extracted by Plant Genomic DNA Extraction Kit (Omega, Norcross, GA, USA) according to the manufacturer's instructions.

4.2. PCR Amplification, Cloning, and Sequencing

PCR primers of DNA barcodes (*rbcL*, *matK*, *psbA-trnH*, *ITS2*, *rpoB*, *rpoC1*, *ycf1b*, *trnL*, *rpl32_trnL(UAG)* and *At103*) were designed based on DNA barcodes [9–11,16], CBOL Plant Working Group (2009), China Bole Group (2011) as well as reported chloroplast and genomic sequences of kiwifruit [40–43]. Primer sequences are listed in Table 1. They were synthesized in BGI (Beijing Genomics Institute, Beijing, China). Polymerase chain reaction made use of the same reaction system. Amplification procedures of ten different primer pairs are shown in Table 1. The PCR products were separated by 1.2~1.5% agarose gel electrophoresis. The remaining PCR products were recovered and purified. Then, the purified products were ligated into pMD18-T vector (Takara, DaLian, China) and transformed into DH-5α *E. coli*. Positive clones were screened and sequenced in Beijing Genomics Institute.

4.3. Sequence Alignments and Analysis

Multiple sequence alignment was performed by using the results of *psbA-trnH*, *rbcL*, *rpoB*, *rpoC1* and *trnL* amplification by http://multalin.toulouse.inra.fr/multalin/. (Supplementary Figure S1). Multi-sequence alignment of *ITS2*, *matK*, *rpl32_trnL*(*UAG*) and *ITS2+matK+rpl32_trnL*(*UAG*) were carried out by ClustalX 1.8 [44] software (Conway Institute UCD, Dublin, Ireland), respectively. Then, the phylogenetic trees of these twelve kiwifruit commercial varieties were reconstructed. Neighbor-joining (NJ) method of MEGA 7.0 is applied to construct the phylogenetic tree. One thousand bootstrap resamplings were performed to assess the reliability. The steps are as follows: Phylogeny/construct/Test Neighbor-joining tree/Nucleotide Sequences/Statistical method: Neighbor-joining, No. of bootstrap repetitions: 1000, Model: Kimura 2-parameter model, Missing Data Treatment: Pairwise deletion/compute. The K2P distances between different genetic resources were calculated by MEGA7.0. Variation of different commercial varieties was analyzed according to the genetic distance of each barcode fragment.

Supplementary Materials:
Figure S1: *psbA-trnH*, *rbcL*, *rpoB*, *rpoC1* and *trnL* sequences alignment of 12 commercial kiwifruits. Figure S2: Phylogenetic analysis of 12 kiwifruit commercial varieties. Table S1: The information of the experimental materials, Table S2: Phenotypic characteristics of twelve commercial varieties of kiwifruit, Table S3: *ITS2* sequence of 72 kiwifruit samples, Table S4: *matK* sequence of 72 kiwifruit samples. Table S5: *rpl32_trnL* (*UAG*) sequence of 72 kiwifruit samples. Table S6: Genetic distance of *ITS2*, *matK*, *rpl32_trnL*(*UAG*) and *ITS2+matK+rpl32_trnL*(*UAG*) sequences of 12 kiwifruit commercial varieties.

Author Contributions: Conceptualization, Q.X., Y.W. and B.Z.; methodology, Q.X. and H.Z.; validation, H.Z., C.Y., S.W. and J.L.; data curation, Q.X. and F.Y.; writing—original draft preparation, Q.X., Y.W., H.Z. and F.Y.; writing—review and editing, C.Y., J.L. and S.W.; visualization, Q.X. and F.Y.; supervision, C.Y., J.L. and S.W.; project administration, Y.W. and B.Z.; funding acquisition, Q.X.

Acknowledgments: In this section you can acknowledge any support given which is not covered by the author contribution or funding sections. This may include administrative and technical support, or donations in kind (e.g., materials used for experiments).

References

1. Hunter, D.C.; Greenwood, J.; Zhang, J.L.; Skinner, M.A. Antioxidant and 'Natural Protective' Properties of Kiwifruit. *Curr. Top. Med. Chem.* **2011**, *11*, 1811–1820. [CrossRef] [PubMed]
2. Analysis of global kiwifruit yield and production area distribution. Available online: http://www.chyxx.com/industry/201602/389018.html (accessed on 22 December 2018).
3. Garcia, C.V.; Quek, S.Y.; Stevenson, R.J.; Winz, R.A. Characterisation of bound volatile compounds of a low flavour kiwifruit species: Actinidia eriantha. *Food Chem.* **2012**, *134*, 655–661. [CrossRef] [PubMed]
4. Nishiyama, I.; Yamashita, Y.; Yamanaka, M.; Shimohashi, A.; Fukuda, T.; Oota, T. Varietal difference in vitamin C content in the fruit of kiwifruit and other Actinidia species. *J. Agric. Food Chem.* **2004**, *52*, 5472–5475. [CrossRef] [PubMed]
5. Hebert, P.D.N.; Cywinska, A.; Ball, S.L.; DeWaard, J.R. Biological identifications through DNA barcodes. *Proc. R. Soc. B-Biol. Sci.* **2003**, *270*, 313–321. [CrossRef] [PubMed]
6. Jiang, Y.; Ding, C.B.; Zhang, L.; Yang, R.W.; Zhou, Y.H.; Tang, L. Identification of the genus Epimedium with DNA barcodes. *J. Med. Plants Res.* **2011**, *5*, 6413–6417. [CrossRef]
7. Liu, J.X.; Shi, L.C.; Han, J.P.; Li, G.; Lu, H.; Hou, J.Y.; Zhou, X.T.; Meng, F.Y.; Downie, S.R. Identification of species in the angiosperm family Apiaceae using DNA barcodes. *Mol. Ecol. Resour.* **2014**, *14*, 1231–1238. [CrossRef] [PubMed]
8. Hollingsworth, P.M.; Forrest, L.L.; Spouge, J.L.; Hajibabaei, M.; Ratnasingham, S.; van der Bank, M.; Chase, M.W.; Cowan, R.S.; Erickson, D.L.; Fazekas, A.J.; et al. A DNA barcode for land plants. *Proc. Natl. Acad. Sci. USA* **2009**, *106*, 12794–12797.
9. Yao, H.; Song, J.Y.; Ma, X.Y.; Liu, C.; Li, Y.; Xu, H.X.; Han, J.P.; Duan, L.S.; Chen, S.L. Identification of

Dendrobium Species by a Candidate DNA Barcode Sequence: The Chloroplast psbA-trnH Intergenic Region. *Planta Med.* **2009**, *75*, 667–669. [CrossRef] [PubMed]

10. Ma, X.Y.; Xie, C.X.; Liu, C.; Song, J.Y.; Yao, H.; Luo, K.; Zhu, Y.J.; Gao, T.; Pang, X.H.; Qian, J.; et al. Species Identification of Medicinal Pteridophytes by a DNA Barcode Marker, the Chloroplast psbA-trnH Intergenic Region. *Biol. Pharm. Bull.* **2010**, *33*, 1919–1924. [CrossRef] [PubMed]
11. Chen, S.; Yao, H.; Han, J.; Liu, C.; Song, J.; Shi, L.; Zhu, Y.; Ma, X.; Gao, T.; Pang, X.; et al. Validation of the ITS2 region as a novel DNA barcode for identifying medicinal plant species. *PLoS ONE* **2010**, *5*, e8613. [CrossRef] [PubMed]
12. Fazekas, A.J.; Burgess, K.S.; Kesanakurti, P.R.; Graham, S.W.; Newmaster, S.G.; Husband, B.C.; Percy, D.M.; Hajibabaei, M.; Barrett, S.C. Multiple multilocus DNA barcodes from the plastid genome discriminate plant species equally well. *PLoS ONE* **2008**, *3*, e2802. [CrossRef] [PubMed]
13. Dong, W.; Xu, C.; Li, C.; Sun, J.; Zuo, Y.; Shi, S.; Cheng, T.; Guo, J.; Zhou, S. ycf1, the most promising plastid DNA barcode of land plants. *Sci. Rep.* **2015**, *5*, 8348. [CrossRef] [PubMed]
14. Yang, Y.; Zhan, A.; Cao, L.; Meng, F.; Xu, W. Selection of a marker gene to construct a reference library for wetland plants, and the application of metabarcoding to analyze the diet of wintering herbivorous waterbirds. *PeerJ* **2016**, *4*, e2345. [CrossRef] [PubMed]
15. Li, M.G.; Wunder, J.; Bissoli, G.; Scarponi, E.; Gazzani, S.; Barbaro, E.; Saedler, H.; Varotto, C. Development of COS genes as universally amplifiable markers for phylogenetic reconstructions of closely related plant species. *Cladistics* **2008**, *24*, 727–745. [CrossRef]
16. Bruni, I.; De Mattia, F.; Galimberti, A.; Galasso, G.; Banfi, E.; Casiraghi, M.; Labra, M. Identification of poisonous plants by DNA barcoding approach. *Int. J. Leg. Med.* **2010**, *124*, 595–603. [CrossRef] [PubMed]
17. Fu, T.; Wang, Z.L.; Lin, L.J.; Lin, L.; Li, W.; Yuan, D.M. Molecular Phylogenetic Analysis of Wild Cerasus Plants in South China. *J. Nucl. Agric. Sci.* **2018**, *32*, 2126–2134.
18. Lee, S.C.; Wang, C.H.; Yen, C.E.; Chang, C. DNA barcode and identification of the varieties and provenances of Taiwan's domestic and imported made teas using ribosomal internal transcribed spacer 2 sequences. *J. Food Drug Anal.* **2017**, *25*, 260–274. [CrossRef] [PubMed]
19. Enan, M.R.; Ahmed, A. Cultivar-level phylogeny using chloroplast DNA barcode psbK-psbI spacers for identification of Emirati date palm (Phoenix dactylifera L.) varieties. *Genet. Mol. Res.* **2016**, *15*. [CrossRef] [PubMed]
20. Jaakola, L.; Suokas, M.; Haggman, H. Novel approaches based on DNA barcoding and high-resolution melting of amplicons for authenticity analyses of berry species. *Food Chem.* **2010**, *123*, 494–500. [CrossRef]
21. He, Y.; Hou, P.; Fan, G.; Song, Z.; Arain, S.; Shu, H.; Tang, C.; Yue, Q.H.; Zhang, Y. Authentication of Angelica anomala Ave-Lall cultivars through DNA barcodes. *Mitochondrial DNA* **2012**, *23*, 100–105. [CrossRef] [PubMed]
22. de Paula Careta, F.; Paneto, G.G. Recent patents on high-throughput single nucleotide polymorphism (SNP) genotyping methods. *Recent Pat. DNA Gene Seq.* **2012**, *6*, 122–126. [CrossRef] [PubMed]
23. Yamamoto, T.; Nagasaki, H.; Yonemaru, J.; Ebana, K.; Nakajima, M.; Shibaya, T.; Yano, M. Fine definition of the pedigree haplotypes of closely related rice cultivars by means of genome-wide discovery of single-nucleotide polymorphisms. *BMC Genom.* **2010**, *11*, 267. [CrossRef] [PubMed]
24. Yuan, Y.; Jiang, C.; Liu, L.; Yu, S.; Cui, Z.; Chen, M.; Lin, S.; Wang, S.; Huang, L. Convenient, sensitive and high-throughput method for screening botanic origin. *Sci. Rep.* **2014**, *4*, 5395. [CrossRef] [PubMed]
25. Wang, H.T.; Kim, M.K.; Kim, Y.J.; Lee, H.N.; Jin, H.Z.; Chen, J.; Yang, D.C. Molecular authentication of the Oriental medicines Pericarpium Citri Reticulatae and Citri Unshius Pericarpium using SNP markers. *Gene* **2012**, *494*, 92–95. [CrossRef] [PubMed]
26. Maiden, M.C.J.; van Rensburg, M.J.J.; Bray, J.E.; Earle, S.G.; Ford, S.A.; Jolley, K.A.; McCarthy, N.D. MLST revisited: The gene-by-gene approach to bacterial genomics. *Nat. Rev. Microbiol.* **2013**, *11*, 728–736. [CrossRef] [PubMed]
27. Kumar, S.; Stecher, G.; Tamura, K. MEGA7: Molecular Evolutionary Genetics Analysis Version 7.0 for Bigger Datasets. *Mol. Biol. Evol.* **2016**, *33*, 1870–1874. [CrossRef] [PubMed]
28. Kress, W.J.; Wurdack, K.J.; Zimmer, E.A.; Weigt, L.A.; Janzen, D.H. Use of DNA barcodes to identify flowering plants. *Proc. Natl. Acad. Sci. USA* **2005**, *102*, 8369–8374. [CrossRef] [PubMed]
29. Ferri, G.; Alu, M.; Corradini, B.; Licata, M.; Beduschi, G. Species Identification Through DNA "Barcodes". *Genet. Test. Mol. Biomark.* **2009**, *13*, 421–426. [CrossRef] [PubMed]

30. Chase, M.W.; Salamin, N.; Wilkinson, M.; Dunwell, J.M.; Kesanakurthi, R.P.; Haider, N.; Savolainen, V. Land plants and DNA barcodes: Short-term and long-term goals. *Philos. Trans. R. Soc. B-Biol. Sci.* **2005**, *360*, 1889–1895. [CrossRef] [PubMed]
31. Pennisi, E. Taxonomy—Wanted: A barcode for plants. *Science* **2007**, *318*, 190–191. [CrossRef] [PubMed]
32. Newmaster, S.G.; Fazekas, A.J.; Steeves, R.A.D.; Janovec, J. Testing candidate plant barcode regions in the Myristicaceae. *Mol. Ecol. Resour.* **2008**, *8*, 480–490. [CrossRef] [PubMed]
33. Chase, M.W.; Cowan, R.S.; Hollingsworth, P.M.; van den Berg, C.; Madrinan, S.; Petersen, G.; Seberg, O.; Jorgsensen, T.; Cameron, K.M.; Carine, M. A proposal for a standardised protocol to barcode all land plants. *Taxon* **2007**, *56*, 295–299. [CrossRef]
34. Kress, W.J.; Erickson, D.L. A two-locus global DNA barcode for land plants: The coding rbcL gene complements the non-coding trnH-psbA spacer region. *PLoS ONE* **2007**, *2*, e508. [CrossRef] [PubMed]
35. Hollingsworth, P.M.; Graham, S.W.; Little, D.P. Choosing and using a plant DNA barcode. *PLoS ONE* **2011**, *6*, e19254. [CrossRef] [PubMed]
36. Feng, S.G.; Jiao, K.L.; Zhu, Y.J.; Wang, H.F.; Jiang, M.Y.; Wang, H.Z. Molecular identification of species of Physalis (Solanaceae) using a candidate DNA barcode: The chloroplast psbA-trnH intergenic region. *Genome* **2018**, *61*, 15–20. [CrossRef] [PubMed]
37. Chen, X.C.; Liao, B.S.; Song, J.Y.; Pang, X.H.; Han, J.P.; Chen, S.L. A fast SNP identification and analysis of intraspecific variation in the medicinal Panax species based on DNA barcoding. *Gene* **2013**, *530*, 39–43. [CrossRef] [PubMed]
38. Huang, Q.; Duan, Z.; Yang, J.; Ma, X.; Zhan, R.; Xu, H.; Chen, W. SNP typing for germplasm identification of Amomum villosum Lour. Based on DNA barcoding markers. *PLoS ONE* **2014**, *9*, e114940. [CrossRef] [PubMed]
39. Murphy, B.P.; Tranel, P.J. Identification and Validation of Amaranthus Species-Specific SNPs within the ITS Region: Applications in Quantitative Species Identification. *Crop. Sci.* **2018**, *58*, 304–311. [CrossRef]
40. Huang, S.X.; Ding, J.; Deng, D.J.; Tang, W.; Sun, H.H.; Liu, D.Y.; Zhang, L.; Niu, X.L.; Zhang, X.; Meng, M. Draft genome of the kiwifruit Actinidia chinensis. *Nat. Commun.* **2013**, *4*, 2640. [CrossRef] [PubMed]
41. Pilkington, S.M.; Crowhurst, R.; Hilario, E.; Nardozza, S.; Fraser, L.; Peng, Y.Y.; Gunaseelan, K.; Simpson, R.; Tahir, J.; Deroles, S.C. A manually annotated Actinidia chinensis var. chinensis (kiwifruit) genome highlights the challenges associated with draft genomes and gene prediction in plants. *BMC Genom.* **2018**, *19*, 257. [CrossRef] [PubMed]
42. Lin, M.; Qi, X.; Chen, J.; Sun, L.; Zhong, Y.; Fang, J.; Hu, C. The complete chloroplast genome sequence of Actinidia arguta using the PacBio RS II platform. *PLoS ONE* **2018**, *13*, e0197393. [CrossRef] [PubMed]
43. Yao, X.; Tang, P.; Li, Z.; Li, D.; Liu, Y.; Huang, H. The First Complete Chloroplast Genome Sequences in Actinidiaceae: Genome Structure and Comparative Analysis. *PLoS ONE* **2015**, *10*, e0129347. [CrossRef] [PubMed]
44. Thompson, J.D.; Gibson, T.J.; Plewniak, F.; Jeanmougin, F.; Higgins, D.G.J.N. The CLUSTAL_X windows interface: Flexible strategies for multiple sequence alignment aided by quality analysis tools. *Nucleic Acids Res.* **1997**, *25*, 4876–4882. [CrossRef] [PubMed]

14

Quantification and Confirmation of Fifteen Carbamate Pesticide Residues by Multiple Reaction Monitoring and Enhanced Product Ion Scan Modes via LC-MS/MS QTRAP System

Ying Zhou, Jian Guan, Weiwei Gao, Shencong Lv and Miaohua Ge *

Jiaxing Center for Disease Control and Prevention, Zhejiang 314050, China; zhouyingand@sina.cn (Y.Z.); jguan001@sina.com (J.G.); dannyday@126.com (W.G.); thestorm2008@126.com (S.L.)
* Correspondence: gemiaohua0823@sina.com

Academic Editor: Alessandra Gentili

Abstract: In this research, fifteen carbamate pesticide residues were systematically analyzed by ultra-high performance liquid chromatography–quadrupole-linear ion trap mass spectrometry on a QTRAP 5500 system in both multiple reaction monitoring (MRM) and enhanced product ion (EPI) scan modes. The carbamate pesticide residues were extracted from a variety of samples by QuEChERS method and separated by a popular reverse phase column (Waters BEH C18). Except for the current conformation criteria including selected ion pairs, retention time and relative intensities from MRM scan mode, the presence of carbamate pesticide residues in diverse samples, especially some doubtful cases, could also be confirmed by the matching of carbamate pesticide spectra via EPI scan mode. Moreover, the fragmentation routes of fifteen carbamates were firstly explained based on the mass spectra obtained by a QTRAP system; the characteristic fragment ion from a neutral loss of CH_3NCO (−57 Da) could be observed. The limits of detection and quantification for fifteen carbamates were 0.2–2.0 μg kg^{-1} and 0.5–5.0 μg kg^{-1}, respectively. For the intra- (n = 3) and inter-day (n = 15) precisions, the recoveries of fifteen carbamates from spiked samples ranged from 88.1% to 118.4%, and the coefficients of variation (CVs) were all below 10%. The method was applied to pesticide residues detection in fruit, vegetable and green tea samples taken from local markets, in which carbamates were extensively detected but all below the standard of maximum residue limit.

Keywords: carbamates; multiple reaction monitoring (MRM); enhanced product ion (EPI); mass fragmentation; confirmatory method; pesticide residues

1. Introduction

Carbamate pesticides, namely, N-methyl carbamate esters not only share with organophosphates the capacity to kill insects by inhibiting the enzyme acetylcholinesterase (AChE) but also show lower toxicity to human being [1–3]. Gradually, they have become one kind of most popular pesticides in agriculture to guarantee food production. However, the improper use of carbamate pesticides have adversely affected food safety g [4,5]. Carbamate pesticide residues are known to be frequently present in fruits, vegetables and green teas on the market at levels close to or below the standard of maximum residue limit [6]. The legitimate content has brought chronic and continuous intake of carbamate residues that could be relevant to multiple ailments and further pose a threat to public health [7–9].

Meeting the challenge of many samples in daily work, developing a best concise, highly specific and sensitive analytical method to detect the carbamate residues in diverse samples from the area of food safety is important.

The challenge to modify this analytical method could be ascribed to two main factors: the existence of a range of carbamate pesticides and samples with complex composition that depend on the instrumental analysis and the sample pretreatment. With the efforts of researchers for half a century, the sample pretreatment of pesticide residues has been developed from Soxhlet extraction, liquid–liquid partition chromatography and column chromatography to solid-phase extraction, solid-phase micro-extraction, molecularly imprinted solid-phase extraction, super critical fluid extraction, matrix solid phase dispersion and QuEChERS (Quick, Easy, Cheap, Effective, Rugged, and Safe) methods [10–12]. Developed in 2003 by Anastassiades, QuEChERS sample preparation has been readily accepted by many pesticide residue analysts over the years [13,14]. In the section of instrumental analysis, a literature survey (Table 1) shows several methods for the detection of carbamate pesticide residues have been developed at low concentration levels in various samples, such as electrochemical detection [15], liquid chromatography [16–24], high-performance thin-layer chromatography [25], liquid chromatography with post-column fluorescence derivatization [26,27] and liquid chromatography with different mass spectrometry [28–32]. Of all the instrumental methods, the LC-MS system that combines the separation ability of liquid chromatography along with the sensitivity and specificity of detection from mass spectrometry and abandoned the additional derivation procedure of whole analysis process has gone mainstream for pesticide residue analysis. Among the different mass spectrometers, the ion trap triple quadrupole mass spectrometer allows for multiple reaction monitoring (MRM) scan modes is the ascendant instrument to quantify the residues of pesticides at trace amount. In MRM mode, the selected precursor ion of relevant compound is pre-screened in the first quadrupole (Q1), dissociated in the second quadrupole (Q2) and identified and quantified in the third quadrupole (Q3). During the analysis procedure, both Q1 and Q3 are set at several specific masses, allowing only those distinct fragment ions from the certain precursor ions to be detected. The setting of certain precursor ions and specific masses results in increased sensitivity and the structural specificity of the analyte [33,34].Compared with traditional triple quadrupole mass spectrometry, the Triple Quadrupole Linear Ion Traps system (QTRAP, AB SCIEX) equips an extra linear ion trap (LIT) located in the third quadrupole set and represents the new generation of LC-MS/MS system with unique scan mode of multiple reaction monitoring–information dependent acquisition–enhanced product ion (MRM-IDA-EPI) scan mode, which could maximize the information of the analytes in a single run, including selected ion pairs, retention time, relative intensities and the mass spectra. In this mode, the ions of analytes enter Q3 and are stopped from leaving this region by the changing of voltages. Then, the trap is allowed to fill with target ions with a set amount of time. The voltages are changed to stop any more ions from entering. Finally, the packet of ions is scanned out from the trap in a controlled manner. The target analytes are quantified as ever, and further identified on the automatically acquired MS/MS spectra while the precursor ions intensity exceeds the fixed area threshold setting. In particular, the linear ion trap could keep on capturing the interested precursor ions and product ions for a few milliseconds. This accumulation signifies a 500-times increase in the scanning sensitivity and brings a high intensity of the mass spectra recorded in EPI mode. It double checks the doubtful samples by comparing the acquired mass spectra with the mass spectral library built from standards [35]. This study developed a new analytical strategy to combine QuEChERS sample preparation and QTRAP LC-MS/MS system to improve the efficiency of quantification and confirmation of carbamate residues while ensuring high degree of reproducibility. In addition, this method should be stable enough to avoid false negative or false positive results.

Table 1. Comparative study between the published analysis methods for carbamates.

Target	Object	Sample Pretreatment	Instrumental Analysis	Analysis Limits	Disadvantage	Ref.
Carbaryl	soil	ionic liquid-dispersive liquid–liquid microextraction	HPLC-FD (fluorescence detector)	0.63–4.0 ng g^{-1}	Ionic liquids are not commercially available No confirmation spectra	[11]
Seven carbamates	gram, wheat, lentil, soybean, fenugreek leaves apple	column chromatography	HPLC-UV	0.08–1.16 mg L^{-1}	Large amount of Solvents Lack of sensitivity No confirmation spectra	[17]
Seven carbamates	/	/	HPLC post-column derivatization and fluorescence detection	0.2–0.7 ng	No application No confirmation spectra	[26]
Eleven carbamates	/	/	HPLC post-column derivatization and fluorescence detection	0.5 ng mL^{-1}	The tempestuously hydrolyzation of standards in dilute hydrochloric acid solution No confirmation spectra	[27]
Fifteen carbamates	corn, cabbage, tomato	QuEChERS method	LC-MS (QDa)	1 μg mL^{-1}	Lack of sensitivity; expensive No confirmation spectra	[28]
Thirteen carbamates	orange, grape, onion, tomatoes	Matrix solid-phase dispersion (concentration)	LC-MS (ESI/APCI)	0.001–0.01 mg kg^{-1}	No confirmation spectra	[29]
Thirteen carbamates	Traditional Chinese Medicine	QuEChERS method	UPLC-MS (MRM)	5.0–10.0 μg kg^{-1}	No confirmation spectra	[30]

2. Results and Discussion

2.1. *Optimization of Sample Preparation*

As shown in Figure 1, the scheme of analysis method, extraction process and purification process are the two main components of the QuEChERS pretreatment. In this study, acetonitrile with different concentrations of formic acid supplement was compared for extraction efficiency. However, acetonitrile and acidified acetonitrile did not show any obvious distinction in the recovery experiment. Thus, it was more convenient to adopt acetonitrile in the extraction process. In the purification process, 1.0 g of sodium chloride and 2.0 g of anhydrous sodium sulfate were applied to separate the aqueous solution and acetonitrile part. In this study, the traditional anhydrous magnesium sulfate was replaced by anhydrous sodium sulfate because of the heat release in the moisture absorbing process. In the purification process, the commonly used cleaning agents are ethylenediamine-N-propylsilane (PSA), graphitized carbon black (GCB) and C18. PSA is pertinent to organic acids, some pigments, and some sugars and fats; GCB has a strong adsorption effect on pigments; and C18 aims to remove non-polar impurities such as fats. Different proportions of PSA, GCB, and C18 were combined to find the best recipe with the best average recoveries. It was proventhat the combination of 50 mg C18 and 150 mg PSA as cleaning agent was most suitable for determining quantitatively carbamate residues in multi-matrices by recovery tests and samples determination (Figure S1).

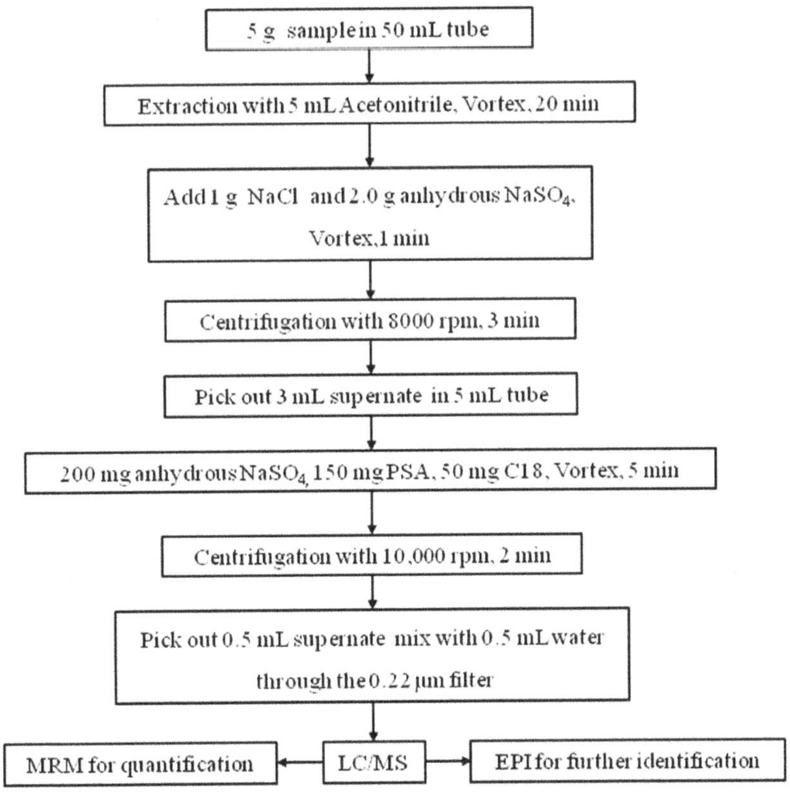

Figure 1. Scheme of analysis for the simultaneous determination and identification of fifteen carbamate pesticide residues in multi-matrices.

2.2. *Optimization of Mass Spectrometry and Chromatographic Conditions*

To confirm the ion pairs for quantification, the mixed standard solution of fifteen carbamate pesticides with a concentration of 100 μg L^{-1} was infused into the QTRAP mass spectrometer at a flow rate of 7 μL min^{-1} to obtain automatic analytes optimization by the ESI positive mode. In the optimal mass spectrometry conditions, all carbamates except tsumacide acquired two pairs of optimum ions for quantification and identification.

For the high sensitivity, symmetric shape of the ionic peaks and minimal retention time, this study examined the elution type, flow rate, gradient and the type of the chromatographic column. Once the most general chromatographic column C18 was decided, the separation and ionization of the analytes were mainly affected by the compositions of the elution type and the elution gradient. Therefore, several classical compositions of the mobile phase were performed including methanol, acetonitrile, water and water with ammonium acetates or formic acid. Finally, water and acetonitrile, both supplemented with 0.1% formic acid, were chosen as the optimal mobile phases. The final gradient elution with the total flow rate of 0.3 mL/min was as follows: 0–2.5 min, 15% A; 2.5–5 min, 15–50% A; 5–7 min, 50% A; 7–8 min, 50–15% A and 8–10 min, 15% A. The column oven was maintained at 40 °C and the injection volume was 5 μL. The representative LC-QTRAP-MS/MS chromatograms were merged (Figure 2). The retention time (RT) and MS information for each analyte including precursor and product ions, DP and CE are shown in Table 2.

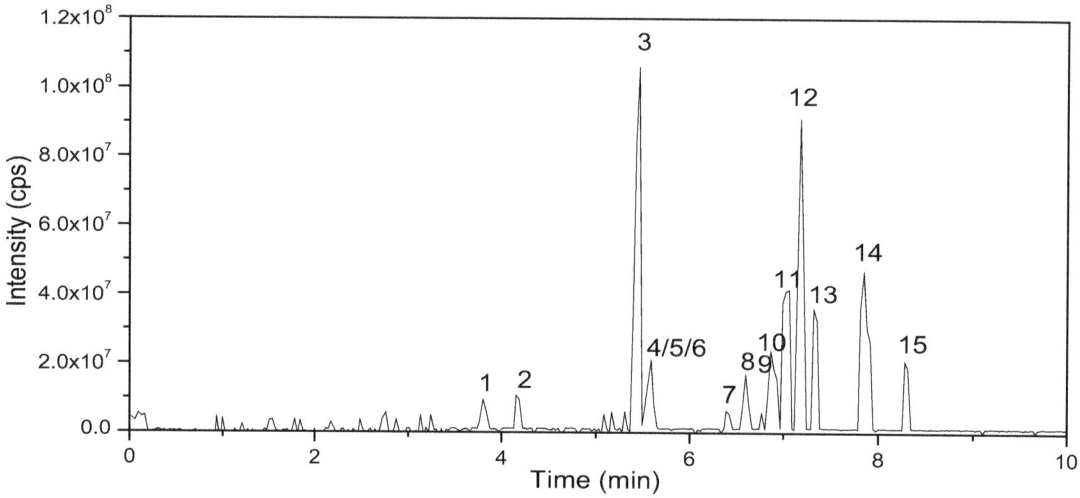

Figure 2. LC-MS/MS chromatograms of fifteen carbamate pesticides (50 ng mL^{-1}).

The EPI survey scans in IDA experiment should be triggered when the ionic intensity exceeded the threshold of 1000 cps. The total scan time (including pauses) was 1.08 s for all MRM transitions. The range of the mass spectra was set between 50 and 300 amu with a scan rate of 10,000 Da s^{-1}. The mass spectra of fifteen carbamates and their characteristic fragmentations are shown in the Supplementary Materials.

Table 2. Retention time and MS parameters of the fifteen carbamates pesticides.

No.	Compound	Retention Time (min)	[1] CAS No.	Precursor Ion (m/z)	Product Ion (m/z)	Declustering Potential (V)	Collision Energy (eV)
1	Aldicarb-sulfoxide	3.81	1646-87-3	206.9	132.0 89.0 *	130 130	8 17
2	Aldicarb-sulfone	4.17	1646-88-3	222.8	166.1 * 148.0	140 140	11 18
3	Pirimicarb	5.45	23103-98-2	239.1	182.1 * 72.0	80 80	20 30
4	Carbofuran-3-hydroxy	5.57	16655-82-6	237.8	181.0* 163.0	150 150	15 19
5	Methomyl	5.58	16752-77-5	162.8	135.0 106.0 *	180 180	26 28
6	Oxamyl	5.59	23135-22-0	220.0	72.0 163.1 *	40 40	18 10
7	Aldicarb	6.40	116-06-3	212.7	89.0 * 156.0	150 150	18 19
8	Tsumacide	6.60	1129-41-5	166.0	109.0 *	50	13
9	Propoxur	6.87	114-26-1	210.0	168.0 * 153.0	60 60	18 10
10	Carbofuran	6.98	1563-66-2	221.7	165.0 * 123.0	120 120	15 27
11	Carbaryl	7.04	63-25-2	202.0	145.0 * 117.0	60 80	12 35
12	Isoprocarb	7.18	2631-40-5	194.0	95.0 137.0 *	100 100	19 11
13	Methiocarb	7.33	2032-65-7	226.0	169.0 * 121.0	55 55	12 24
14	Fenobucarb	7.84	3766-81-2	208.0	95.0 * 152.0	70 70	19 11
15	Banol	8.29	671-04-5	214.1	157.1 * 121.2	60 60	12 16

[1] CAS: chemical abstracts service; *: quantitative ion.

2.3. *Fragmentation Manner of Carbamate Pesticides*

Based on the structure, the fifteen carbamates can be divided into three kinds: N-methyl amino formic acid aromatic ester, N-methyl amino formic acid oxime ester and heterocyclic N-methyl amino formic acid ester. The former two include carbaryl, tsumacide, methiocarb and methomyl; and aldicarb sulfone, aldicarb sulfoxide, etc., respectively, while the latter includes pirimicarb, isolan (highly regulated and unavailable), etc. Generally speaking, in the positive mode of electrospray ionization mass spectrometry, the proton first attaches to the protonation site, and then triggers the cleavages by migrating to the reactive center [36]. With the common structure of amino nitrogen and carbonyl oxygen, the protons could gravitate to these preferred protonation sites, and then trigger the dissociation reaction by migrating to the reactive center. Therefore, the oxygen of ester near the carbonyl carbon is a suitable active site to induce this fragmentation of carbamates. For example, 3-hydroxyl carbofuran belonging to N-methyl amino formic acid aromatic ester, could generate a stable fragment ion m/z 181.1 by a neutral loss of CH_3NCO (−57 Da). The resulting spectrum is shown in Figure 3. According to this fragment mechanism, all fifteen carbamate pesticides were investigated and the vast majority of carbamate pesticides, except pirimicarb, could dissolve at the same active site and obtain the same characteristic loss of 57 Da (Figures 4 and 5).

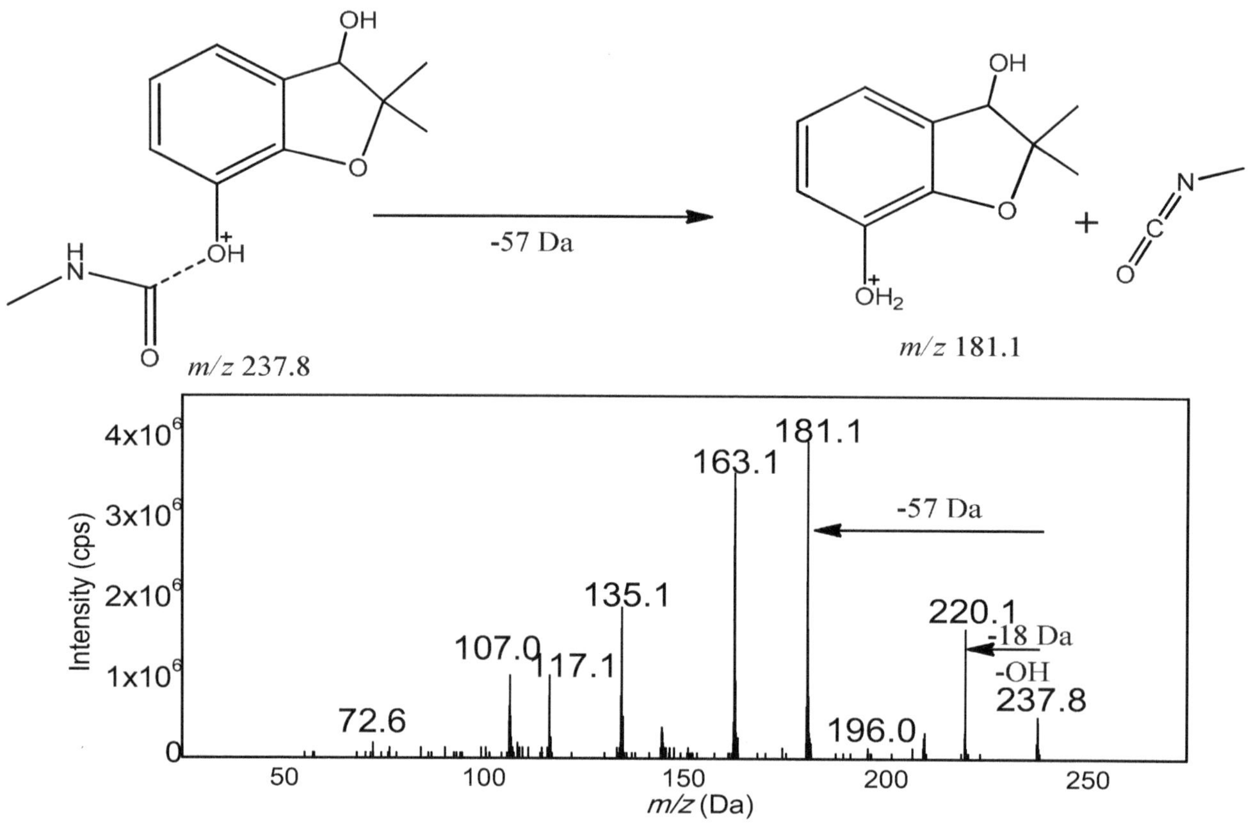

Figure 3. Product ion spectrum and probable fragmentation routes of Carbofuran-3-hydroxy.

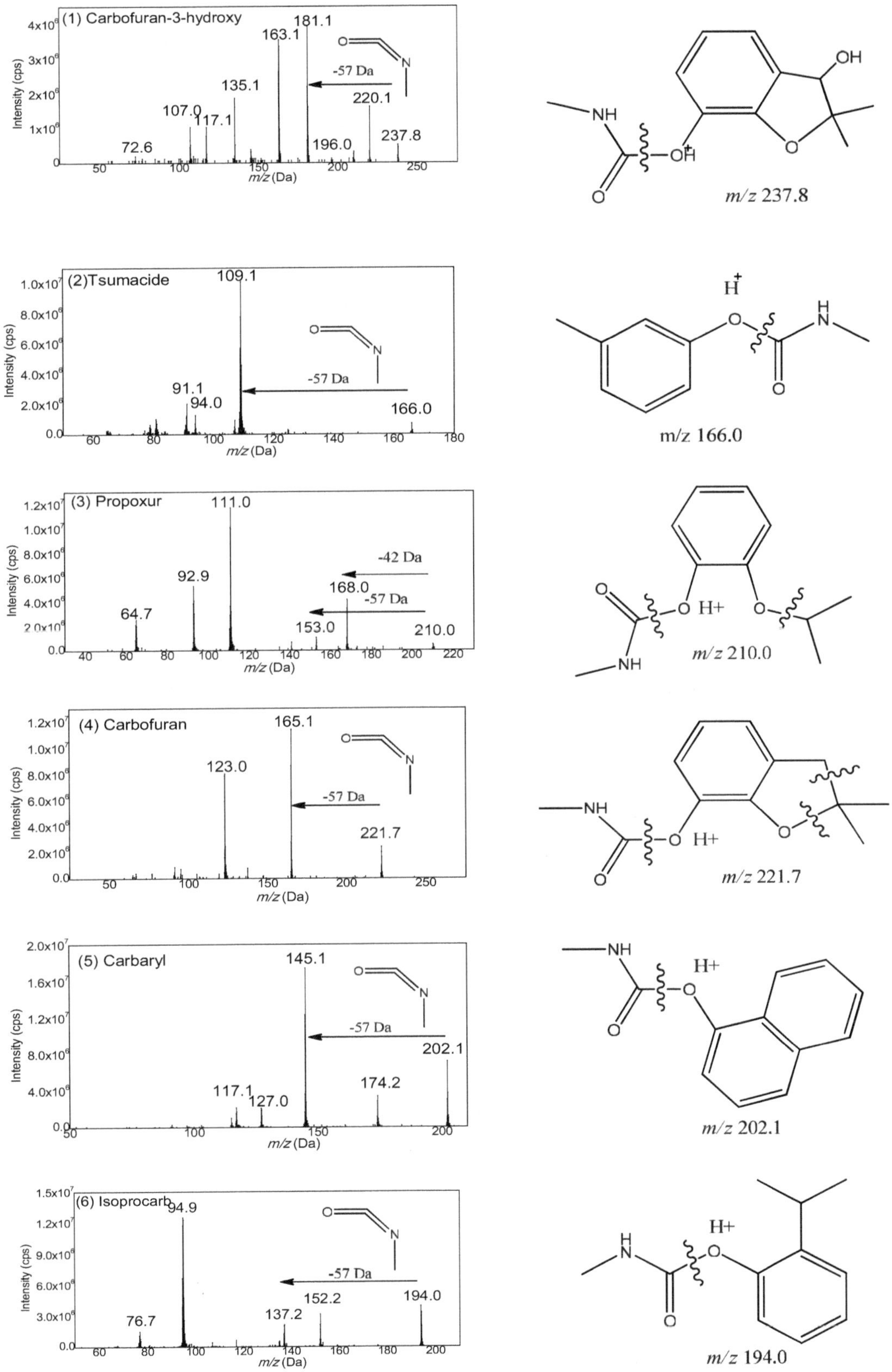

Figure 4. *Cont.*

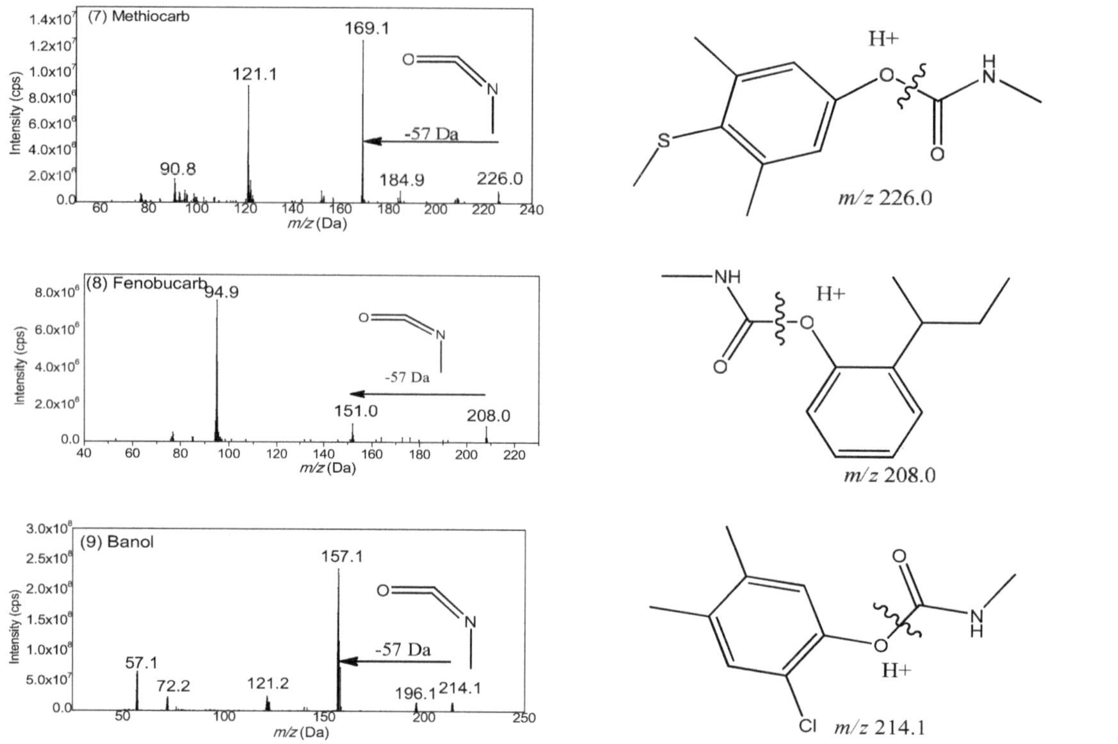

Figure 4. Product ion spectra and proposed fragmentation pathway of nine N-methyl amino formic acid aromatic ester: (**1**) Carbofuran-3-hydroxy; (**2**) Tsumacide; (**3**) Propoxur; (**4**) Carbofuran; (**5**) Carbaryl; (**6**) Isoprocarb; (**7**) Methiocarb; (**8**) Fenobucarb; and (**9**) Banol.

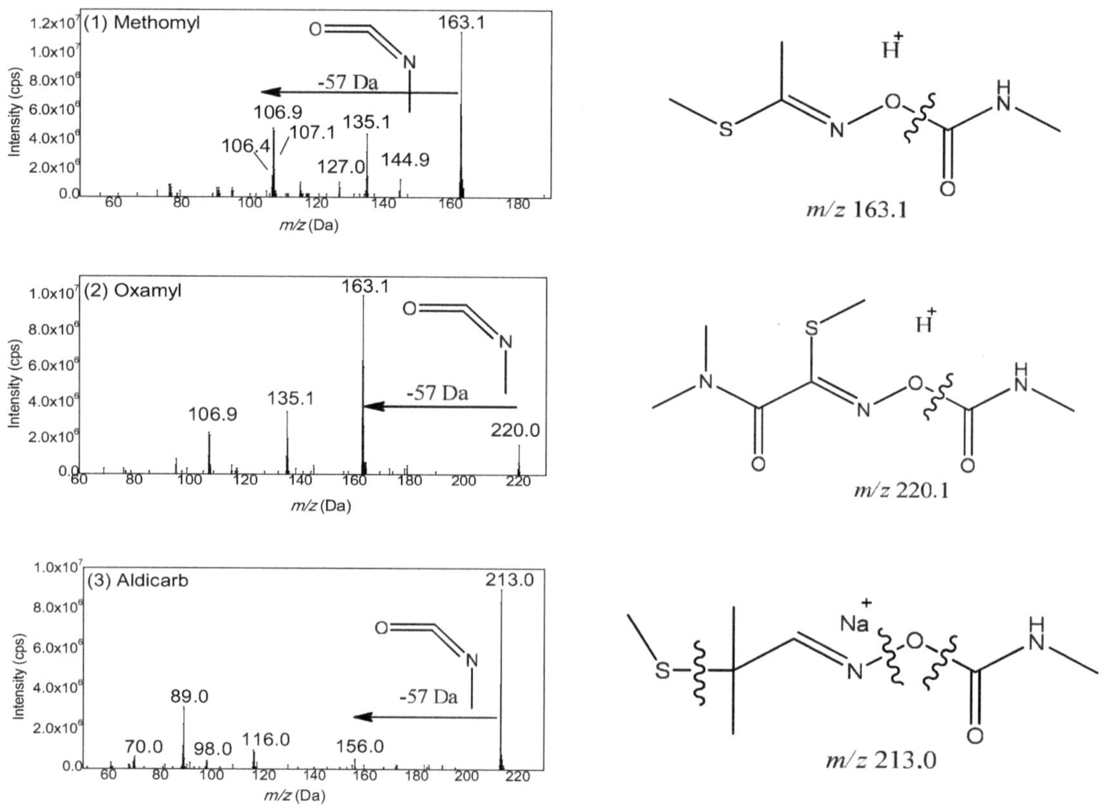

Figure 5. *Cont.*

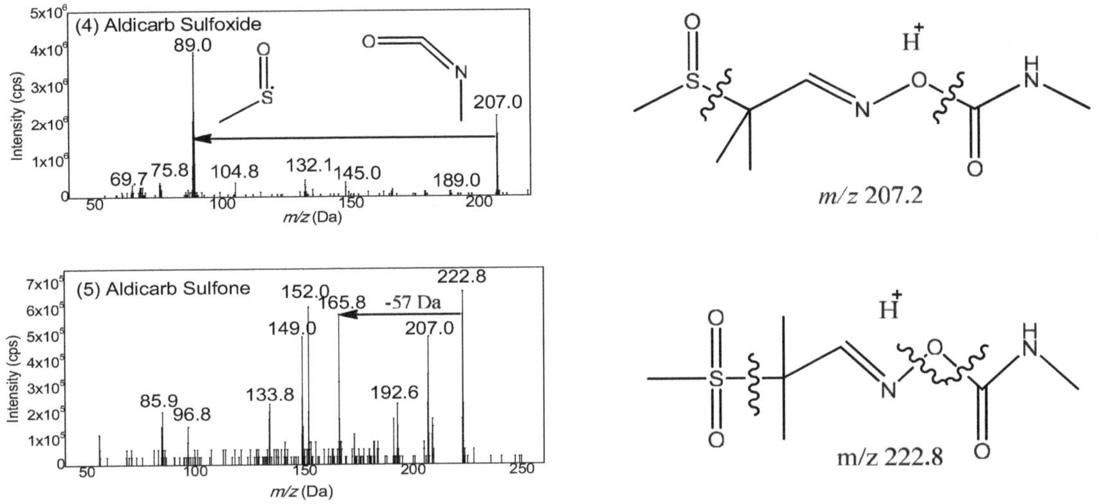

Figure 5. Product ion spectra and proposed fragmentation pathway of three N-methyl amino formic acid oxime ester: (**1**) Methomyl; (**2**) Oxamyl; (**3**) Aldicarb; (**4**) Aldicarb-sulfoxide; and (**5**) Aldicarb-sulfone.

Unexpectedly, 57 Da was lost in pirimicarb. Structurally speaking, it has two methyl groups linked to the amino group and it should produce a loss of 72 Da. In fact, the mass spectrum of pirimicarb indicated an unexpected loss of 57 Da rather than 72 Da, as shown in Figure 6. This is because of the nitrogen-containing pyrimidine heterocyclic ring adjoined the ester oxygen that could induce a methyl transfer reaction. According to a literature survey [37–39], the proton is localized at the pyrimidine nitrogen first; and then the charge-remote become the reaction center that forming an ion-neutral complex and induce an alkyl cation transfer.

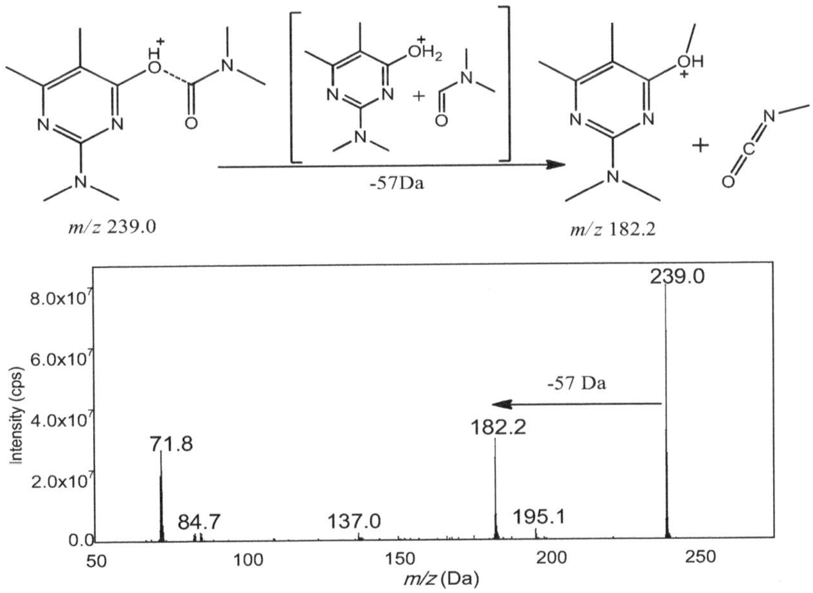

Figure 6. Product ion spectra and probable fragmentation routes of Pirimicarb.

2.4. Method Validation

2.4.1. Matrix Effect

As shown in Figure 7, the complex composition of diverse matrices might have different effects on the accuracy of identification and quantification. It was necessary to correct the matrix effects in LC-MS detection. Eight distinct types of organic vegetable samples or pesticide-free samples were used as matrix-matched blank to evaluate the matrix effects on the ionization of fifteen carbamate residues. The evaluation of matrix effects was calculated using the following equation [40]:

$$ME = \frac{A_{Matrix}}{A_S} \times 100\%$$

where A_{Matrix} is the peak area of the standard solution with the matrix-matched blank and A_S is the peak area of the standard solution with initial mobile phase. As shown in Table 3, the percentages of the matrix effects of 15 carbamate pesticides at three different concentrations (2, 20, and 200 ng mL^{-1}) ranged 95.4–111.2% (see Table 2 for details). Since there was no obvious ion suppression or ion enhancement at the chosen levels of quantification for 15 carbamate pesticides in eight distinct types of samples, the matrix effect can be ignored.

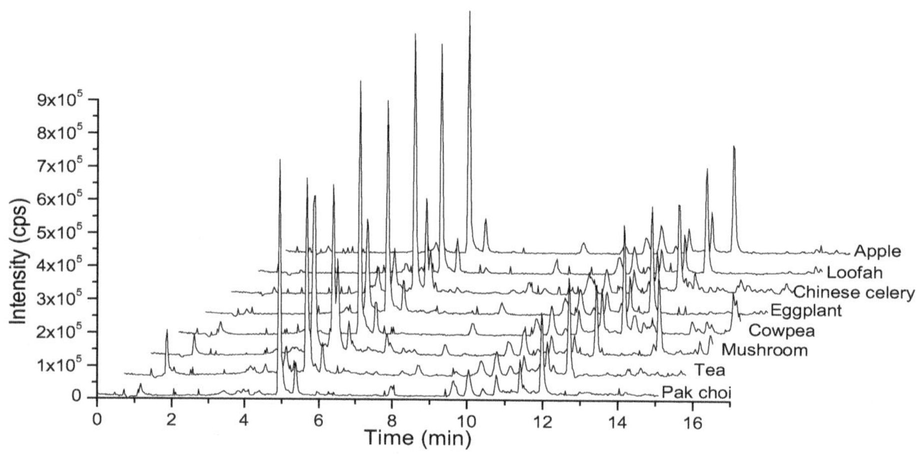

Figure 7. Typical total ion chromatograms of fifteen carbamates at 2 μg/kg spiked in eight matrices (cleaning by 50 mg C18 and 150 mg PSA).

Table 3. Matrix effects of the fifteen carbamates pesticides in distinct samples.

Compound	Matrix Effects/% (n = 3; 2, 20, 200 ng mL^{-1})							
	Pak Choi	Chinese Celery	Loofah	Eggplant	Cowpea	Apple	Mushroom	Tea
Aldicarb-sulfoxide	98.6	98.5	96.3	98.8	96.0	95.4	96.6	97.6
Aldicarb-sulfone	97.8	96.5	98.3	99.1	95.7	96.2	98.4	96.9
Pirimicarb	105.1	102.0	98.0	98.7	96.3	98.9	101.2	99.2
Carbofuran-3-hydroxy	102.3	101.3	103.5	107.5	101.2	102.5	101.0	105.4
Methomyl	110.2	108.6	107.2	108.1	105.4	111.2	105.7	109.2
Oxamyl	102.5	98.6	105.2	106.3	102.0	102.1	105.2	106.2
Aldicarb	107.5	102.1	100.2	101.3	100.8	103.2	100.5	102.3
Tsumacide	105.0	100.9	102.3	98.6	97.9	99.0	102.6	102.7
Propoxur	102.3	105.4	102.7	104.9	101.8	102.4	103.1	101.5
Carbofuran	102.5	105.9	106.2	103.1	98.7	99.2	97.4	98.5
Carbaryl	102.3	104.1	106.5	104.5	101.7	102.4	106.9	107.4
Isoprocarb	102.3	104.6	102.5	104.8	101.2	107.5	108.4	101.8
Methiocarb	105.9	104.1	102.4	106.2	107.1	105.1	106.3	105.1
Fenobucarb	102.1	105.2	102.6	107.8	110.0	108.4	107.6	106.8
Banol	105.6	104.2	102.9	104.1	103.2	102.5	102.3	101.1

2.4.2. Linearity and Analytical Limits

Satisfactory correlation coefficients (R > 0.996) were obtained for fifteen carbamate residues in matrix-matched blank over the concentration range of 0.05–200.0 ng mL^{-1}. The analytical limits of the proposed method were measured by the limit of detection (LOD) and the limit of quantification (LOQ) using the following equations [41]:

$$LOD = C_S \frac{3}{S/N}$$

$$LOQ = C_S \frac{10}{S/N}$$

where S/N is the average signal to noise ratio and C_S is the concentration of the specific pesticide. The estimated values were tested with a suitable number and kind of samples containing the 15 carbamate pesticides at the corresponding concentrations. All related parameters of the proposed method are summarized in Table 4. The LOD and LOQ were 0.2–2.0 μg kg^{-1} and 0.5–5.0 μg kg^{-1}, respectively, which demonstrated a good analytical limit of this method for carbamate residues.

Table 4. Linear ranges, linear equations, correlation coefficients (R) limits of detection (LODs) and limits of quantification (LOQs) of the fifteen carbamate pesticides.

Compound	Linear Range (ng mL^{-1})	Linear Equation	R ($1/X^2$)	LOD (μg kg^{-1})	LOQ (μg kg^{-1})
Aldicarb-sulfoxide	0.05–200	$Y = 2.42 \times 10^4X + 812.9$	0.9963	0.2	0.5
Aldicarb-sulfone	0.50–200	$Y = 1.44 \times 10^4X + 10{,}476$	0.9972	2.0	5.0
Pirimicarb	0.05–200	$Y = 5.49 \times 10^5X + 15{,}110.0$	0.9968	0.2	0.5
Carbofuran-3-hydroxy	0.10–200	$Y = 1.75 \times 10^4X + 457.2$	0.9964	0.3	1.0
Methomyl	0.05–200	$Y = 2.91 \times 10^4X + 102{,}771$	0.9962	0.2	0.5
Oxamyl	0.05–200	$Y = 8.38 \times 10^4X + 2287.1$	0.9965	0.2	0.5
Aldicarb	0.05–200	$Y = 1.96 \times 10^4X + 7283.2$	0.9973	0.2	0.5
Tsumacide	0.05–200	$Y = 6.64 \times 10^4X + 3330.4$	0.9967	0.2	0.5
Propoxur	0.05–200	$Y = 1.27 \times 10^5X + 3441.9$	0.9963	0.2	0.5
Carbofuran	0.20–200	$Y = 6.31 \times 10^4X + 543.2$	0.9982	0.3	1.0
Carbaryl	0.05–200	$Y = 6.01 \times 10^4X + 4445.6$	0.9968	0.2	0.5
Isoprocarb	0.05–200	$Y = 9.33 \times 10^4X + 15{,}304$	0.9957	0.2	0.5
Methiocarb	0.05–200	$Y = 9.86 \times 10^4X + 6565.0$	0.9965	0.2	0.5
Fenobucarb	0.05–200	$Y = 1.24 \times 10^5X + 7436.7$	0.9962	0.2	0.5
Banol	0.05–200	$Y = 1.01 \times 10^5X + 2436.7$	0.9971	0.2	0.5

Y, peak area; X, mass concentration, ng mL^{-1}; $1/X^2$, least square method.

2.4.3. Accuracy and Precision

The accuracy and precision of the method were measured by the recoveries and coefficients of variation (CVs) for intra- and inter-day. Therefore, the standard mixtures solution of fifteen carbamate residues was spiked into distinct types of samples including greengrocery, Chinese celery, loofah, eggplant, cowpea, apple, mushroom and tea leaves, and 24 spiked samples (eight types at three concentrations of 2, 20, and 200 μg kg^{-1}) were obtained. All spiked samples were detected three times a day on five different days following the process of Figure 1. The results are shown in the Supplementary Materials. The recovery of fifteen carbamate residues (88.1–118.4%) fell within the recommended Eurachem guidelines of 80–120% [42]. The analysis precision, measured as the coefficient of variation percentage (% CV) of the recovery (3.2–9.8%), was well under the criteria of 10% [42].

2.5. *Sample Analyses*

After validation of the analyti cal methodology through above experimentation, it was used to detect numerous types of samples including apple, carrot, chili pepper, Chinese celery, coriander, cowpea, crown daisy, eggplant, garlic sprout, grape, ginger, leek, lettuce, loofah, mushroom, needle mushroom, orange, pakchoi, radish, shiitake, spinach and tea leaves. Carbofuran and its metabolite Carbofuran-3-hydroxy were the predominant detected residues of the fifteen carbamates. Meanwhile, of the 26 samples (Table S3), eggplant contributed the detected residues of maximum frequency, with concentrations of carbofuran at 85.5, 77.4, 68.4, 35.6, 25.6, 17.3, 16.5 and 11.8 μg kg^{-1}, respectively. Pokchoi, spinach, and Chinese celery all had considerable detection rate and concentration of carbofuran. Among all samples, although green tea had the second highest detection rate, the concentration of carbamate residues were localized at relative low concentration of 6.4, 1.5, 1.2, 0.8, 0.6, 0.6, 0.4, 0.3, 0.2 and 0.2 μg kg^{-1}, respectively. This was mainly due to the further degradation of carbamate residues in the drying process of tea manufacturing. Cowpea occasionally had the highest concentration of carbofuran at 180.2 μg kg^{-1}.

3. Materials and Methods

3.1. Chemicals and Reagents

Acetonitrile and Methanol were of HPLC grade and obtained from Merck (Darmstadt, Germany). Formic acid (≥98%) was of HPLC grade and purchased from Aladdin (Shanghai, China). High purity water was obtained using a Milli-Q water purification system from Millipore (Bedford, MA, USA). Cleanert® Pesticarb, Cleanert® PSA and Cleanert® C18 for QuEChERS were purchased from Agela Technologies (Beijing, China). Millipore filters of polytetrafluoroethylene (0.22 μm) were purchased from ANPEL Lab (Shanghai, China).

Standards of fifteen carbamates were purchased from Dr. Ehrenstorfer GmbH (Augsburg, Germany): Aldicarb-sulfoxide, Aldicarb-sulfone, Pirimicarb, Carbofuran-3-hydroxy, Methomyl, Oxamyl, Banol, Aldicarb, Metolcarb, Propoxur, Carbofuran, Carbaryl, Isoprocarb, Mercaptodimethur and Fenobucarb. Individual standard solutions were prepared in HPLC-grade methanol at a concentration of 1.0 mg/mL and stored at −28 °C temporarily.

Samples were collected from local supermarkets, including vegetables (leeks, pakchoi, crowndaisy, coriander, spinach, lettuce, Chinese celery, garlic sprout, loofah, chili pepper, eggplants, cowpea, radish, and turnip), fruits (apples, kiwis, grapes, and oranges), tea (green tea, black tea, and flower tea), and edible fungus (shiitake, mushroom, and needle mushroom). Each fresh sample was crushed and mixed into homogenate, and stored in plastic bottles at −20 °C until analysis.

3.2. Instrumentation

HPLC analysis was performed on a Shimadzu LC-30AD system (Shimadzu, Kyoto, Japan) which consist of two interconnected pump units: one with an integrated degasser and the other with a mixer, and is comprised of a UHPLC gradient system, a refrigerated autosampler, and a column oven compartment. Mass spectrometric detection was performed on an AB SCIEX QTRAP® 5500 (AB SCIEX instruments, Foster, CA, USA) in MRM mode and EPI mode. A Turbo V™ Ion Source (ESI) interface in positive ionization mode was used. Both the UHPLC and mass spectrometer were controlled remotely using Analyst® software v. 1.6.2 (AB SCIEX instruments, Foster City, CA, USA). A Waters BEH C18 column (1.7 μm 2.1 mm × 100 mm, Waters, Milford, MA, USA) was applied for analysis. The mass spectrometer was equipped with an electrospray ionization source and spectra were acquired in the positive ion multiple reaction monitoring (MRM) mode and enhanced product ion (EPI) scan modes. MS was optimized using a capillary voltage of 5.50 kV and desolvation temperature of 500 °C. The cone gas pressure and desolvation gas pressure were 50 psi. Nitrogen was used as the cone and collision gasses, respectively. The raw data were analyzed using an Analyst 1.6.2 workstation (AB SCIEX, Foster, CA, USA).

3.3. Sample Extraction

Five grams of the homogenized sample was weighed in 50 mL Corning extraction tubes, and 5 mL of acetonitrile was added. The tubes were vibrated vigorously for 20 min in batches by a Multi Reax Vortexer (Heidolph, Schwabach, Germany). After the vortex, 1.0 g of sodium chloride and 2.0 g of sodium sulfate were separately added to each tube for the elimination of moisture, and the tubes were immediately manually shaken vigorously for 1 min to prevent sodium sulfate agglomeration. The process was followed by centrifugation at 8000 rpm for 3 min at 4 °C in a Sigma 2–16 K. centrifuge (Sartorius AG, Göttingen, Germany). The supernatant was collected in 10 mL extraction tubes and subjected to QuEChERS dispersive SPE cleaning. The dSPE agent contained 100 mg of primary secondary amine (PSA), 50 mg of C18EC, and 200 mg of magnesium sulfate. The tubes were vibrated for 30 s and centrifuged in the Sigma 2–16 K centrifuge at 13,000 rpm for 2 min at 20 °C. The supernatants were collected and filtered through 0.22 μm Anpel Syringe Hydrophilic PTFE filters (Anpel, Shanghai, China) before LC-MS/MS analysis.

Organic vegetables were cultivated by Ying Zhou's mother from her garden. The samples were homogenized with an Ultra-Turrax T 25 homogeniser (IKA® Werke, Staufen, Germany) and extracted using the QuEChERS method.

4. Conclusions

In this paper, a quick and credible method to quantify and confirm fifteen carbamate pesticide residues by modified QuEChERS sample pretreatment combined ultra-high performance liquid chromatography–quadrupole-linear ion trap mass spectrometry was systematically developed. For modified QuEChERS part, the composition of the cleaning agents was optimized to adapt the general samples. For the instrumental analysis, based on traditional conformation criteria including selected ion pairs, retention time and relative intensities between transitions from MRM scan mode, the presence of carbamate pesticides in doubtful samples was further confirmed by the matching of carbamate pesticide spectra via EPI scan mode. Then, the fragmentation routes of all the carbamates were preliminarily explained based on the mass spectra that the carbamate could generate a stable characteristic fragment ion by a neutral loss of CH_3NCO (−57 Da). The linearity, matrix effect, analysis limits, accuracy and repeatability were validated in diverse samples. All fifteen carbamates have good correlation coefficients above 0.996 at the range of 0.05–200 ng mL^{-1}. The recoveries of intra- and inter-day experiments were in the range of 80–120% at three concentrations with coefficients of variation all better than 10%. Moreover, the whole process of one dozen samples from pretreatment to final report did not exceed 2 h, which is shorter and exacter than traditional methods. Finally, the method was applied to carbamate residues detection in 26 kinds of samples from local markets, in which carbamates were extensively detected but all below the standard of maximum residue limit.

Author Contributions: Y.Z. conceived and designed the study; M.G., J.G., S.L. and W.G. performed the experiments; and Y.Z. analyzed the data and wrote the paper.

Acknowledgments: This work was supported by all colleagues of Jiaxing Center for Disease Control and Prevention. Special thanks are given to Ying Zhou's mother, Yafen Wang, for the organic vegetables and fruits from her garden.

References

1. Carbamate. Available online: https://en.wikipedia.org/wiki/Carbamate#Carbamate_insecticides (accessed on 15 August 2018).
2. Fukuto, T.R. Mechanism of action of organophosphorus and carbamate insecticides. *Environ. Health Persp.* **1990**, *87*, 245–254. [CrossRef] [PubMed]
3. Richard, D. Insecticides. Action and Metabolism. *Bull. Entomol. Soc. Ameri.* **1968**, *14*, 258–259. [CrossRef]
4. Geoffrey, M.C.; Wayne, T.S.; Margot, B.; Jerome, M.B.; Louise, N.M. Surveillance of pesticide-related illness and injury in humans. In *Handbook of Pesticide Toxicology*, 2nd ed.; Robert, K., Ed.; Academic Press: Cambridge, MA, USA, 2001; Volume 1, pp. 603–641.
5. Michal, R.; Shlomo, A. Evaluation of the decarbamylation process of cholinesterase during assay of enzyme activity. *Clin. Chim. Acta* **1995**, *240*, 107–116. [CrossRef]
6. Announcement of Food Safety Inspection. Qinhai Food and Drug Administration. No. 68, 2018. Available online: http://www.sdaqh.gov.cn/html/201896/n140026114.html (accessed on 10 September 2018).

7. Tsiplakou, E.; Anagnostopoulos, C.J.; Liapis, K.; Haroutounian, S.A.; Zervas, G. Pesticides residues in milks and feedstuff of farm animals drawn from Greece. *Chemosphere* **2010**, *80*, 504–512. [CrossRef] [PubMed]
8. Daphne, B.M. Public health impacts of organophosphate and carbamates. In *Toxicology of Organophosphate & Carbamate Compounds*; Gupta, R.C., Ed.; Academic Press: London, UK, 2005; pp. 599–601.
9. Marina, P.G.; Margarita, L.D.; Rajnarayanan, R.V. Carbamate Insecticides Target Human Melatonin Receptors. *Chem. Res. Toxicol.* **2017**, *30*, 574–582. [CrossRef]
10. Pinto, M.I.; Gerhard, S.; Bernardino, R.J.; Noronha, J.P. Pesticides in water and the performance of the liquid-phase microextraction based techniques. A review. *Microchem. J.* **2010**, *96*, 225–237. [CrossRef]
11. María, A.R.; Javier, H.B.; Teresa, M.B.M.; Miguel, Á.R.D. Ionic liquid-dispersive liquid-liquid microextraction for the simultaneous determination of pesticides and metabolites in soils using high-performance liquid chromatography and fluorescence detection. *J. Chromatogr. A* **2011**, *1218*, 4808–4816. [CrossRef]
12. Wang, J.; Zhang, K.; Tech, K.; Brown, D. Multiresidue Pesticide Analysis of Ginseng Powders Using Acetonitrile- or Acetone-Based Extraction, Solid-Phase Extraction Cleanup, and Gas Chromatography-Mass Spectrometry/Selective Ion Monitoring (GC-MS/SIM) or -Tandem Mass Spectrometry (GC-MS/MS). *J. Agric. Food Chem.* **2010**, *58*, 5884–5896. [CrossRef] [PubMed]
13. Anastassiades, M.; Scherbaum, E.; Tasdelen, B.; Stajnbaher, D. Recent Developments in QuEChERS Methodology for Pesticide Multiresidue Analysis. In *Pesticide Chemistry: Crop Protection, Public Health, Environmental Safety*; Ohkawa, H., Miyagawa, H., Lee, P.W., Eds.; Wiley-VCH: Weinheim, Germany, 2007; pp. 439–458.
14. Wilkowska, A.; Biziuk, M. Determination of pesticides residues in food matrices using the QuEChERS methology. *Food Chem.* **2011**, *125*, 803–812. [CrossRef]
15. Rao, T.N.; Loo, B.H.; Sarada, B.V.; Terashima, C.; Fujishima, A. Electrochemical Detection of Carbamate Pesticides at Conductive Diamond Electrodes. *Anal. Chem.* **2002**, *74*, 1578–1583. [CrossRef] [PubMed]
16. Tao, L.; Su, H.; Qu, X.; Ju, P.; Lin, C.; Ai, S. Acetylcholinesterase biosensor based on 3-carboxyphenylboronic acid/reduced graphene oxide-gold nanocomposites modified electrode for amperometric detection of organophosphorus and carbamate pesticides. *Sens. Actuators B Chem.* **2011**, *160*, 1255–1261. [CrossRef]
17. Rajendra, P.; Niraj, U.; Vijay, K. Simultaneous determination of seven carbamate pesticide residues in gram, wheat, lentil, soybean, fenugreek leaves and apple matrices. *Microchem. J.* **2013**, *111*, 91–96. [CrossRef]
18. Sanchez, B.C.; Albero, B.; Tadeo, J.L. High-performance liquid chromatography multiresidue method for the determination of N-methyl carbamates in fruit and vegetable juices. *J. Food Prot.* **2004**, *67*, 2565–2569. [CrossRef]
19. Tsumura, Y.; Ujita, K.; Tonogai, Y.; Ito, Y. Simultaneous determination of aldicarb, ethiofencarb, methiocarb and their oxidized metabolites in grains, fruits and vegetables by high performance liquid chromatography. *J. Food Prot.* **1995**, *58*, 217–222. [CrossRef]
20. Bernal, J.L.; Nozal, M.J.; Torlblo, L.; Jlmenez, J.J.; Atienza, J. High-performance liquid chromatographic determination of benomyl and carbendazim residues in apiarian samples. *J. Chromatogr. A* **1997**, *787*, 129–136. [CrossRef]
21. Sandahl, M.; Mathiasson, L.; Jonsson, J.A. Determination of thiophanate-methyl and its metabolites at trace level in spiked natural water using the supported liquid membrane extraction and the microporous membrane liquid-liquid extraction techniques combined on-line with high-performance liquid chromatography. *J. Chromatogr. A* **2000**, *893*, 123–131. [CrossRef] [PubMed]
22. Vandecasteele, K.; Gaus, I.; Debreuck, W.; Walraevens, K. Identification and quantification of 77 pesticides in groundwater using solid phase coupled to liquid–liquid microextraction and reversed-phase liquid chromatography. *Anal. Chem.* **2000**, *72*, 3093–3101. [CrossRef] [PubMed]
23. Sanchez, B.C.; Albero, B.; Tadeo, J.L. Multiresidue analysis of carbamate pesticides in soil by sonication-assisted extraction in small columns and liquid chromatography. *J. Chromatogr. A* **2003**, *1007*, 85–91. [CrossRef]
24. Gou, Y.; Eisert, R.; Pawliszyn, J. In-tube solid-phase microextraction coupled to capillary LC for carbamate analysis in water samples. *Anal. Chem.* **2000**, *72*, 2774–2779. [CrossRef] [PubMed]
25. Rami, A.; Wolfgang, S. Effect of bromine oxidation on high-performance thin-layer chromatography multi-enzyme inhibition assay detection of organophosphates and carbamate insecticides. *J. Chromatogr. A* **2011**, *1218*, 2775–2784. [CrossRef]
26. Sabala, A.; Portillo, J.L.; Broto-Puig, F.; Comellas, L. Development of a new high-performance liquid

chromatography method to analyse N-methylcarbamate insecticides by a simple post-column derivatization system and fluorescence detection. *J. Chromatogr. A* **1997**, *778*, 103–110. [CrossRef]

27. Waters Technologies Ltd. *Waters ACQUITY UPLC H-Class FLR System for Carbamate Analysis Method Guide*; Waters Technologies Ltd.: Shanghai, China, 2010.
28. Huang, D.F. *Quantitative Analysis of 15 Carbamates in Vegetables Using DisQuE Cleanup and UHPLC with Mass Detection*; Waters Technologies Ltd.: Shanghai, China, 2008.
29. Fernández, M.; Picó, Y.; Mañes, J. Determination of carbamate residues in fruits and vegetables by matrix solid-phase dispersion and liquid chromatography-mass spectrometry. *J. Chromatogr. A* **2000**, *871*, 43–56. [CrossRef]
30. Chen, L.; Song, F.; Liu, Z.; Zheng, Z.; Xing, J.; Liu, S. Multi-residue method for fast determination of pesticide residues in plants used in traditional Chinese medicine by ultra-high-performance liquid chromatography coupled to tandem mass spectrometry. *J. Chromatogr. A* **2012**, *1225*, 132–140. [CrossRef] [PubMed]
31. Romero, R.G.; Frenich, A.G.; Vidal, J.L.M. Multiresidue method for fast determination of pesticides in fruit juices by ultra performance liquid chromatography coupled to tandem mass spectrometry. *Talanta* **2008**, *76*, 211–225. [CrossRef] [PubMed]
32. Didier, O.; Patrick, E.; Claude, C. Multiresidue analysis of 74 pesticides in fruits and vegetables by liquid chromatography-electrospray-tandem mass spectrometry. *Anal. Chim. Acta* **2004**, *520*, 33–45. [CrossRef]
33. Anderson, N.L.; Hunter, C.L. Quantitative Mass Spectrometric Multiple Reaction Monitoring Assays for Major Plasma Proteins. *Mol. Cell Proteom.* **2006**, *5*, 573–588. [CrossRef] [PubMed]
34. Multiple Quadrupoles, Hybrids and Variations. Available online: https://en.wikipedia.org/wiki/Quadrupole_mass_analyzer#Multiple_quadrupoles,_hybrids_and_variations (accessed on 28 August 2018).
35. Iwona, M.Z.; Barbara, W.; Andrzej, P. Comparison of the Multiple Reaction Monitoring and Enhanced Product Ion Scan Modes for Confirmation of Stilbenes in Bovine Urine Samples Using LC–MS/MS QTRAP® System. *Chromatographia* **2016**, *79*, 1003–1012. [CrossRef]
36. Sun, H.; Chai, Y.; Pan, Y. Dissociative Benzyl Cation Transfer versus Proton Transfer: Loss of Benzene from Protonated N-Benzylaniline. *J. Org. Chem.* **2012**, *77*, 7098–7102. [CrossRef] [PubMed]
37. Guo, C.; Yue, L.; Guo, M.; Pan, Y. Elimination of Benzene from Protonated N-Benzylindoline: Benzyl Cation/Proton Transfer or Direct Proton Transfer? *J. Am. Soc. Mass Spectrom* **2013**, *24*, 381–387. [CrossRef] [PubMed]
38. Li, F.; Zhang, X.; Zhang, H.; Jiang, K. Gas-phase fragmentation of the protonated benzyl ester of proline: Intramolecular electrophilic substitution versus hydride transfer. *J. Mass Spectrom.* **2013**, *48*, 423–429. [CrossRef] [PubMed]
39. Yue, L.; Guo, C.; Chai, Y.; Yin, X.; Pan, Y. Gas-phase reaction: Alkyl cation transfer in the dissociation of protonated pyridyl carbamates in mass spectrometry. *Tetrahedron* **2014**, *70*, 9500–9505. [CrossRef]
40. Bo, H.B.; Wang, J.H.; Guo, C.H. Determination of Strobilurin Fungicide Residues in Food by Gas Chromatography-Mass Spectrometry. *Chin. J. Anal. Chem.* **2008**, *36*, 1471–1475.
41. Chen, F.; Huber, C.; May, R.; Schröder, P. Metabolism of oxybenzone in a hairy root culture: Perspectives for phytoremediation of a widely used sunscreen agent. *J. Hazard. Mater.* **2016**, *306*, 230–236. [CrossRef] [PubMed]
42. Commission of the European Communities. *Commission Decision 2002/657/EC Implementing Council Directive 96/23/EC Concerning the Performance of Analytical Methods and the Interpretation of Results*; Commission of the European: Brussels, Belgium, 2002.

15

Flavor Profile Evolution of Bottle Aged Rosé and White Wines Sealed with Different Closures

Meng-Qi Ling [1,2], Han Xie [1,2], Yu-Bo Hua [3], Jian Cai [4], Si-Yu Li [1,2], Yi-Bin Lan [1,2], Ruo-Nan Li [1,2], Chang-Qing Duan [1,2] and Ying Shi [1,2,*]

1 Center for Viticulture & Enology, College of Food Science and Nutritional Engineering, China Agricultural University, Beijing 100083, China; knighhtt@cau.edu.cn (M.-Q.L.); m18800135515@163.com (H.X.); siyuli@cau.edu.cn (S.-Y.L.); clanyibin@gmail.com (Y.-B.L.); liruonan0011@163.com (R.-N.L.); chqduan@cau.edu.cn (C.-Q.D.)
2 Key Laboratory of Viticulture and Enology, Ministry of Agriculture and Rural Affairs, Beijing 100083, China
3 Shandong Taila Winery Co., Ltd., Weihai 264500, China; huayubo-111@163.com
4 College of Biological Resources and Food Engineering, Qujing Normal University, Qujing 655011, China; caijian928@outlook.com
* Correspondence: shiy@cau.edu.cn

Abstract: Bottle aging is the final stage before wines are drunk, and is considered as a maturation time when many chemical changes occur. To get a better understanding of the evolution of wines' flavor profile, the flavor compounds (phenolic and volatile compounds), dissolved oxygen (DO), and flavor characters (OAVs and chromatic parameters) of rosé and dry white wines bottled with different closures were determined after 18 months' bottle aging. The results showed the main phenolic change trends of rosé wines were decreasing while the trends of white wines were increasing, which could be the reason for their unique DO changing behaviors. Volatile compounds could be clustered into fluctuating, increasing, and decreasing groups using k-means algorithm. Most volatile compounds, especially some long-chain aliphatic acid esters (octanoates and decanoates), exhibited a lower decrease rate in rosé wines sealed with natural corks and white wines with screw caps. After 18 months of bottle aging, wines treated with natural corks and their alternatives could be distinguished into two groups based on flavor compounds via PLS-DA. As for flavor characters, the total intensity of aroma declined obviously compared with their initial counterparts. Rosé wines exhibit visual difference in color, whereas such a phenomenon was not observed in white wines.

Keywords: rosé wines; white wines; bottle aging; flavor profile; closures

1. Introduction

Wine flavor is composed of a wide variety of compounds with different organoleptic properties, which will slightly evolve during bottle aging due to the limited quantities of oxygen penetrating through the closures [1–4]. The increase of dissolved oxygen (DO) in wines means the replenishment of oxygen is higher than the consumption of oxygen by antioxidants (such as phenolic compounds), while the decrease of DO represents a relatively higher consumption of oxygen [5]. Different types of closures exhibit different abilities in preventing oxygen penetration due to their structural differences [6,7]. Natural corks are the traditional choice of closure in the wine industry, but other types of closures are also used by wine producers. Agglomerated corks and technical corks are made of offcuts of oak wood, thus can be named oak-based corks as opposed to natural corks. In comparison with oak-based corks, polymer synthetic plugs and screw caps are more economical and less dependent on the raw material limitation.

Bottle aging is an important period when wine's flavor characters must be preserved as much as possible. The main function of closures is to ensure a good seal and to prevent any organoleptic deterioration of wine during storage. However, during bottle aging, various reactions may occur, such as oxidation, hydrolysis, and reactions caused by charge transfer and formation of covalent bonds, which will influence wine flavor evolution [6]. Wine aroma quality like the fruity and floral perception usually decreases due to the diminishment of critical aroma compounds including long-chain aliphatic acid ethyl esters, terpenes, and norisoprenoids [8,9]. Wine astringency also decreases because of a decline in the mean polymerization degree of tannins [10], while the hue and color stability usually increase due to the formation of stable orange-yellow pigments such as pyranoanthocyanins [11]. So far, most research observing wine flavor changes during bottle aging has focused on dry red wines [1,12], or is only concerned about the flavor quality, determined either by volatile compounds or non-volatile compounds [3,4,13,14]; comprehensive investigations monitoring the evolution of rosé and dry white wines' flavor profiles and oxidation patterns during bottle aging are quite limited.

The objectives of our research were: (1) understanding the evolution of flavor compounds of rosé and white wines during an 18-month bottle aging; (2) observing the oxidation pattern differences of rosé and white wines during an 18-month bottle aging and finding out the possible reasons that caused the differences; (3) comparing the effects of natural cork and its alternatives on the flavor profiles of rosé and white wines after an 18-month bottle aging.

2. Results and Discussion

2.1. *Evolution of Phenolic Compounds and Dissolved Oxygen during Bottle Aging*

Seventeen phenolic compounds were quantified in rosé wines (Figure 1a) and only four non-anthocyanin compounds were quantified, which might be due to the short maceration time during winemaking. Phenolic compounds in all rosé wine samples were clustered into two groups using k-means algorithm after normalization process (dividing the concentration by the maximum value of each compounds among all samples). A boxplot based on the normalized data was carried out to exhibit the change trend of phenolic compounds of each cluster, the fluctuating group (Cluster 1) and the decreasing group (Cluster 2) (Figure 1b). The majority of phenolic compounds belonged to Cluster 2, which was similar to previous reports on red wines [11,15]. Considering the limited non-anthocyanin phenolics detected in rosé wines, the decrease of major anthocyanins should not be strongly correlated with the formation of polymeric pigments [10]. Therefore, these compounds should participate in reactions such as oxidation and degradation [16]. Only caffeic acid and 4-hydroxycinnamic acid were detected in white wines, which increased from 1.07 ± 0.17 mg/L to 1.98 ± 0.26 mg/L and 0.97 ± 0.04 mg/L to 1.6 ± 0.11 mg/L, respectively (see Table S1), due to hydrolysis of their tartaric esters [10].

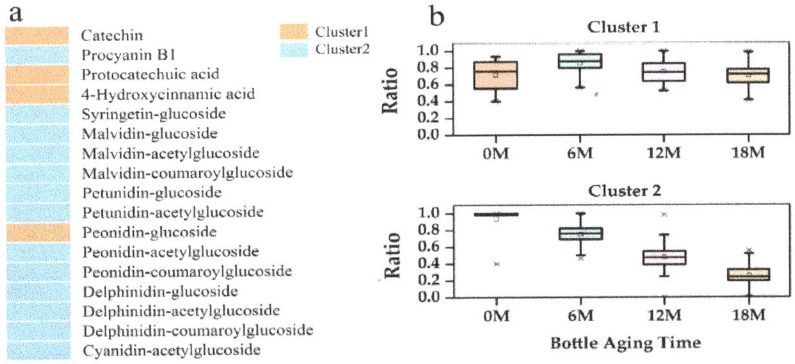

Figure 1. Statistical clustering for change trends of phenolic compounds in all rosé wines during an 18-month bottle aging. (**a**) General phenolic profile of rosé wines; (**b**) phenolic compounds' change trends.

The level of dissolved oxygen (DO) in wines mainly depends on oxygen concentration in the headspace at bottling and the ingress rate of oxygen into the bottle through closures [9]. During bottling, wines are exposed to air and thus have a chance to absorb oxygen. The oxidation substrates, generally phenolic compounds in wines, are positively correlated to wines' oxygen absorption capacity [6]. Rosé and white wines have unique DO evolution behaviors (Figure 2). In all rosé wines, the concentrations of DO decreased constantly during the first 10 months of bottle aging (Figure 2a), possibly because their major antioxidants, the phenolic compounds, were decreasing due to oxidation (Figure 1). However, in all white wines, the change trends of DO showed a slight decrease in the first two months, and then a drastic fluctuation (Figure 2b). The accumulation of DO meant the replenishment of oxygen was higher than the consumption. The oxygen dissolved in white wines probably came from the headspace oxygen, which was much higher in white wines (5.12 ± 0.76 mg/L) than in rosé wines (1.57 ± 0.73 mg/L). Moreover, caffeic acid and 4-hydroxycinnamic acid in white wines are good antioxidants. However, they might react slowly or not react at all until their concentrations reach a certain degree due to the hydrolysis of their tartaric esters, just like the formation of pinotins, a group of pyranoanthocyanins slowly formed by reactions between hydroxycinnamic acids and anthocyanins during wine aging [17]. After a four-month bottle aging, the accumulation of caffeic acid and 4-hydroxycinnamic acid was sufficient to trigger oxidation reactions, leading to a drastic decrease in DO (Figure 2b). Furthermore, white wines sealed with a screw cap had the lowest concentration of DO during bottle aging, which means a screw cap might provide better preservation for white wines. The DO concentrations were below the detection limit after 10 months of bottle aging for all rosé and white wines, indicating that the oxygen replenished from outside of the closures was far less than the oxygen consumption potential of wines. According to the results, we believe that although closures act as an oxygen barrier [18], DO's change trends during preservation depend more on the wine type and also the initial oxygen dissolved during the bottling process.

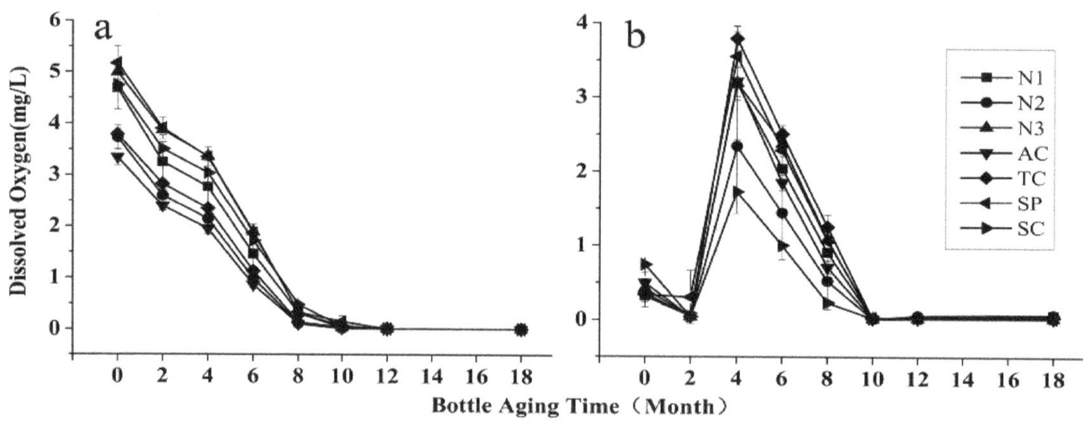

Figure 2. Change trends of DO in rosé (**a**) and white wines (**b**) during an 18-month bottle aging. (Symbols and abbreviations: N1: natural cork-1; N2: natural cork-2; N3: natural cork-3; AC: agglomerated cork; TC: '1+1' technical cork; SP: polymer synthetic plug; SC: screw cap.)

2.2. Evolution of Volatile Compounds during Bottle Aging

There were 72 volatile compounds quantified in this study, including 15 alcohols, 32 esters, five aliphatic acids, seven terpenes, three norisoprenoids and 10 other volatile compounds (Figure 3a). Detailed information is shown in Table S2. These two types of wines shared similar aromatic profiles, except there were more terpenes only quantified in whites (terpinolene, α-terpineol, and β-farnesene) to represent floral notes (Figure 3a). (E)-2-hexen-1-ol, 1-heptanol, benzyl alcohol, heptyl acetate, ethyl 2-hydroxy-4-methylpentanoate, benzaldehyde, and guaiacol were only quantified in rosé wines. Some volatile compounds, such as ethyl decanoate, n-decanoic acid, and dodecanoic acid, were reported to be associated with the oxidation of wines [8]. Wine oxidation was responsible for some unpleasant sensory descriptors ('green apple,' 'cooked potato,' and 'curry'), and could also interact with remaining pleasant aroma compounds, leading to the suppression of certain positive attributes [19]. The evolution of volatile compounds in wines during bottle aging was analyzed by k-means algorithm. It was clear that all compounds could be classified into three groups, namely, Cluster 1, Cluster 2, and Cluster 3, shown in the boxplot based on the normalized data from k-means algorithm (Figure 3b,c). Cluster 1 compounds stayed stable and were mostly alcohols, some ethyl esters, aliphatic acids, and a few terpenes (linalool and terpinen-4-ol). Compounds in Cluster 2 and 3 exhibited an increasing and decreasing trend, respectively, which indicated that they were the main changing compounds. Overall, more volatile compounds had a decreasing trend than an increasing trend during bottle aging. Almost half of the esters detected in our study were found in Cluster 3, which corroborates previous studies' finding that esters decreased during ageing due to hydrolysis [2,20]. By calculating the absolute values of the differences between final concentrations (after an 18-month bottle aging) and initial concentrations of those changing compounds, we used clustering analysis to differentiate wines sealed with natural corks and their alternatives (Figure 4).

For rosé wines, wines treated with similar closures (natural corks or alternatives of natural corks) did not show much consistency based on Cluster 2 compounds (Figure 4a), while Cluster 3 compounds could precisely distinguish wines with natural corks from their alternatives (Figure 4b). In Figure 4b, the difference between final concentration and initial concentration of Cluster 3 compounds showed less of a heat response in rosé wines sealed with natural corks than those sealed with alternative closures, which meant natural corks could better prevent aroma loss, especially for acetates (heptyl acetate, hexyl acetate, 3-hexen-1-ol, phenethyl acetate). β-Damascenone and (6E)-nerolidol decreased more in rosé wines sealed with a '1+1' technical cork and agglomerated cork, while the compounds in wines with a polymer synthetic plug seemed to have the largest decrease rate during bottle aging.

As for white wines, agglomerated cork and '1+1' technical cork were mixed with natural corks in a clustering analysis for compounds in Cluster 2 and 3, mainly because they were all oak-based closures (Figure 4c,d). In Cluster 2, α-ionone and TDN had a relatively lower increase rate in wines with natural corks compared to their alternatives, which might be due to the 'scalping phenomenon' of corks. Some studies have showed that cork and synthetic closures may scalp several aroma compounds from wine, such as TDN and methoxypyrazines [18] (Figure 4c). In Figure 4d, white wines with a screw cap had the most compounds that exhibited less heat response among all samples, which meant the screw cap was better ay preserving aroma quality in white wines. This might be due to there being the lowest level of DO concentration in white wines sealed with a screw cap (Figure 2b). It was those long-chain aliphatic acid esters that decreased less in both rosé and white wines sealed with a screw cap, such as ethyl decanoate, ethyl 9-decenoate, ethyl dodecanoate, ethyl hexadecanoate, isoamyl decanoate, isoamyl octanoate, and isobutyl octanoate, which could differentiate a screw cap from other closures. The result corresponded to a previous study that found that long-chain aliphatic acid esters would decrease with a higher oxidation level [8], while a screw cap seemed to be the best choice in terms of minimizing wine oxidation [1,9]. On the contrary, agglomerated cork and polymer synthetic plug could lead to more loss in the aroma compounds (Figure 4d).

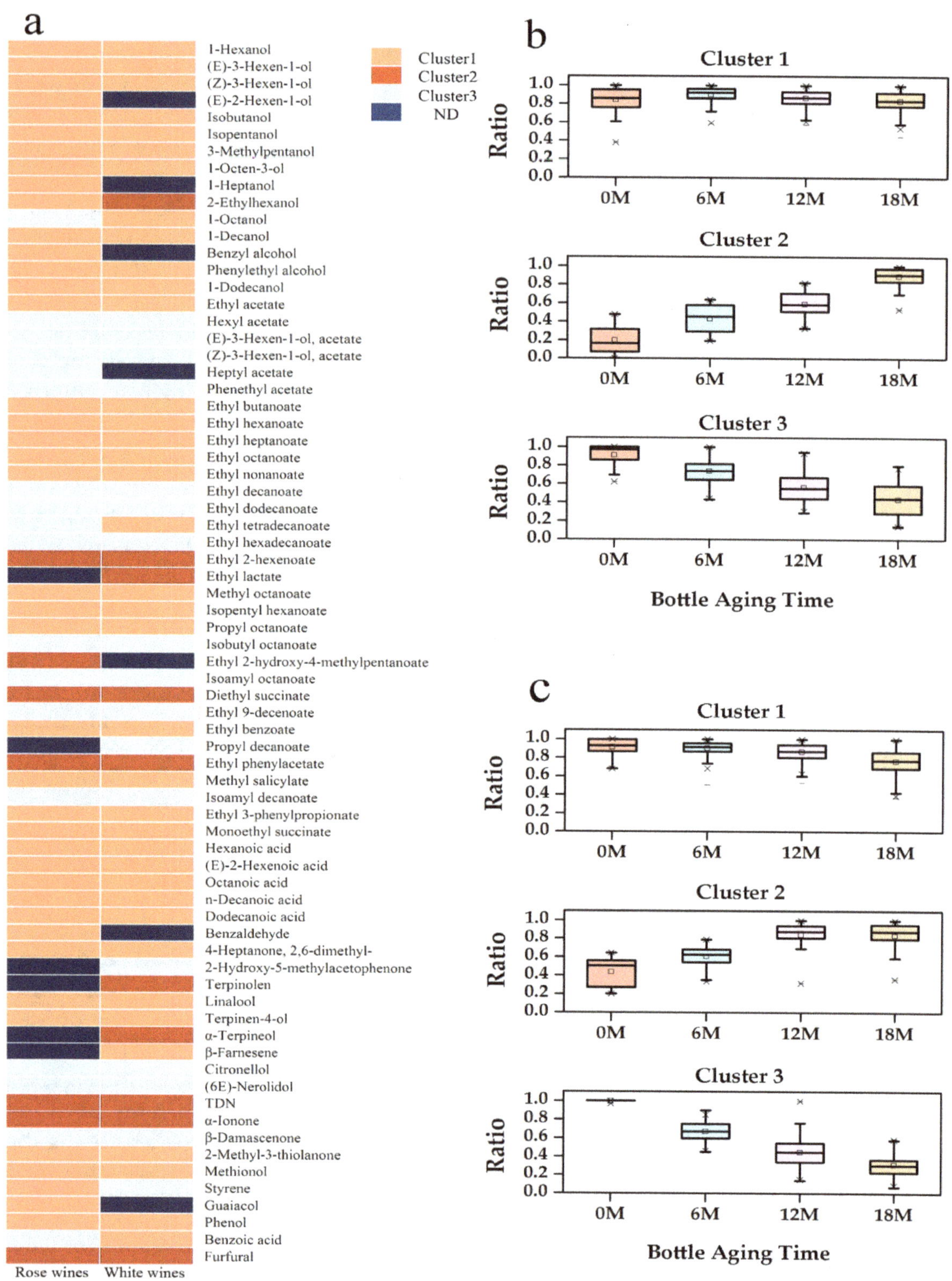

Figure 3. Change trends of volatile compounds during an 18-month bottle aging. (**a**) General volatile profile of rosé and white wines; (**b**) volatile compounds' change trends in rosé wines; (**c**) volatile compounds' change trends in white wines.

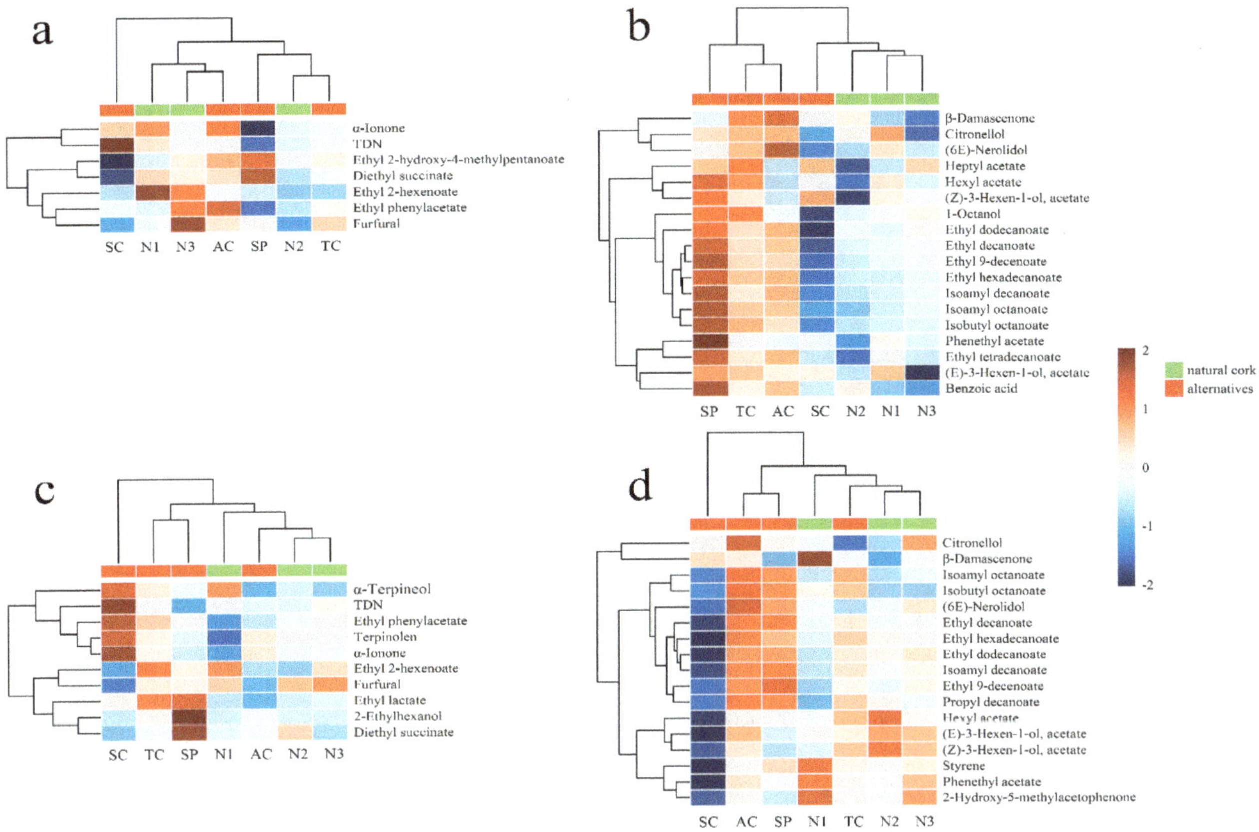

Figure 4. Clustering analysis of wines sealed with natural corks and their alternatives based on absolute values of the differences between final concentrations and initial concentrations. (**a**) Difference value of increasing volatile compounds in rosé wines; (**b**) difference value of decreasing volatile compounds in rosé wines; (**c**) difference value of increasing volatile compounds in white wines; (**d**) difference value of decreasing volatile compounds in white wines.

2.3. Flavor Profile Analysis of Wines after an 18-Month Bottle Aging

During bottle aging, wines are expected to maintain their original organoleptic characters as much as possible. In this section, partial least squares-discriminate analysis (PLS-DA) of different flavor compounds (phenolic and volatile compounds) was conducted in wines after an 18-month bottle aging. Wines sealed with three natural corks (N1, N2, N3) were set up as a group, while wines sealed with other closures (AC, TC, SP, SC) were set up as a single group of natural cork alternatives to see the effect of natural corks and their alternatives on wines' flavor profile (Figure 5). The use and interpretation of the VIP (variable importance in projection) values obtained from the PLS-DA model make it possible to determine potential flavor markers in the classes selected. Besides calculating the VIP values in PLS-DA, a one-way ANOVA test was also applied to check the statistical differences between different flavor compounds and guarantee a more statistically reliable selection of critical flavor markers (Table 1). Moreover, by calculating OAVs and chromatic parameters in all initial wines (prior to bottle aging) and final wines (after an 18-month bottle aging), the aromatic and chromatic characters of these wines were also compared.

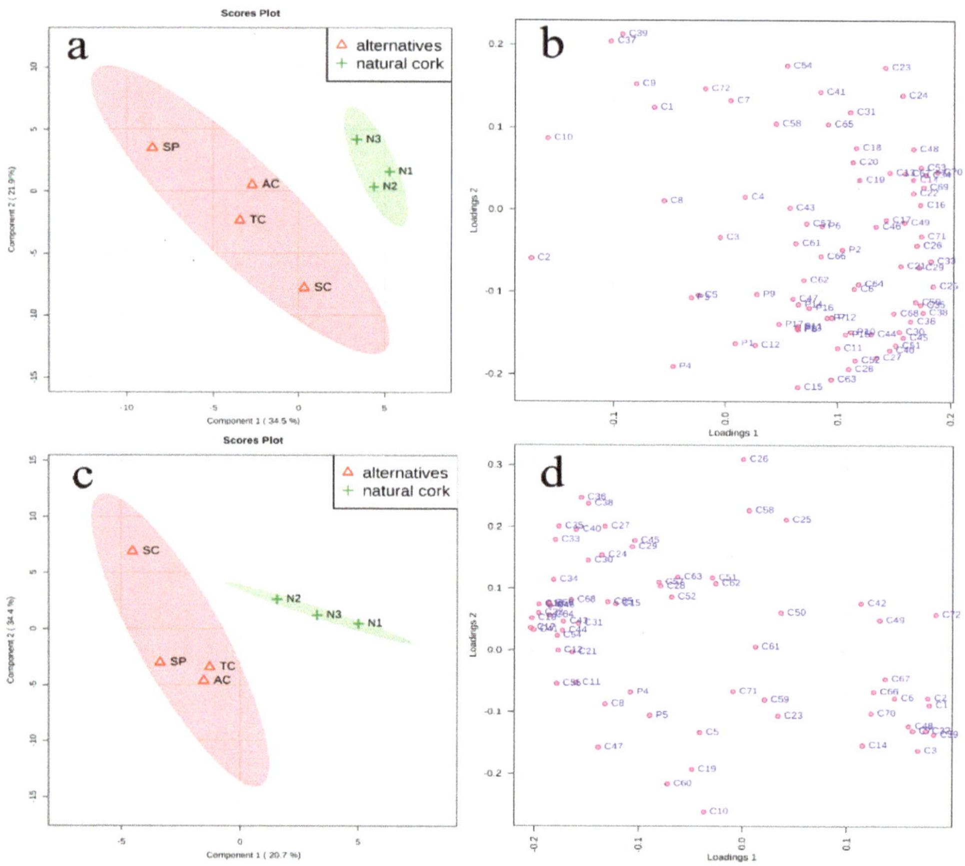

Figure 5. PLS-DA of flavor compounds in 18-month bottle-aged wines treated with natural corks and their alternatives. (**a**) PLS-DA model for rosé wine differentiation; (**b**) scattering plot of PLS-DA model for rosé wines; (**c**) PLS-DA model for white wine differentiation; (**d**) scattering plot of PLS-DA model for white wines. Note: Flavor compounds′ numbers in (**b**,**d**) are provided in Tables S1,S2.

Table 1. Differentiated flavor compounds in rosé and white wines (VIP > 1, p < 0.05).

No. [a]	CAS	Compounds [b]	'Cabernet Sauvignon' Rosé Wines			'Chardonnay' Dry White Wines		
			VIP1	VIP2	p values	VIP1	VIP2	p values
C10	104-76-7	2-Ethylhexanol	1.145	1.055	**0.024**	1.045	1.229	**7.07×10^{-7}**
C12	112-30-1	1-Decanol	0.286	0.617	0.001	1.443	1.022	**0.035**
C24	106-30-9	Ethyl heptanoate	1.876	1.720	**0.022**	0.348	1.025	0.992
C29	124-06-1	Ethyl tetradecanoate	1.404	1.245	**0.022**	0.155	0.944	0.029
C33	111-11-5	Methyl octanoate	1.266	1.187	**0.021**	0.711	1.121	0.954
C38	2035-99-6	Isoamyl octanoate *	1.136	1.111	**0.002**	0.117	1.44	0.678
C48	142-62-1	Hexanoic acid **	1.655	1.464	**0.022**	0.717	0.917	0.975
C55	1450-72-2	2-Hydroxy-5-methylacetophenone	Trace	Trace	Trace	1.673	1.218	**0.013**
C60	18794-84-8	β-Farnesene	Trace	Trace	Trace	1.053	0.963	**0.000**
C64	8013-90-9	α-Ionone *	0.341	0.851	0.003	1.464	1.039	**0.037**
C65	23726-93-4	β-Damascenone **	1.301	1.274	**0.014**	0.428	0.883	0.749
C68	100-42-5	Styrene	0.691	0.934	0.001	1.121	0.847	**0.010**
C69	90-05-1	Guaiacol *	1.777	1.575	**0.04**	Trace	Trace	Trace

The data in bold were the most differentiated compounds (VIP>1, p < 0.05); [a.] The No. was in accordance with Table S2; [b.] * OAV > 0.1, ** OAV > 1.0.

2.3.1. Flavor Compounds

In rosé wines, principal component 1 (PC1) and principal component 2 (PC2) accounted for 56.4% of the total variance (Figure 5a). According to variable importance in projection values (VIP > 1) of the first two PCs from the PLS-DA model and P values (p < 0.05) from a one-way ANOVA test, differentiated flavor compounds in rosé wines sealed with natural corks and their alternatives were selected (Table 1). Closures seemed to have more effect on volatile compounds than phenolic compounds because those differentiated flavor compounds were all volatile compounds. Only rosé wines with natural corks were in the quadrant where PC1 and PC2 were both positive. PC1, which accounted for 34.5% of the total variance, could distinguish wines sealed with natural corks from those with a polymer synthetic plug (SP) (Figure 5a). As shown in Figure 5b, the positive direction of PC1 was mostly driven by esters, such as ethyl octanoate (C25), methyl octanoate (C33), isopentyl hexanoate (C34), and isoamyl octanoate (C38). Except for 2-ethylhexanol (C10), the other differentiated compounds all lay in the positive part in PC1. Ethyl heptanoate (C24), hexanoic acid (C48), β-damascenone (C65), and guaiacol (C69) had a higher concentration in wines with natural corks (see Table S2). All phenolic compounds, especially 4-hydroxycinnamic acid (P4) and catechin (P1), drove PC2 in the negative direction, where wines with a screw cap lay.

As for white wines, PC1 and PC2 accounted for 55.1% of the total variance (Figure 5c). Since phenolic compounds are quite limited in white wines, aroma properties play a more important role from an organoleptic perspective. According to the loading plot, the quadrant where PC2 was positive and PC1 was negative contained the most types of volatile compounds (alcohols, esters, aliphatic acids, terpenes, and norisoprenoids) (Figure 5d). Wines sealed with a screw cap lay in this quadrant. The most differentiated compounds (VIP > 1, $P < 0.05$) in white wines were 2-ethylhexanol (C10), 1-decanol (C12), 2-hydroxy-5-methylacetophenone (C55), β-farnesene (C60), α-ionone (C64), and styrene (C68). Those volatile compounds characterized the negative direction of PC1 (Table 1, Figure 5d) to distinguish wines sealed with a screw cap from those with natural corks. Wines sealed with agglomerated cork (AC), '1+1' technical cork (TC), and polymer synthetic plug (SP) were together in the third quadrant, close to the 'zero' of plot score.

Comparing the flavor compounds in rosé and white wines, wines with natural corks could be distinguished from their alternatives by the results of PLS-DA. The only differentiated flavor compound rosé and white wines had in common was 2-ethylhexanol, which was related to the citrus odor. The quadrant where rosé wines with natural corks lay contained more compounds, indicating that natural corks could perform better in preventing flavor compounds from loss during bottle aging in rosé wines. Whereas, in white wines, it was wines with a screw cap that stood out with more volatile compounds, which might be due to having the best sealing environment. This result was also in accordance with a previous study in Semillon wines that bottling with a screw cap tends to lead to a higher fruity, citrus sensory score and less of an oxidized aroma [9].

2.3.2. Flavor Characters

To learn more about the aroma profile in rosé and white wines, Duncan's multiple range tests based on OAVs were applied to identify which types of closures had a different effect on wines after an 18-month bottle aging (Table 2). Overall, the rosé and white wines were both fruity, floral, and sweet in odor, which makes sense given that most of the volatile compounds quantified in this study were esters and the thresholds of terpenes and norisoprenoids were usually low. White wines exhibited more 'berry' and 'sweet' odors, mainly due to the high level of ethyl esters, such as ethyl acetate, ethyl octanoate, ethyl decanoate, and ethyl dodecanoate (Table S2). During an 18-month bottle aging, the total intensity of aroma in both types of wines declined prominently, especially 'tropical fruity,' 'floral,' 'berry,' and 'sweet' aroma descriptions. The results indicated that the period of bottle aging was likely to contribute to the loss of aroma properties. The decrease was more obvious in white wines, as the studies showed that white wines were very unstable and lost their desirable fresh and fruity characters over time [20]. According to the results of Duncan's multiple range tests, the difference between different closures mainly concerned the 'berry' and 'sweet' odors. Ethyl acetate, ethyl 2-hydroxy-4-methylpentanoate, ethyl decanoate, and β-damascenone were representative compounds of those odorant series. Wine samples sealed with a screw cap contained a higher concentration of long-chain aliphatic acid esters, which might be the reason why it performed better in preventing 'berry' and 'sweet' odors from declining. There was no significant difference between rosé wines sealed with three natural corks (N1, N2, N3) and a screw cap (SC), except for the 'berry' aroma description.

On the contrary, OAVs in rosé wines sealed with a polymer synthetic plug (SP) were significantly lower than those sealed with other closures after an 18-month bottle aging. White wines with agglomerated cork (AC) and polymer synthetic plug (SP) had the lowest OAVs among other types of closures, indicating that those two types of closures were not appropriate for the preservation of white wines during bottle aging.

Table 2. OAVs in all initial wines (0M) and final wines (after an 18-month bottle aging).

Aroma Description	0M	N1	N2	N3	TC	AC	SP	SC
	'Cabernet Sauvignon' rosé wines							
Tropical Fruity	56.01 ± 5.02	33.11 ± 1.07ab	31.9 ± 1.13ab	33.68 ± 2.62a	29.96 ± 0.65b	30.51 ± 0.11ab	30.26 ± 0.67b	31.77 ± 1.49ab
Floral	35.17 ± 3.08	12.98 ± 0.06ab	12.18 ± 0.5cd	13.48 ± 0.2a	11.49 ± 0.08de	11.28 ± 0.32e	12.44 ± 0.2bc	12.34 ± 0.37bc
Berry	14.86 ± 1.33	10.95 ± 0.65b	11.15 ± 0.8b	10.68 ± 1.08bc	9.74 ± 0.37bc	9.34 ± 0.21c	7.51 ± 0.2d	13.57 ± 0.62a
Herbaceous/Vegetal	1.27 ± 0.07	1.27 ± 0.05ab	1.31 ± 0.06a	1.27 ± 0.08ab	1.24 ± 0.03ab	1.24 ± 0.01ab	1.18 ± 0.02b	1.3 ± 0.06a
Chemical	8.57 ± 0.8	8.94 ± 0.43ab	8.47 ± 0.31ab	8.69 ± 0.94ab	7.83 ± 0.2b	8.09 ± 0.19b	6.65 ± 0.18c	9.32 ± 0.5a
Fatty	9.66 ± 0.27	9.01 ± 0.32a	9.04 ± 0.58a	9.01 ± 0.36a	8.8 ± 0.14a	8.58 ± 0.16a	7.24 ± 0.32b	8.99 ± 0.19a
Sweet	26.05 ± 2.21	21.02 ± 1.18ab	21.14 ± 1.16ab	20.97 ± 2.41ab	19.07 ± 0.65bc	19.15 ± 0.45bc	17.02 ± 0.43c	23.01 ± 1.12a
	'Chardonnay' dry white wines							
Tropical Fruity	44.8 ± 1.19	28.13 ± 0.21a	26.77 ± 1.12a	28.44 ± 1.13a	26.62 ± 2.71a	27.02 ± 0.5a	27.86 ± 2.97a	28.61 ± 1.03a
Floral	19.81 ± 0.35	8.93 ± 0.21b	9.66 ± 0.04a	9.41 ± 0.11a	9.47 ± 0.27a	9.43 ± 0a	9.6 ± 0.24a	9.57 ± 0.17a
Berry	21.12 ± 0.49	8.73 ± 0.39ab	8.13 ± 0.12bc	8.11 ± 0.61bc	7.56 ± 0.65bc	6.66 ± 0.42c	6.72 ± 1.16c	10.16 ± 0.54a
Herbaceous/Vegetal	1.3 ± 0.02	1.01 ± 0.17a	0.89 ± 0.03a	0.88 ± 0.04a	0.85 ± 0.04a	0.83 ± 0.02a	0.88 ± 0.1a	0.93 ± 0.05a
Chemical	10.68 ± 0.4	6.43 ± 0.02ab	5.99 ± 0.3ab	6.43 ± 0.52ab	5.82 ± 0.73b	5.68 ± 0.3b	5.33 ± 0.95b	7.14 ± 0.32a
Fatty	12.39 ± 0.36	11.74 ± 0.62ab	11.45 ± 0.48ab	10.77 ± 0.16b	11.46 ± 0.61ab	11.14 ± 0.34b	10.4 ± 0.74b	12.9 ± 0.94a
Sweet	34.94 ± 0.81	20.58 ± 0.15ab	18.68 ± 0.94ab	20.01 ± 1.09ab	18.34 ± 2.24ab	17.87 ± 0.59b	18.58 ± 2.87ab	22.09 ± 1.11a

Odor activity values (OAVs) were shown through average ± standard error. The odor thresholds were taken from [21–27]. The details are shown in Table S2. Different letters in the same row indicate significant differences at $p < 0.05$ by Duncan's multiple-range test.

As for chromatic characters, due to the decrease of anthocyanins in rosé wines during bottle aging, a^* values were much lower, while b^* values were higher in final wines than in initial wines (Figure 6a), indicating that the hue tended to undergo a yellowing phenomenon after 18 months of bottle aging. Although the ΔE values of rosé wines in our research ranged from 1.62 to 2.37 and the chromatic difference in wines was perceivable by human eyes when the ΔE value was above 2.8 [28], judging from the direct comparison in Figure 6a, visual differences existed in all rosé wines after an 18-month bottle aging compared to the initial wines. This may be due to the extreme brightness caused by low color substances in rosé wines, since it was reported that an ΔE value >1.0 in model solutions was visually perceivable by the human eye [28]. The ΔE value of white wines ranged from 0.08 to 0.33, corresponding to no visual differences in all white wines after an 18-month bottle aging (Figure 6b).

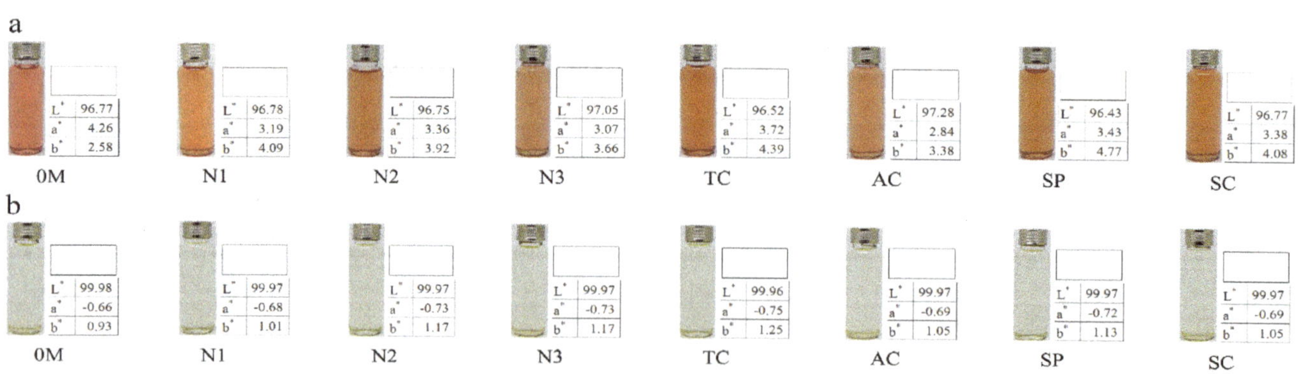

Figure 6. Sample photos and chromatic parameters of rosé (**a**) and white wines (**b**) prior to and after an 18-month bottle aging.

3. Materials and Methods

3.1. Wine Samples

'Cabernet Sauvignon' rosé wines and 'Chardonnay' dry white wines were made strictly according to the local published winemaking standards in October 2014, at Shandong Taila Winery Co., Ltd.

(Rushan, Shandong, China). Detailed information about the two original wine samples is given in Table S3. Bottling was performed at the packaging line of this winery to guarantee a 750 mL volume of each bottle. Seven types of closures were used for treatment, including three different natural corks [named natural cork-1 (N1), natural cork-2 (N2), and natural cork-3 (N3)] and four natural cork alternatives ['1+1' technical cork (TC), agglomerated cork (AC), polymer synthetic plug (SP) and screw cap (SC)]. The physical indexes of these closures are provided in Table S4. For each closure treatment, 30 bottles of wine were collected as samples. After bottling, wine samples were stored in a cellar, with an average temperature of 16 ± 1 °C and a relative humidity of 65 ± 5%. The length of the experiment was 18 months.

3.2. Oxygen Measurements

The dissolved oxygen (DO) levels were recorded by noninvasive oxygen sensors (5 mm sensor spots PSt3, NomaSense O_2 P6000, Yantai Vinventions Co., Ltd., Yantai, China). For each closure treatment, three bottles of wine were selected randomly to equip one sensor positioned at mid height and one in the neck of the bottle. Wine samples were analyzed every two months for oxygen data acquisition.

3.3. Flavor Compounds Detection

Phenolic compounds were detected via high-performance liquid chromatography/triple-quadrupole tandem mass spectrometry (HPLC-QqQ-MS/MS) method using an Agilent series 1200 instrument fitted with a Poroshell 120 EC-C18 column (150 × 2.1 mm, 2.7 μm, Agilent Technologies, Santa Clara, CA, USA). The method was established and validated in our earlier work, and is reliable for detecting and quantifying 45 non-anthocyanin and 95 anthocyanin compounds [29,30]. Volatile compounds were detected using headspace solid phase micro-extraction (HS-SPME) with a 2 cm DVB/CAR/PDMS 50/30 μm SPME fiber (Supelco, Bellefonte, PA, USA) and an Agilent 7890 gas chromatography kit equipped with an Agilent 5975 mass spectrometer (GC–MS). The details were given in our previous study [31]. Retention indices (RI) [adjected by C7–C24 n-alkane series (Supelco)] were compared with those in the NIST11 database via an Automatic Mass Spectral Deconvolution & Identification System (AMDIS) for identification. The ChemStation software (Agilent Technologies, Inc.) was used to calculated the peak areas and quantification was conducted on the basis of the calibration curve of volatile compound standards purchased from Sigma-Aldrich (St. Louis, MO, USA).

Three bottles of wine samples from each closure treatment were collected randomly every six months for flavor compound detection. Each sample was analyzed in duplicate.

3.4. Flavor Analysis

By calculating the ratio of the volatile compound concentration to its odor perception threshold, the odor activity value (OAV) was obtained. This analysis was used to assess the potential contribution of individual volatile compounds to wine aroma.

Chromatic parameters were measured in triplicate using a spectrophotometer Shimadzu UV-Vis 2600 (Shimadzu, Kyoto, Japan) [30] and analyzed via CIELab system to detect the lightness (L^*), reddish attribute (a^*), and yellowish attribute (b^*) of each wine [32]. The visual change in wine, described as ΔE, was calculated using the following equation [28]:

$$\Delta E* = [(\Delta L^*)^2 + (\Delta a^*)^2 + (\Delta b^*)^2]^{1/2} \quad (1)$$

3.5. Statistical Analysis

K-means algorithm and clustering analysis were conducted respectively by 'kmeans' function and 'pheatmap' function in R environment (3.4.0) (http://www.r-project.org/). Partial least squares-discriminate analysis (PLS-DA) was conducted by MetaboAnalyst 4.0 (http://www.

metaboanalyst.ca/). One-way ANOVA tests and Duncan's multiple range tests were carried out using SPSS software version 20.0 for Windows (SPSS Inc., Chicago IL, USA).

4. Conclusions

In conclusion, by assessing the dissolved oxygen (DO), phenolic and volatile compounds, OAV and chromatic parameters, a comprehensive understanding of the evolution of oxidation patterns and flavor profiles in rosé and dry white wines during an 18-month bottle aging was established. Those two types of wines had different change trends of DO because of their unique phenolic compositions and the initial concentration of headspace oxygen, which makes it important for winemakers to control the bottling process. Many volatile compounds exhibited a decreasing trend during bottle aging and the total intensity of aroma declined obviously compared with their counterparts prior to bottling in all wines. Rosé wines sealed with natural corks and screw caps exhibited no significant difference after an 18-month bottle aging, except for in the 'berry' aroma description, while a polymer synthetic plug could lead to a greater loss of aroma quality in rosé wines during bottle aging. Agglomerated corks and polymer synthetic plugs were not appropriate for the preservation of white wines, while screw caps behaved better than any other closures in maintaining aroma compounds in white wines. After an 18-month bottle aging, rosé wines exhibit a difference in color, whereas this phenomenon was not observed in white wines. Our study has extended the research into the evolution of flavor profiles in rosé and white wines; further attention should be given to the wines' flavor chemistry and quality control during bottle aging.

Author Contributions: C.-Q.D. and Y.S. designed the study. M.-Q.L., H.X., Y.-B.H., J.C., S.-Y.L., Y.-B.L. and R.-N.L. conducted the experiments and collected the data. M.-Q.L. analyzed the data and drafted the manuscript.

References

1. Kwiatkowski, M.J.; Skouroumounis, G.K.; Lattey, K.A.; Waters, E.J. The impact of closures, including screw cap with three different headspace volumes, on the composition, colour and sensory properties of a Cabernet Sauvignon wine during two years' storage. *Aust. J. Grape Wine R* **2007**, *13*, 81–94. [CrossRef]
2. Guaita, M.; Petrozziello, M.; Motta, S.; Bonello, F.; Cravero, M.C.; Marulli, C.; Bosso, A. Effect of the closure type on the evolution of the physical-chemical and sensory characteristics of a Montepulciano d'Abruzzo *Rosé* wine. *J. Food Sci.* **2013**, *78*, C160–C169. [CrossRef] [PubMed]
3. Liu, D.; Xing, R.-R.; Li, Z.; Yang, D.-M.; Pan, Q.-H. Evolution of volatile compounds, aroma attributes, and sensory perception in bottle-aged red wines and their correlation. *Eur. Food Res. Technol.* **2016**, *242*, 1937–1948. [CrossRef]
4. Gao, Y.; Tian, Y.; Liu, D.; Li, Z.; Zhang, X.X.; Li, J.M.; Huang, J.H.; Wang, J.; Pan, Q.H. Evolution of phenolic compounds and sensory in bottled red wines and their co-development. *Food Chem.* **2015**, *172*, 565–574. [CrossRef] [PubMed]
5. Ugliano, M. Oxygen contribution to wine aroma evolution during bottle aging. *J. Agric. Food Chem.* **2013**, *61*, 6125–6136. [CrossRef] [PubMed]
6. Karbowiak, T.; Gougeon, R.D.; Alinc, J.-B.; Brachais, L.; Debeaufort, F.; Voilley, A.; Chassagne, D. Wine oxidation and the role of cork. *Crit. Rev. Food Sci. Nutr.* **2009**, *50*, 20–52. [CrossRef]
7. Lopes, P.; Silva, M.A.; Pons, A.; Tominaga, T.; Lavigne, V.; Saucier, C.; Darriet, P.; Teissedre, P.L.; Dubourdieu, D. Impact of oxygen dissolved at bottling and transmitted through closures on the composition and sensory properties of a Sauvignon Blanc wine during bottle storage. *J. Agric. Food Chem.* **2009**, *57*, 10261–10270. [CrossRef] [PubMed]

8. Lee, D.H.; Kang, B.S.; Park, H.J. Effect of oxygen on volatile and sensory characteristics of Cabernet Sauvignon during secondary shelf life. *J. Agric. Food Chem.* **2011**, *59*, 11657–11666. [CrossRef] [PubMed]
9. Godden, P.; Francis, L.; Field, J.; Gishen, M.; Coulter, A.; Valente, P.; HØJ, P.; Robinson, E. Wine bottle closures: Physical characteristics and effect on composition and sensory properties of a Semillon wine 1. Performance up to 20 months post-bottling. *Aust. J. Grape Wine R* **2001**, *7*, 64–105. [CrossRef]
10. Wirth, J.; Morel-Salmi, C.; Souquet, J.; Dieval, J.; Aagaard, O.; Vidal, S.; Fulcrand, H.; Cheynier, V. The impact of oxygen exposure before and after bottling on the polyphenolic composition of red wines. *Food Chem.* **2010**, *123*, 107–116. [CrossRef]
11. Vazallo-Valleumbrocio, G.; Medel-Maraboli, M.; Peña-Neira, Á.; López-Solís, R.; Obreque-Slier, E. Commercial enological tannins: Characterization and their relative impact on the phenolic and sensory composition of Carménère wine during bottle aging. *LWT-Food Sci. Technol.* **2017**, *83*, 172–183. [CrossRef]
12. Han, G.; Ugliano, M.; Currie, B.; Vidal, S.; Dieval, J.B.; Waterhouse, A.L. Influence of closure, phenolic levels and microoxygenation on Cabernet Sauvignon wine composition after 5 years' bottle storage. *J. Sci. Food Agric.* **2015**, *95*, 36–43. [CrossRef] [PubMed]
13. He, J.; Zhou, Q.; Peck, J.; Soles, R.; Qian, M.C. The effect of wine closures on volatile sulfur and other compounds during post-bottle ageing. *Flavour Frag. J.* **2013**, *28*, 118–128. [CrossRef]
14. Skouroumounis, G.K.; Kwiatkowski, M.J.; Francis, I.L.; Oakey, H.; Capone, D.L.; Duncan, B.; Waters, E.J. The impact of closure type and storage conditions on the composition, colour and flavour properties of a Riesling and a wooded Chardonnay wine during five years' storage. *Aust. J. Grape Wine R* **2005**, *11*, 369–377. [CrossRef]
15. Puértolas, E.; Saldaña, G.; Condón, S.; Álvarez, I.; Raso, J. Evolution of polyphenolic compounds in red wine from Cabernet Sauvignon grapes processed by pulsed electric fields during aging in bottle. *Food Chem.* **2010**, *119*, 1063–1070. [CrossRef]
16. Waterhouse, A.L.; Laurie, V.F. Oxidation of Wine Phenolics: A Critical Evaluation and Hypotheses. *Am. J. Enol. Viticult.* **2006**, *57*, 306–313.
17. Ivanova-Petropulos, V.; Hermosín-Gutiérrez, I.; Boros, B.; Stefova, M.; Stafilov, T.; Vojnoski, B.; Dörnyei, Á.; Kilár, F. Phenolic compounds and antioxidant activity of Macedonian red wines. *J. Food Compos. Anal.* **2015**, *41*, 1–14. [CrossRef]
18. Silva, M.A.; Julien, M.; Jourdes, M.; Teissedre, P.-L. Impact of closures on wine post-bottling development: A review. *Eur. Food Res. Technol.* **2011**, *233*, 905–914. [CrossRef]
19. Coetzee, C.; Van Wyngaard, E.; Suklje, K.; Silva Ferreira, A.C.; du Toit, W.J. Chemical and sensory study on the evolution of aromatic and nonaromatic compounds during the progressive oxidative storage of a Sauvignon blanc wine. *J. Agric. Food Chem.* **2016**, *64*, 7979–7993. [CrossRef] [PubMed]
20. Makhotkina, O.; Pineau, B.; Kilmartin, P.A. Effect of storage temperature on the chemical composition and sensory profile of Sauvignon Blanc wines. *Aust. J. Grape Wine R* **2012**, *18*, 91–99. [CrossRef]
21. Peinado, R.A.; Mauricio, J.C.; Moreno, J. Aromatic series in sherry wines with gluconic acid subjected to different biological aging conditions by *Saccharomyces cerevisiae* var. capensis. *Food Chem.* **2006**, *94*, 232–239. [CrossRef]
22. Franco, M.; Peinado, R.A.; Medina, M.; Moreno, J. Off-vine grape drying effect on volatile compounds and aromatic series in must from Pedro Ximénez grape variety. *J. Agric. Food Chem.* **2004**, *52*, 3905–3910. [CrossRef] [PubMed]
23. Ferreira, V.; López, R.; Cacho, J.F. Quantitative determination of the odorants of young red wines from different grape varieties. *J. Sci. Food Agric.* **2000**, *80*, 1659–1667. [CrossRef]
24. Bao, J.; Zhenwen, Z. Volatile compounds of young wines from Cabernet Sauvignon, Cabernet Gernischet and Chardonnay varieties grown in the Loess Plateau Region of China. *Molecules* **2010**, *15*, 9184–9196. [CrossRef]
25. Sacks, G.L.; Gates, M.J.; Ferry, F.X.; Lavin, E.H.; Kurtz, A.J.; Acree, T.E. Sensory threshold of 1,1,6-trimethyl-1,2-dihydronaphthalene (TDN) and concentrations in young Riesling and non-Riesling Wines. *J. Agric. Food Chem.* **2012**, *60*, 2998–3004. [CrossRef] [PubMed]
26. Li, H.; Tao, Y.-S.; Wang, H.; Zhang, L. Impact odorants of Chardonnay dry white wine from Changli County (China). *Eur. Food Res. Technol.* **2007**, *227*, 287–292. [CrossRef]
27. Zea, L.; Moyano, L.; Moreno, J.; Cortes, B.; Medina, M. Discrimination of the aroma fraction of Sherry wines obtained by oxidative and biological ageing. *Food Chem.* **2001**, *75*, 79–84. [CrossRef]
28. Habekost, M. Which color differencing equation should be used. *Int. Circ. Graph. Educ. Res.* **2013**, *6*, 20–33.

29. Li, S.-Y.; He, F.; Zhu, B.-Q.; Xing, R.-R.; Reeves, M.J.; Duan, C.-Q. A systematic analysis strategy for accurate detection of anthocyanin pigments in red wines. *Rapid Commun. Mass Spectrom.* **2016**, *30*, 1619–1626. [CrossRef] [PubMed]
30. Li, S.-Y.; He, F.; Zhu, B.-Q.; Wang, J.; Duan, C.-Q. Comparison of phenolic and chromatic characteristics of dry red wines made from native Chinese grape species and *Vitis vinifera*. *Int. J. Food Prop.* **2017**, *20*, 2134–2146. [CrossRef]
31. Lan, Y.B.; Qian, X.; Yang, Z.J.; Xiang, X.F.; Yang, W.X.; Liu, T.; Zhu, B.Q.; Pan, Q.H.; Duan, C.Q. Striking changes in volatile profiles at sub-zero temperatures during over-ripening of 'Beibinghong' grapes in Northeastern China. *Food Chem.* **2016**, *212*, 172–182. [CrossRef] [PubMed]
32. Ayala, F.; Echávarri, J.F.; Negueruela, A.I. A New Simplified Method for Measuring the Color of Wines. III. All Wines and Brandies. *Am. J. Enol. Vitic.* **1999**, *50*, 359–363.

16

Analysis of Enantiomers in Products of Food Interest

Chiara Fanali [1,*], Giovanni D'Orazio [2], Alessandra Gentili [3] and Salvatore Fanali [4]

1 Department of Medicine, University Campus Bio-Medico of Rome, Via Alvaro del Portillo 21, 00128 Rome, Italy
2 Istituto per I Sistemi Biologici, Consiglio Nazionale delle Ricerche, Via Salaria km 29, 300-00015 Monterotondo, Italy; giovanni.dorazio@cnr.it
3 Department of Chemistry, University of Rome "La Sapienza", Piazzale Aldo Moro 5, P.O. Box 34, Posta 62, 00185 Roma, Italy; alessandra.gentili@uniroma1.it
4 Teaching Committee of Ph.D. School in Natural Science and Engineering, University of Verona, 37134 Verona, Italy; salvatore.fanali@gmail.com
* Correspondence: c.fanali@unicampus.it

Abstract: The separation of enantiomers has been started in the past and continues to be a topic of great interest in various fields of research, mainly because these compounds could be involved in biological processes such as, for example, those related to human health. Great attention has been devoted to studies for the analysis of enantiomers present in food products in order to assess authenticity and safety. The separation of these compounds can be carried out utilizing analytical techniques such as gas chromatography, high-performance liquid chromatography, supercritical fluid chromatography, and other methods. The separation is performed mainly employing chromatographic columns containing particles modified with chiral selectors (CS). Among the CS used, modified polysaccharides, glycopeptide antibiotics, and cyclodextrins are currently applied.

Keywords: chiral; chiral stationary phases; enantiomers; food; review

1. Introduction

In recent years, in the field of separation science, there has been great attention to analyzing products of food interest. All this has been done to meet the needs of both the industries operating in the field and the control laboratories belonging to the various national and international agencies. The use of modern and reliable analytical methods permits knowing the chemical composition of these products allowing, e.g., to determine their quality, to trace problems related to production, and storage processes, etc. In addition, the presence of dangerous pollutants such as pesticides present in foodstuff and/or additives can also be quantified.

A large number of compounds present in food products have been successfully examined, e.g., amino acids, proteins, carbohydrates, vitamins, lipids, mycotoxins, colorants, preservatives, herbicides, ionic compounds, fungicides, and enantiomers, etc.

Among all the classes of compounds analyzed so far, particular attention has been paid to the separation and determination of chiral compounds. Two enantiomers have the same chemical composition with, e.g., one asymmetric center, and due to the different spatial orientation of the substituent groups, they exhibit quite similar physical-chemical properties. However, due to the chirality, they can participate, with a different effect, to the various biochemical processes. In these processes, enantiomers could react differently with, e.g., enzymes, proteins, and peptides, etc., present in humans determining beneficial or dangerous effects. In addition, natural products must contain only one enantiomer as in the case of fruit juices where only the *L*-amino acid must be present. The existence of the antipode can indicate either an adulteration (addition of a racemate) or other problems related to poor storage. Therefore, it is very important to have reliable analytical methods able to determine qualitatively and quantitatively the enantiomeric forms present in samples of food interest.

Analytical techniques so far used for the analysis of enantiomers include: Gas chromatography (GC) [1], supercritical fluid chromatography (SFC) [2,3], thin layer chromatography (TLC) [4], high-performance liquid chromatography (HPLC) [5], capillary electrophoresis (CE), and capillary-/nano-liquid chromatography (CLC/nano-LC) [6,7], etc.

In this review, general principles of enantiomers separation utilizing chromatographic techniques and the most applied chiral stationary phases (CSPs) are reported and discussed. In addition, the most important applications of some analytical techniques to the analysis of chiral compounds in food matrices published in the last two years are presented.

2. General Separation Methods in Chiral Resolution

As previously mentioned, the very similar physical-chemical properties of two enantiomers are the main reasons for the difficulty in obtaining their resolution by using conventional methods such as, for example, reversed phase one in liquid chromatography. However, their separation can be easily achieved utilizing a chiral background.

Generally, two resolution methods can be used, namely indirect and direct ones. In the first procedure, the two compounds react with a chiral selector (CS) on forming stable diastereoisomers where strong bonds are formed. The two new compounds can be separated employing a non-chiral stationary phase (SP). The separation method offers some benefits, e.g., the introduction of additional groups useful for i) further interactions and ii) increase the sensitivity when UV and mass spectrometry (MS) detectors are applied. However, a high reagent purity is necessary and it is time-consuming [8].

The direct resolution method is the most applied in separation science. It is based on the use of a CSP where the two analytes interact continuously on forming diastereoisomeric complexes, and involving weak bonds (hydrogen, π-π, hydrophobic etc.). While some models have been proposed for chiral recognition, the "Three point" interaction could be considered in order to achieve compounds separation. However, enantiomers resolution can also be obtained when repulsion, in addition to interactions, are involved in the stereoselective mechanism [9].

Figure 1 shows a scheme of the "Three-point" interaction model.

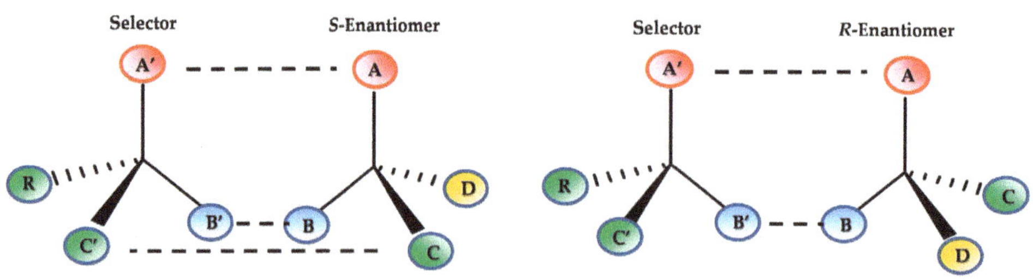

Figure 1. Scheme of "Three-point" interaction model.

Even if the number of CSs, applied to the resolution of enantiomers is quite large and varied, it is noteworthy to mention that it is not possible to find a universal one. They have been utilized employing different analytical techniques, e.g., GC, SFC, HPLC, CLC/nano-LC, and CE etc. [8,10–13].

Among all the chiral selectors used so far with the different analytical techniques, those that deserve special attention are peptides, chiral amino acids, cyclodextrins, polysaccharide derivatives, quinine-based, glycopeptides antibiotics, and chiral crown-ethers, etc.

3. The Most Used Chiral Stationary Phases for Enantiomers Separation

In this section, the most applied CSPs to the separation of enantiomers utilizing the analytical techniques above mentioned will be summarized. Some of them have been applied to the analysis of compounds of food interest. Over the years, a large number of CSPs has been prepared and studied for enantiomers separation. Several of them are commercially available under different trade name also containing the same CS. This is the case of Lux or Chiralpak by Phenomenex (USA) and Chiral Technologies (USA), respectively. They make use of modified polysaccharides either coated or immobilized on silica particles. In order to improve the enantioresolution capacity of this CS type, cellulose or amylose have been modified with phenylcarbamate groups containing substituents such as methyl, chloro, and bromo, etc., alone or in combination. The presence of these substituents with electron-withdrawing or electron-donating modifies the properties of the phenyl promoting different interactions with the enantiomers to be separated [14]. Concerning the interactions involved in the chiral recognition when polysaccharides are employed, hydrogen, π-π bonding have to be mentioned. The inert support is usually silica of different dimensions (sub-2 μm to 5μm) porous or recently core-shell [15,16]. Monolithic material has also been used [17,18]. These CSPs have been largely employed in HPLC [5,19], SFC [11,20], capillary electrochromatography (CEC) [21,22], and nano-LC [23].

Another CS, widely applied to enantiomers separation, deserving attention, includes cyclodextrins (CDs) or their derivatives. It can be either bonded to silica particles or monolithic material or rarely added to the mobile phase. In addition it can be coated on the capillary wall, as used in GC [24]. CDs or their derivatives have been widely used in capillary electrophoresis added to the background electrolyte [12]. CDs are oligosaccharides with a shape like a truncated cone with a hydrophobic cavity and a hydrophilic outside (presence of hydroxyl groups). The recognition mechanism is based on inclusion complexation where the two enantiomers fit into the cavity. Steresoselective interactions with hydroxyl or substituent groups (e.g., methyl, ethyl, and hydroxypropyl, etc.) take place. However, adsorption interactions could also be involved, especially if organic solvents are employed in the mobile phase. Schurig has showed a unified enantioselective approach for enantiomers separation. Here the same capillary column (50 μm i.d.) containing a permethylated-β-CD, thermally coated, has been employed in both open-GC, -SFC, -LC, and –CEC [25].

Among other CS used for enantiomers separation, chiral crown-ethers have a certain interest. These CSP type are commercially available, e.g., crownpakcr(+) and cr(−) (Chiral technologies, USA) or Larihc CF6- P (AZYP, USA). The second column contains cyclofructans groups forming a basket structure based on crown-ether. Therefore, enantiomers enter into the basket and on the contrary of CDs, the hydrophilic groups interact with the cavity. This CSP type has been used for the enantiomers separation of compounds containing amino groups in their structure (amino acids, amines, and derivatives). Recently Armstrong's group reported the screening of 119 primary amine-containing compounds achieving enantioresolution for 92% of analytes [26]. Finally, it is noteworthy to mention the use of ion-exchange type CSP [27–30] and those containing glycopeptides antibiotics (vancomycin, teicoplanin, etc.). All of them are commercially available columns offering good/high enantioselectivity towards a large number and type of compounds [31–35].

Glycopeptide antibiotics contain in their chemical structure a consistent number of asymmetric centers, amino, amide, aromatic, carboxylic groups offering different interaction types with the analyzed enantiomers. The recognition mechanism is based on the affinity interaction strongly influenced by the type of mobile phase, pH, ionic strength, organic solvent type, and concentration, etc. They have been used bonded to either porous or core-shell particles (1.7–5 μm diameter) [23,36].

4. Some Selected Applications to Enantiomers Separation in Food Chemistry

This section describes the recent data present in the literature reported in 2017–2018 and related to the enantiomers separation in food samples. Data on this topic, related to previous years, can be found in previous reviews [23,37].

4.1. Supercritical Fluid Chromatography

Although SFC is not a recent analytical technique, it has enormous potential for both analytical and preparative purposes. Among its main features, we can mention: Fast balance of the columns, high efficiency, and use of non-hazardous solvents. Making use of stationary phases, developed for HPLC applications, SFC has also been applied to the separation of enantiomers. Among these CSPs, those based on polysaccharides, cyclodextrins, vancomycin, and Pirkle-type are the most employed ones. The mobile phase contains carbon dioxide often modified with some organic solvents such as methanol, ethanol, isopropanol, and acetonitrile at relatively low concentrations. In addition to enhancing chiral resolution, other additives, e.g., formic acid (FA), acetic acid (HAc), trifluoracetic acid (TFA), ammonia (NH_3), and diethylamine (DEA) have been used [37].

As can be observed in Table 1 six papers, dealing with enantiomers separation by SFC in samples of food interest, appeared in the considered last two years. In the developed analytical methods, chiral resolution was obtained utilizing CSP based on vancomycin and amylose or cellulose derivative.

Prothioconazole is a triazole fungicide widely used for its curative and protective action. Its enantiomers have been separated and analyzed by Jiang et al. [38] by using SFC. Analytes were extracted in a food sample (tomatoes) applying the Quick Easy Cheap Effective Rugged Safe (QuEChERS) sample preparation method and analyzed in a silica column coated with a modified cellulose derivative. Chiral separation was achieved in less than four minutes with good precision, accuracy, and recovery. In this study, method optimization was done by investigating some experimental parameters, e.g., the different type of polysaccharide columns, the composition of the mobile phase (carbon dioxide with additives such as methanol, ethanol, isopropanol-IPA). Best results were obtained using IPA as an additive at 30% (*v*/*v*) concentration. Another CS, namely amylose tris-(3,5-dimethylphenylcarbamate), coated on silica particles was used for the chiral separation of fenbuconazol and its metabolites by SFC. Analytes were also characterized by using MS/MS. Fenbuconazole is a compound with fungicidal activity, currently used for the treatment of fruits and vegetables. Two of its degradation lactone metabolites could contaminate groundwater and therefore their determination is of great interest. Very good enantiomers separation was obtained by SFC using a silica column coated with amylose tris-(3,5-dimethylphenylcarbamate) trough CO_2/ethanol applying a gradient elution mode. Sample preparation was done employing a QuEChERS method. The six stereoisomers were separated in less than four min. In order to increase the MS ionization, 0.1% (*v*/*v*) formic acid/methanol compensation solution was added. The optimized method was validated with various samples, e.g., tomatoes, cucumbers, apples, peaches, rice, and wheat. Application to a real sample (greenhouse cucumbers) revealed the presence of only the two enantiomers of fenbuconazol [39]. Figure 2 reports the chemical structure of the two enantiomers of fenbuconazol and its diastereoisomers.

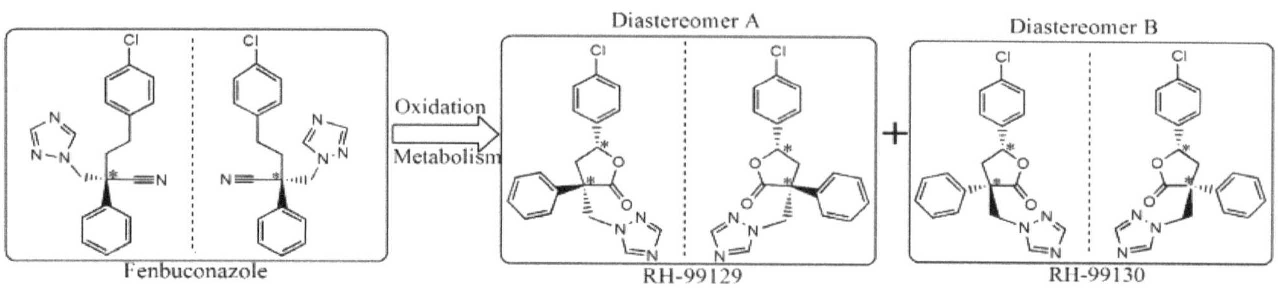

Figure 2. Chemical structure of fenbuconazol enantiomers and its metabolite diastereoisomers. Reproduced with permission of Elsevier from ref. [39].

Another polysaccharide based column (tris(3,5-dimethylphenylcarbamoyl) cellulose) was used for analysis of triticonazole enantiomers in cucumbers and tomatoes with the same method. The separation was achieved in three min (in HPLC, eighteen min) eluting with CO_2-ethanol (80:20, *v*/*v*) [40]. Four stereoisomers of propiconazole, a widely used triazole fungicide, were separated by SFC using different polysaccharide SPs. Among them, Chiralpak AD3 and Chiralpak IA3 allowed the separation of all stereoisomers in less than five min [41]. The two SPs contained amylose modified with tris(3,5-dimethylphenylcarbamate) coated and immobilized, respectively. Retention order was assessed by measuring the optical rotation of the four studied compounds. The optimized method was validated and applied to extract samples of wheat, grape, and soil matrices with good results concerning recovery.

4.2. High-Performance Liquid Chromatography

Selected applications to the analysis of chiral compounds utilizing HPLC are reported in Table 2 and described in this section.

Clenbuterol is a β_2 agonist that is used in the treatment of respiratory disorders in humans. However, this drug, in some countries, even if forbidden, is also administered to the animals. The two enantiomers have been recently separated utilizing both HPLC and SFC employing columns containing vancomicin or teicoplanin [42]. Baseline separation of the two enantiomers was obtained with both techniques, however, a shorter analysis time was observed utilizing SFC (3.5 and 6 min, respectively). The methods were applied to the analysis of meat samples and urines of humans after eating the meat. Cattle meat contained enriched *R*-(−)-clenbuterol. More recently, the two analytical techniques have also been used by Chen and Zhang [43] for the chiral resolution of some isobutylhydroxyamides. Different commercial columns, amylose-based have been employed. While analyzed compounds belonging to the same class, two different techniques and different columns had to be used. These compounds are quite important in the field of phytochemistry and traditional Chinese medicine; they are present in Sichuan pepper. In this study, authors separated and characterized several isobutylhydroxyamides enantiomers also studying protective effects (PC12 Cells), reporting that compounds with *R* configuration resulted protective on the contrary of the *S* form.

Vegetables such as spinach, tomatoes, and cucumber have been analyzed for the determination of *R* and *S*-titriconazol using HPLC. This compound belongs to the family of fungicides used in agriculture. In this study, the single enantiomers and the racemic mixture were used and the bioactivity was investigated. The analyzed vegetables contained a high concentration of titriconazol after foliar spraying. However, the concentration of the two enantiomers decreased along the time. The *R*-isomer was dissipated more than its antipode. Therefore, it can be concluded that using the single enantiomer (*R*-) would reduce potential health risk [44]. A different fungicide, pyrisoxazole was studied by Yang et al. [45]. Enantiomers were separated with a commercial cellulose tris(4-methylbenzoate) column by HPLC in less than 10 min. This fungicide contains two asymmetric centers and therefore two couples of enantiomers. A similar study was carried out by Wang et al. [46] to study the degradation of isofenphos-methyl, a pesticide, after treating cowpea, cucumber, and pepper.

Table 1. Selected applications of supercritical fluid chromatography (SFC) to Food analysis.

Samples	Matrix	Sample Preparation	Technique	Column and Chiral Stationary Phase	Mobile Phase	Detection	References
(*R*) and (S)-Prothioconazole	Tomatoes	QuEChERS	SFC	cellulose tris(3,5-dimethylphenylcarbamate)-silica coated, EnantioPak (China) (150 × 4.6 mm, 5 μm)	CO_2/2-propanol (80:20, *v*/*v*); 2.5 mL/min	UV, 254 nm	[38]
Fenbuconazole and metabolites	tomatoes, cucumbers, apples, peaches, rice and wheat,	QuEChERS	SFC	amylose tris-(3,5-dimethylphenylcarbamate)-coated chiral column	CO_2/ethanol; 1.8 mL/min	MS/MS	[39]
Triticonazole	cucumbers and tomatoes	QuEChERS	SFC	tris(3,5-dimethylphenylcarbamoyl) cellulose-coated silica gel EnantioPak OD column (China) (150 mm × 4.6 mm, 5 μm)	CO_2/ethanol (80:20, *v*/*v*)	-	[40]
Propiconazole	Wheat	Solid phase	SFC	Amylose tris(3,5-dimethylphenylcarbamate) and immobilized Amylose tris(3,5-dimethylphenylcarbamate), Chiralpak AD-3 and IA3, respectively (Daicel Chemical Industries, Japan) (150 mm × 4.6 mm i.d., 3 μm particle size)	CO_2/ethanol (93:7, *v*/*v*)	MS/MS	[41]
Clenbuterol	Meat	Liquid-liquid extractionand Solid phase	SFC or HPLC	Vancomycin, Astecchirobiotic V2 (150 × 4.6 mm or 2.1 mm, particle size 5 μm); Teicoplanin, Astecchirobiotic T (150 × 4.6 mm or 2.1 mm, particlesize 5 μm);	CO_2/ammonia or formic acid in SFC and MeOH (with water or ACN or 2-propanol) in HPLC	MS/MS	[42]
Isobutylhydroxyamides ZP-amide A and ZP-amide B	Pepper	Liquid-liquid extraction	SFC or HPLC	Amylose tris-(3,5-dimethylphenylcarbamate) coated, Chiralpak AD-H column (chiral column 1); Amylose tris(5-chloro-2-methylphenylcarbamate) coated, Chiralpak AY-H column (chiral column 2) (15 cm × 3 cm i.d., 5 μm; 15 cm × 3 cm i.d., 5 μm,); Chiralpak AD-H column (chiral column 3); (25 cm × 0.46 cm i.d., 5 μm, 15 cm × 0.46 cm i.d.)	SFC, (40% MeOH, 3 mL/min); HPLC different MP (ethanol/diethylamine (100:0.1 *v*/*v*); 0.2 mL/min) or n-hexane/isopropanol/diethylamine(85:15:0.1 *v*/*v*/*v*); 1 mL/min or n-hexane/ethanol/diethylamine (85:15:0.1 *v*/*v*/*v*); 1 mL/min)	MS	[43]

Table 2. Selected applications of HPLC to Food analysis.

Samples	Matrix	Sample Preparation	Technique	Column and Chiral Stationary Phase	Mobile Phase	Detection	References
(*R,S*)-triticonazole	Vegetables (tomatos, spinach, cucumber)	Liquid-liquid extraction and Solid phase	HPLC	cellulose tris (3-chloro-4-methylphenylcarbamate) (Lux Cellulose-2); 3-μm particles (250 mm or 4.6 mm i.d.)	-	UV	[44]
Isofenphos-methyl, (*R,S*)-*O*-methyl-*O*-(2-isopropoxcarbonyl)-phenyl-*N*-isopropylphosphora -midothioate	cowpea, cucumber, and pepper	QuEChERS	HPLC-MS/MS	Lux cellulose-3 chiral, column (250 mm × 4.6 mm i.d., 5 μm,	Isocratic, 0.80 mL/min	MS	[46]
Metconazole	Flour	QuEChERS	HPLC	Cellulose 3, 5-dimethylphenylcarbamate Enantiopak OD column	n-hexane-ethanol mixture (97:3, *v*/*v*) at the flow rate of 1.0 mL/min	UV, 220 nm	[47]
Thiols 3-sulfanylhexan-1-ol and its O-acetate, 3-sulfanylhexyl acetate	Wine	Solid phase	HPLC	Lux Amilose 1 and 2, cellulose 1	Gradient, 5 mM aqueous ammonium bicarbonate (A, pH 8.7) and acetonitrile (B)	MS/MS	[48]
8-*O*-4′ type neolignans	Raspberry	-	HPLC	Chiralpak AD-H 250 × 4.6 mm, 5 μm	2-propanol-n-hexane (various ratios)	Polarimeter	[49]
Fungicide pyrisoxazole	Pakchoi, pepper, cabbage	QuEChERS	HPLC	cellulose tris(4-methylbenzoate)-Lux Cellulose-3, 150 mm × 2.0 mm, 3 μm	methanol and water (70:30 *v*/*v*), 0.35 mL/min	MS/MS	[45]
Chloramphenicol	Honey	Liquid-liquid extraction	HPLC	Chiralpak AGP, 3 × 5 mm, 5 μm	Gradient, water with 0.01% acetic acid (A) and methanol with 0.01% acetic acid (B)	MS/MS	[50]
α- and γ-Hexabromocyclododecanes	egg	Soxhlet extraction	HPLC	Permethylated-β-cyclodextrin-Chiral column Nucleosil β-PM Macherey-Nagel,(GmbH & Co., Düren, Germany), 20 cm × 4 mm × 5 μm	-	MS	[51]
Fluazifop-butyl and fluazifop	tomato, cucumber, pakchoi, rape	Liquid-liquid extraction	HPLC	Cellulose tris(3,5-dichlorophenylcarbamate) Chiralpak IC, 2504.6 mm I.D., 5μm particles	Not reported	MS/MS	[52]
Triacylglycerol	Chicken yolk and meat	Liquid-liquid extraction	HPLC	CHIRALCEL OD-3R, (Daicel Corporation, Tokyo, Japan) 4.6 mm i.d. ×150 mm,	Methanol, 0.5 mL min^{-1}	MS	[53]
Amino acids several	Vinegar, milk, kimchi, yogurt	Liquid-liquid extraction	HPLC	CROWNPAK CRIchiral column.	Acetonitrile/ethanol/water/TFA (80/15/5/0.5, *v*/*v*/*v*/*v*)	TOFMS	[54]
Amino acids	Chimchi (fermented vegetables)	Liquid-liquid extraction	HPLC	CROWNPAK CR-I(+) column, 3.0 mm i.d. 150 mm; particle size, 5 μm	Acetonitrile, ethanol, water, and TFA (80:15:5:0.5, *v*/*v*/*v*/*v*)	TOFMS	[55]

Table 2. *Cont.*

Samples	Matrix	Sample Preparation	Technique	Column and Chiral Stationary Phase	Mobile Phase	Detection	References
Amino acids	Vinegars	Liquid-liquid extraction	HPLC	CROWNPAK CR-I(+) and CR-I(−)(Daicel CPI, Osaka, Japan) (3.0 mm i.d. 150 mm, 5 μm)	acetonitrile, ethanol, water and TFA (80:15:5:0.5, *v/v/v/v*)	MS/MS	[56]
Triazole fungicide (paclobutrazol, myclobutanil, diniconazole, epoxiconazole)	Honey	Solid phase	HPLC	Chiralcel OD-RH column (150 mm × 4.6 mm, 5 μm, daicel, Japan)	ACN/2mM ammonium acetate, 50:42 (*v/v*)	MS/MS	[57]
Pesticides	cucumber, tomato, cabbage, grape, mulberry, apple and pear	Magnetic solid phase extraction	HPLC	Chiralpak IG column (250 mm × 4.6 mm, i.d. 5 μm, Daicel, Japan)	ACN/ water containing 5 mmol L^{-1} ammonium acetate and 0.1% (*v/v*) formic acid (65:35, *v/v*)	MS/MS	[58]
Fungicide prothioconazole and metabolites	Cucumber, pear	QuEChERS and Solid phase	UHPLC and HPLC	Cellulose-tris(4-methylbenzoate) Lux Cellulose-3, 2 or 3, 250 mm × 4.6 mm i.d., 5 μm and 150 mm × 2 mm i.d., 3 μm,	Acetonitrile:water	MS/MS	[59]
(*R*,*S*)-zoxamide	wine	-	UHPLC	Lux Amylose-2 chiral column (150 mm × 2 mm, 3 μm particle	acetonitrile and water (70:30, *v/v*), 0.5 mL/min	MS/MS	[60]
Zoxamide	Vegetable (tomato, cucumber), pepper, potato, grape, strawberry	-	UHPLC	Lux Amylose-2, 150 mm × 2 mm, 3 μm particle size	acetonitrile/water (70:30 *v/v*), 0.5 mL/min	MS/MS	[61]
Amino acids (derivatized with fluorescein isothiocyanate, FITC) (glutamic acid, aspartic acid, isoleucine, tryptophan, phenylalanine, tyrosine, histidine, proline)	Apple juice	-	Nano-LC (open tubular)	polymerization of 3-chloro-2-hydroxypropylmethacrylate (HPMA-Cl) and ethylene dimethacrylate (EDMA) with bonded β-cyclodextrin; 15 cm and i.d. 75 μm	acetonitrile:methanol:H2O at 0.1% *v/v* TFA (85:10:5, *v/v/v*); flow rate; 800 nL/min	UV, 214 nm	[62]

The two enantiomers were analyzed by HPLC and detected with MS/MS utilizing a commercial cellulose-based column (modified with 3-chloro-4-methylphenylcarbamate). Based on the presented results, the type of vegetable strongly influenced the degradation process, e.g., (*R*)-(−)-isofenphos-methyl was faster than (*S*)-(+)-isofenphos-methyl in cowpea and cucumber, while the opposite was observed in pepper. While showing interesting results, the authors did not further report about the reasons for the differences found.

Recently, an interesting HPLC method was studied utilizing cellulose or amylose 3, 5-dimethylphenylcarbamate for the separation of the four isomers of metconazole. An optimum resolution was achieved, after studying various mobile phases, employing the cellulose-based column and eluting with an organic solvent (hexane) with 3%, *v*/*v* of ethanol. The method was validated and applied to the analysis of the four compounds in flour. While the proposed method was carefully optimized studying the effect of various parameters on the compounds separation, the real sample was spiked and there was no finding of the transformation of the enantiomers/diastereoisomers [47].

The use of an HPLC method, employing polysaccharide-based columns (cellulose or amylose derivatives) allowed the chiral resolution of Thiols 3-sulfanylhexan-1-ol (**1**) and its O-acetate, 3-sulfanylhexyl acetate (**2**) after derivatization with 4-thiopyridine. The two compounds are quite important in assessing wine quality because of their potent aroma properties. In order to increase the MS signal and to have additional interaction groups for stereoselective interactions with the SP, the derivatization was helpful. Concerning enantiomers, the authors found that dry wines contained the two enantiomers of **1** and **2** at the same concentration, while in botrysed wines, elevated concentrations of *S*-form were found [48].

Chiralpak AD-H was used for semipreparative purposes for the enantiomers separation of some neolignans present in raspberry. A quantity of 2.5 mg of each enantiomer was obtained and the compounds investigated as potential inhibitors of β-amyloid aggregation. The study could be interesting in the field of nutraceutical research [49]. As reported above, all studies have been carried out using CSP polysaccharides bases, however, other CSs have also been used.

The enantioselective dissipation of an herbicide (fluazofop-butyl) was also studied in some vegetables (tomato, cucumber, pakchoi, and rape). The compound was degraded to fluazifop. Enantiomers separation was carried out in a commercial column packed with silica with immobilized cellulose tris(3,5-dichlorophenylcarbamate). As a result, the authors reported that the type of vegetable influenced the different behavior of the two chiral herbicides *S*-fluazifop-butyl dissipated faster than its antipode, while the contrary was observed in pakchoi, rape. The different behavior accounted for the presence of enzymes and to some chiral endogenous substances present in the plants [52].

α-, β-, and γ-hexabromocyclododecanes (HBCDs) are compounds used for different purposes, e.g., polystyrene foams, thermal insulation buildings, electric insulators, etc. Therefore they can be found in the environment and consequently in animals such as fishes, birds, chickens, etc. Enantiomers and diastereisomers, related to these compounds, have been separated by HPLC and the method applied to study the bioaccumulation in chicken tissues and eggs. Chiral compounds were separated using a permethylated-β-cyclodextrin silica column after recovering the different fractions subjected to sample preparation (soxhlet and gel permeation chromatography). It was observed that in adult chicken tissues (−)-α-HBCD and (+)-γ-HBCD were present at a higher concentration than the corresponding antipode [51].

A different chiral column containing α_1-acid glycoprotein (AGP) has been employed for the enantiomers and diastereoisomers separation of chloramphenicol (CAP) in honey by HPLC. The analysis of this antibiotic is very important because it has been found in some foodstuff causing health problems. Due to the presence of two asymmetric centers, four stereoisomers are exhibited (*RR*- and *SS*-CAP; *RS*- and *SR*-CAP) [50].

Usually, amino acid enantiomers present in the natural product are abundant in their L-form, however, in some fermented food, various D-amino acids could be present. Therefore, the enantiomers analysis is very important in order to assess food quality and properties. Amino acid chiral separation has been carried out utilizing a chiral crown-ether column in chimchi (Chinese cabbage) [55], in black vinegar and yogurt [54], in vinegar [56]. The chiral separation of a large number of amino acids has been obtained without derivatization. Finally, it has been found that the concentration of D-amino acids increased during the storage.

A few papers reported the enantiomers separation utilizing ultra high-performance liquid chromatography (UHPLC). This is a powerful methodology where compounds are separated in columns packed with particles of small diameter (<2–3 μm). High selectivity, high chromatographic efficiency, and short analysis time are the main advantages of this technique. However, it is noteworthy mentioning that increased back-pressure is obtained (high-pressure pumps are necessary) [59–61]. This methodology has been applied to the separation and analysis of a fungicide compound (prothioconazole) and its metabolite prothioconazole-desthio. The analytes have been base-line resolved in their enantiomers using a cellulose- tris(4-methylbenzoate) column by HPLC, while no baseline resolution was observed with UHPLC [59].

Among other recently developed analytical technique, one of them, namely nano-liquid chromatography (nano-LC open tubular) has been proposed for the enantiomers separation of some amino acids in apple juice. Amino acids have been derivatized with fluorescein isothiocyanate (FITC) and separated in a fused silica capillary containing, on the wall, a thin layer of polymeric material (monolithic) modified with a chiral selector (β-cyclodextrin). The optimized method was applied to the analysis of amino acid enantiomers in apple juice. While the technique can offer some advantages over conventional ones, there are some limitations especially considering the sample loading and the limited amount of chiral selector present in the capillary [62]. Table 2 summarizes the data related to the use of HPLC for enantiomers separation in food products. Some representative chiral separations of compounds in food matrices by HPLC-MS are reported in Figure 3.

4.3. Gas Chromatography

Table 3 reports the application to the analysis of enantiomers in food products achieved by gas chromatography.

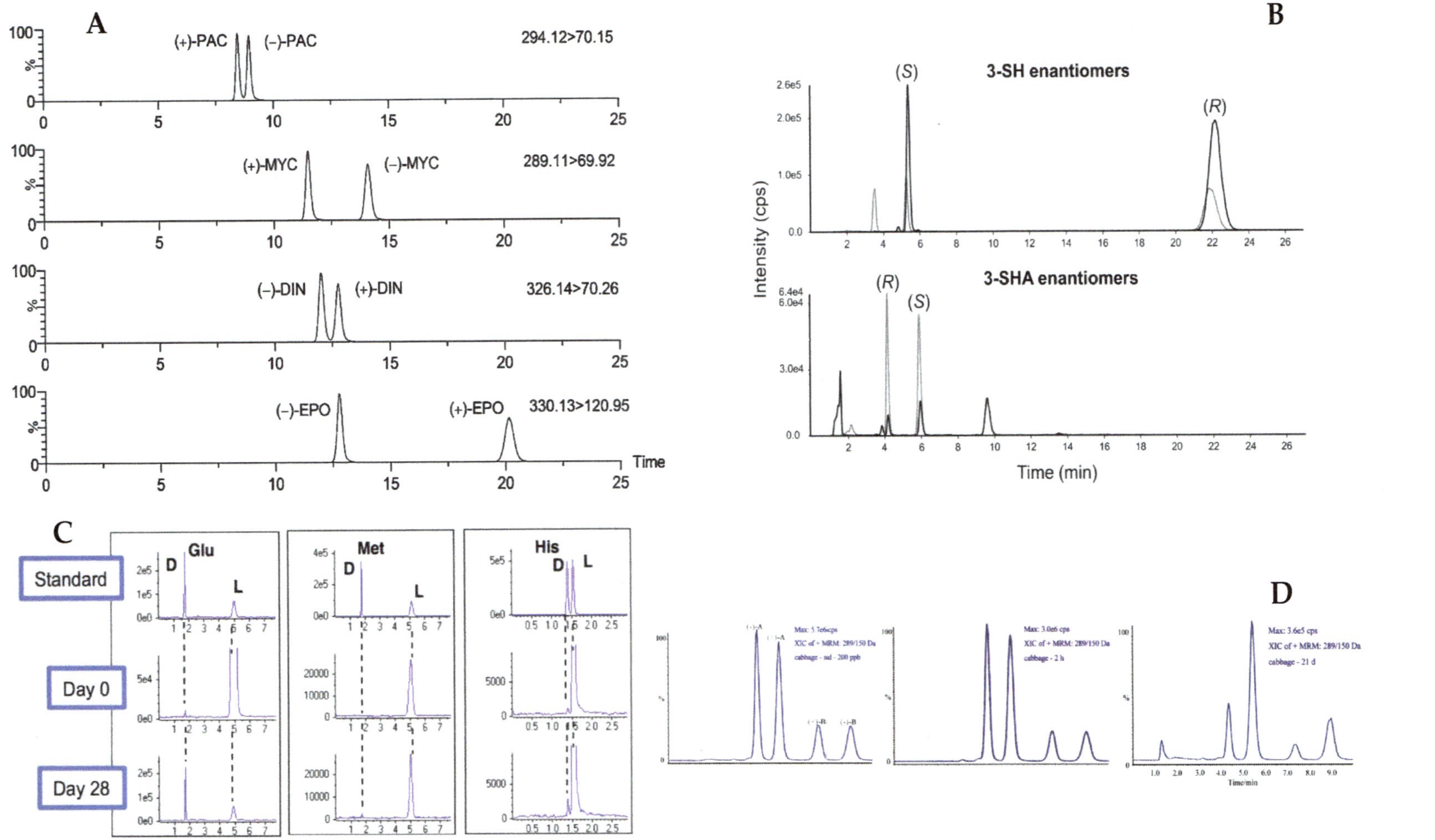

Figure 3. Typical enantioselective LC–MS/MS chromatograms: (**A**) chiral triazole fungicides in blank honey spiked using a Chiralcel OD-RH column modified with permission from [57]; (**B**) 3-sulfanylhexan-1-ol (3-SH 1) and 3-sulfanylhexyl acetate (3-SHA 2) isolated from a Sauvignon blanc wine on a CSP -Amylose-1 column; modified with permission from [48]. (**C**) Enantiomers resolution of some amino acids in kimchi stored for 25/28 days analyzed on a CSP- CROWNPAK CR-I by LC-TOF; modified with permission from [55]; and (**D**) Enatiomeric separation of fungicide pyrisoxazole by LC-MS on a CSP- Lux Cellulose-3; modified with permission from Reference [45].

Table 3. Selected applications of GC to Food analysis.

Samples	Matrix	Sample Preparation	Technique	Column and Chiral Stationary Phase	Mobile Phase	Detection	References
TBECH enantiomers 1,2-dibromo-4-(1,2-dibromoethyl)-cyclohexane (α, β, γ, δ) HBCD 1,2,5,6,9,10-hexabromocyclododecane (α, β, γ)	Marine organisms, including 5 mollusk species, 6 crustacean species, and 19 fish species	pressurized fluid extraction	GC	CHIRALDEX B-TA capillary column (30 m × 0.25 mm i.d., 0.12 μm film thickness)	Carrier:40% methane in a helium carrier gas at a constant flow of 1.2 mL/min.	MS	[63]
α-pinene, β-pinene, borneol, camphene, carvone, linalool, limonene, α-terpineol, α-ionene, terpinen-4-ol	Juices (apple, pear, peach, carrot, lemon flesh, orange flesh, orange peel, tangerine flesh), and tangerine peel	Headspace Solid-Phase Microextraction (HS-SPME)	GC	30% 2,3-di-*O*-ethyl-6-*O*-*tert*-butyldimethylsilyl-β-cyclodextrin (diEt-CD) and 30% 2,6-dimethyl-3-*O*-pentyl-β-cyclodextrin (Pentyl-CD) coated	Hydrogen as carrier gas (1.25 mL/min)	MS	[24]
α-pinene, limonene, linalool, β-caryophyllene	Essential oil Thyme	Steam distillation	GC	non-bonded 2,3-di-*O*-methyl-6-tbutyl silyl derivative of β-cyclodextrin AstecChiraldex B-DM column (30 m length 0.25 mm internal, diameter 0.12 μm film thickness)	hydrogen as carrier gas (constant flow of 2.5 mL/min, 8 psistarting column head pressure	Electronic impact ionization and MS	[64]
Pyrethroid insecticide (α-cypermethrin)	tomato, cucumber, rape, cabbage, and pepper	Liquid-liquidextraction	GC	BGB-172 chiral column	N_2 100–220 °C	ECD	[65]
Limonene, linalool, α-terpineol and 4-terpineol	Tea	Solid-Phase Microextraction	2D-GC	DB-WAX column (30 m × 0.25 mm i.d., 0.5 μm film thickness) (first) Cyclosil-B column (30 m × 0.25 mm i.d., 0.25 μm film thickness) (second)	helium at 1.2 mL/min	MS	[66]

Finally, gas chromatography (GC), widely used for enantiomers resolution, can offer high chromatographic efficiency. It is applied mainly to volatile compounds. The separation is performed in a capillary of thin i.d. where a chiral selector is coated or bonded on the wall. In the publications appeared during this time, polymeric material, β-cyclodextrin-based has been used as a chiral selector. Pollutants such as 1,2,5,6,9,10-hexabromocyclododecane (HBCD) (six diastereisomeric couple of enantiomers) were studied by GC-MS in order to verify the bioaccumulation in fishes [63]. In order to verify whether the processing would influence the enantiomeric ratio, chiral terpenoids present in juice industry by-products have been analyzed with GC-MS by Marsol-Vall et al. [24]. Among the results, authors observed that monoterpene alcohols (terpinen-4- ol and α-terpineol) exhibited a change on the enantiomeric ratio concluding that this was probably due to the heat applied during the drying process. The same technique has also been applied to the analysis of enantiomers present in essential oil from thyme. In this study, authors found the presence of (−)-linalool, (−)-borneol and (+)-limonene concluding that these compounds could be useful to assess origin and authenticity of the product [64]. A more advanced analytical technique, namely two dimensional-GC-MS(2D-GC-MS) has been applied to the analysis of chiral compounds present in tea samples [66]. The study aimed to investigate the chiral stability of lemon-flavored hard tea during storage. Samples were firstly analyzed by GC in a DB-Wax column and then target compounds heart-cutted and online submitted to a chiral column (cyclodextrin-based) for chiral analysis. Authors found that *R*/*S* ratio of limonene did not change during storage, while at high-temperature, *S*-limonene increased. Linalool enantiomers analysis, at the beginning of storage, revealed the presence of a higher amount of *R*-enantiomer. During storage, the *R*/*S* ratio decreased due to the conversion of *S*-isomer to *R*- one. Similar results have been obtained for α-terpineol.

5. Conclusions

The separation of chiral compounds has received great attention also in the last few years in different application fields including food chemistry. HPLC, SFC, and GC are the most utilized techniques because their features employing essentially the direct resolution method. Chiral columns, containing particles modified with selected CSs allowed the enantiomeric resolution of a large number of compounds. The majority of the enantioseparations have been obtained using polysaccharide-type CSs because of their very high chiral selectivity. The presented applications in the field of food analysis document the need of powerful analytical methods to assess the role of enantiomers for safety and product origin. Further studies are needed in order to (i) develop new CSPs (ii) study modern analytical techniques such as UHPLC and nano-LC offering high efficiency and reducing analysis time.

References

1. Gil-Av, E. Present status of enantiomeric analysis by gas chromatography. *J. Mol. Evol.* **1975**, *6*, 131–144. [CrossRef] [PubMed]
2. Novell, A.; Méndez, A.; Minguillón, C. Effects of supercritical fluid chromatography conditions on enantioselectivity and performance of polyproline-derived chiral stationary phases. *J. Chromatogr. A* **2015**, *1403*, 138–143. [CrossRef] [PubMed]
3. Albals, D.; Vander Heyden, Y.; Schmid, M.G.; Chankvetadze, B.; Mangelings, D. Chiral separations of cathinone and amphetamine-derivatives: Comparative study between capillary electrochromatography, supercritical fluid chromatography and three liquid chromatographic modes. *J. Pharm. Biomed. Anal.* **2016**, *121*, 232–243. [CrossRef] [PubMed]
4. Armstrong, D.W.; Faulkner, J.R.J.; Han, S.M. Use of hydroxypropyl and hydroxyethyl-derivatized -cyclodextrins for the thin-layer chromatographic separation of enantiomers and diastereomers. *J. Chromatogr.* **1988**, *452*, 323–330. [CrossRef]

5. Khundadze, N.; Pantsulaia, S.; Fanali, C.; Farkas, T.; Chankvetadze, B. On our way to sub-second separations of enantiomers in high-performance liquid chromatography. *J. Chromatogr. A* **2018**, *1572*, 37–43. [CrossRef] [PubMed]
6. Si-Ahmed, K.; Aturki, Z.; Chankvetadze, B.; Fanali, S. Evaluation of novel amylose and cellulose-based chiral stationary phases for the stereoisomer separation of flavanones by means of nano-liquid chromatography. *Anal. Chim. Acta* **2012**, *738*, 85–94. [CrossRef]
7. Aturki, Z.; Rocco, A.; Rocchi, S.; Fanali, S. Current applications of miniaturized chromatographic and electrophoretic techniques in drug analysis. *J. Pharm. Biomed. Anal.* **2014**, *101*, 194–220. [CrossRef]
8. Fanali, S. Nano-liquid chromatography applied to enantiomers separation. *J. Chromatogr. A* **2017**, *1486*, 20–34. [CrossRef]
9. Davankov, V.A. The nature of chiral recognition:is it a three-point interaction? *Chirality* **1997**, *9*, 99–102. [CrossRef]
10. Cagliero, C.; Sgorbini, B.; Cordero, C.; Liberato, E.; Rubiolo, P.; Bicchi, C. Enantioselective Gas Chromatography with DerivatizedCyclodextrins in the Flavour and Fragrance Field. *Isr. J. Chem.* **2016**, *56*, 925–939. [CrossRef]
11. West, C. Enantioselective separations with supercritical fluids—Review. *Curr. Anal. Chem.* **2014**, *10*, 99–120. [CrossRef]
12. Chankvetadze, B. *Capillary Electrophoresis in Chiral Analysis*, 1st ed.; John Wiley & Sons: Chichester, UK, 1997; pp. 1–555.
13. Chankvetadze, B. Contemporary theory of enantioseparations in capillary electrophoresis. *J. Chromatogr. A* **2018**, *1567*, 2–25. [CrossRef] [PubMed]
14. Chankvetadze, B. Liquid chromatographic separation of enantiomers. In *Liquid Chromatography*, 2nd ed.; Fanali, S., Haddad, P.R., Poole, C.F., Riekkola, M.-L., Eds.; Elsevier Inc.: Amsterdam, The Netherlands, 2017; Volume 2, pp. 69–86.
15. Patel, D.C.; Waha, M.F.; Armstrong, D.W.; Breitbach, Z.S. Advances in high-throughput and high-efficiency chiral liquid chromatographic separations. *J. Chromatogr. A* **2016**, *1467*, 2–18. [CrossRef] [PubMed]
16. Lomsadze, K.; Jibuti, G.; Farkas, T.; Chankvetadze, B. Comparative high-performance liquid chromatography enantioseparations on polysaccharide based chiral stationary phases prepared by coating totally porous and core-shell silica particles. *J. Chromatogr. A* **2012**, *1234*, 50–55. [CrossRef] [PubMed]
17. Chankvetadze, B.; Yamamoto, C.; Okamoto, Y. Very Fast Enantioseparation in High-performance Liquid Chromatography Using Cellulose Tris(3,5-dimethylphenylcarbamate) Coated on Monolithic Silica Support. *Chem. Lett.* **2003**, *32*, 850–851. [CrossRef]
18. Chankvetadze, B.; Yamamoto, C.; Kamigaito, M.; Tanaka, N.; Nakanishi, K.; Okamoto, Y. High-performance liquid chromatographic enantioseparations on capillary columns containing monolithic silica modified with amylose tris(3,5- dimethylphenylcarbamate). *J. Chromatogr. A* **2006**, *1110*, 46–52. [CrossRef] [PubMed]
19. Khatiashvili, T.; Kakava, R.; Matarashvili, I.; Tabani, H.; Fanali, C.; Volonterio, A.; Farkas, T.; Chankvetadze, B. Separation of enantiomers of selected chiral sulfoxides with cellulose tris(4-chloro-3-methylphenylcarbamate)-based chiral columns in high-performance liquid chromatography with very high separation factor. *J. Chromatogr. A* **2018**, *1545*, 59–66. [CrossRef] [PubMed]
20. Pan, X.; Dong, F.; Xu, J.; Liu, X.; Chen, Z.; Zheng, Y. Stereoselective analysis of novel chiral fungicide pyrisoxazole in cucumber, tomato and soil under different application methods with supercritical fluid chromatography/tandem mass spectrometry. *J. Hazard. Mater.* **2016**, *311*, 115–124. [CrossRef] [PubMed]
21. D'Orazio, G.; Kakava, R.; Volonterio, A.; Fanali, S.; Chankvetadze, B. An attempt for fast separation of enantiomers in nano-liquid chromatography and capillary electrochromatography. *Electrophoresis* **2017**, *38*, 1932–1938. [CrossRef]
22. D'Orazio, G.; Fanali, C.; Karchkhadze, M.; Chankvetadze, B.; Fanali, S. Enantiomeric separation of some chiral analytes using amylose 3,5-dimethylphenylcarbamate covalently immobilized on silica by nano-liquid chromatography and capillary electrochromatography. *J. Chromatogr. A* **2017**, *1520*, 127–134. [CrossRef] [PubMed]
23. D'Orazio, G.; Fanali, C.; Asensio-Ramos, M.; Fanali, S. Chiral separations in food analysis. *TrAC Trends Anal. Chem.* **2017**, *96*, 151–171. [CrossRef]
24. Marsol-Vall, A.; Sgorbini, B.; Cagliero, C.; Bicchi, C.; Eras, J.; Balcells, M. Volatile composition and enantioselective analysis of chiral terpenoids of nine fruit and vegetable fibres resulting from juice industry

by-products. *J. Chem.* **2017**, *2017*, 8675014. [CrossRef]
25. Schurig, V.; Mayer, S. Separation of enantiomers by open capillary electrochromatography on polysiloxane-bonded permethyl-beta-cyclodextrin. *J. Biochem. Biophys. Methods* **2001**, *48*, 117–141. [CrossRef]
26. Sun, P.; Armstrong, D.W. Effective enantiomeric separations of racemic primary amines by the isopropyl carbamate-cyclofructan6 chiral stationary phase. *J. Chromatogr. A* **2010**, *1217*, 4904–4918. [CrossRef] [PubMed]
27. Lajkó, G.; Grecsó, N.; Tóth, G.; Fülöp, F.; Lindner, W.; Ilisz, I.; Péter, A. Liquid and subcritical fluid chromatographic enantioseparation of Nα-Fmoc proteinogenic amino acids on Quinidine-based zwitterionic and anion-exchanger type chiral stationary phases. A comparative study. *Chirality* **2017**, *29*, 225–238. [CrossRef] [PubMed]
28. Ilisz, I.; Bajtai, A.; Lindner, W.; Péter, A. Liquid chromatographic enantiomer separations applying chiral ion-exchangers based on Cinchona alkaloids. *J. Pharm. Biomed. Anal.* **2018**, *159*, 127–152. [CrossRef]
29. Schmitt, K.; Woiwode, U.; Kohout, M.; Zhang, T.; Lindner, W.; Lämmerhofer, M. Comparison of small size fully porous particles and superficially porous particles of chiral anion-exchange type stationary phases in ultra-high performance liquid chromatography: Effect of particle and pore size on chromatographic efficiency and kinetic performance. *J. Chromatogr. A* **2018**, *1569*, 149–159. [CrossRef]
30. Patel, D.C.; Breitbach, Z.S.; Yu, J.; Nguyen, K.A.; Armstrong, D.W. Quinine bonded to superficially porous particles for high-efficiency and ultrafast liquid and supercritical fluid chromatography. *Anal. Chim. Acta* **2017**, *963*, 164–174. [CrossRef]
31. Wahab, M.F.; Wimalasinghe, R.M.; Wang, Y.; Barhate, C.L.; Patel, D.C.; Armstrong, D.W. Salient Sub-Second Separations. *Anal. Chem.* **2016**, *88*, 8821–8826. [CrossRef]
32. Barhate, C.L.; Regalado, E.L.; Contrella, N.D.; Lee, J.; Jo, J.; Makarov, A.A.; Armstrong, D.W.; Welch, C.J. Ultrafast Chiral Chromatography as the Second Dimension in Two-Dimensional Liquid Chromatography Experiments. *Anal. Chem.* **2017**, *89*, 3545–3553. [CrossRef]
33. Orosz, T.; Grecsó, N.; Lajkó, G.; Szakonyi, Z.; Fülöp, F.; Armstrong, D.W.; Ilisz, I.; Péter, A. Liquid chromatographic enantioseparation of carbocyclic β-amino acids possessing limonene skeleton on macrocyclic glycopeptide-based chiral stationary phases. *J. Pharm. Biomed. Anal.* **2017**, *145*, 119–126. [CrossRef] [PubMed]
34. Ismail, O.H.; Ciogli, A.; Villani, C.; De Martino, M.; Pierini, M.; Cavazzini, A.; Bell, D.S.; Gasparrini, F. Ultra-fast high-efficiency enantioseparations by means of a teicoplanin-based chiral stationary phase made on sub-2μm totally porous silica particles of narrow size distribution. *J. Chromatogr. A* **2016**, *1427*, 55–68. [CrossRef] [PubMed]
35. Ismail, O.H.; Antonelli, M.; Ciogli, A.; Villani, C.; Cavazzini, A.; Catani, M.; Felletti, S.; Bell, D.S.; Gasparrini, F. Future perspectives in high efficient and ultrafast chiral liquid chromatography through zwitterionic teicoplanin-based 2-μm superficially porous particles. *J. Chromatogr. A* **2017**, *1520*, 91–102. [CrossRef]
36. Rocchi, S.; Fanali, C.; Fanali, S. Use of a Novel Sub-2 μm Silica Hydride Vancomycin Stationary Phase in Nano-Liquid Chromatography. II. Separation of Derivatized Amino Acid Enantiomers. *Chirality* **2015**, *27*, 767–772. [CrossRef] [PubMed]
37. Harps, L.C.; Joseph, J.F.; Parr, M.K. SFC for chiral separations in bioanalysis. *J. Pharm. Biomed. Anal.* **2019**, *162*, 47–59. [CrossRef] [PubMed]
38. Jiang, Y.; Fan, J.; He, R.; Guo, D.; Wang, T.; Zhang, H.; Zhang, W. High-fast enantioselective determination of prothioconazole in different matrices by supercritical fluid chromatography and vibrational circular dichroism spectroscopic study. *Talanta* **2018**, *187*, 40–46. [CrossRef]
39. Tao, Y.; Zheng, Z.; Yu, Y.; Xu, J.; Liu, X.; Wu, X.; Dong, F.; Zheng, Y. Supercritical fluid chromatography–tandem mass spectrometry-assisted methodology for rapid enantiomeric analysis of fenbuconazole and its chiral metabolites in fruits, vegetables, cereals, and soil. *Food Chem.* **2018**, *241*, 32–39. [CrossRef] [PubMed]
40. Tan, Q.; Fan, J.; Gao, R.; He, R.; Wang, T.; Zhang, Y.; Zhang, W. Stereoselective quantification of triticonazole in vegetables by supercritical fluid chromatography. *Talanta* **2017**, *164*, 362–367. [CrossRef]
41. Cheng, Y.; Zheng, Y.; Dong, F.; Li, J.; Zhang, Y.; Sun, S.; Li, N.; Cui, X.; Wang, Y.; Pan, X.; et al. Stereoselective analysis & dissipation of propiconazole in wheat, grapes, & soil by supercritical fluid chromatography-tandem mass spectrometry. *J. Agric. Food Chem.* **2017**, *65*, 234–243. [CrossRef] [PubMed]
42. Parr, M.K.; Blokland, M.H.; Liebetrau, F.; Schmidt, A.H.; Meijer, T.; Stanic, M.; Kwiatkowska, D.; Waraksa, E.; Sterk, S.S. Distinction of clenbuterol intake from drug or contaminated food of animal origin in a controlled

administration trial–the potential of enantiomeric separation for doping control analysis. *Food Addit. Contam. Part A* **2017**, *34*, 525–535. [CrossRef] [PubMed]

43. Chen, J.; Zhang, T.; Zhang, Q.; Liu, Y.; Li, L.; Si, J.; Zou, Z.; Hua, H. Isobutylhydroxyamides from Sichuan Pepper and Their Protective Activity on PC12 Cells Damaged by Corticosterone. *J. Agric. Food Chem.* **2018**, *66*, 3408–3416. [CrossRef]
44. Zhang, Q.; Zhang, Z.; Tang, B.; Gao, B.; Tian, M.; Sanganyado, E.; Shi, H.; Wang, M. Mechanistic Insights into Stereospecific Bioactivity and Dissipation of Chiral Fungicide Triticonazole in Agricultural Management. *J. Agric. Food Chem.* **2018**, *66*, 7286–7293. [CrossRef]
45. Yang, X.; Qi, P.; Wang, X.; Wang, Z.; Sun, Y.; Wang, L.; Xu, X.; Xu, H.; Wang, Q.; Wang, X.; et al. Stereoselective Analysis and Degradation of Pyrisoxazole in Cabbage, Pakchoi, and Pepper by Liquid Chromatography Tandem Mass Spectrometry. *J. Agric. Food Chem.* **2017**, *65*, 8295–8301. [CrossRef] [PubMed]
46. Wang, L.; Wang, X.; Di, S.; Qi, P.; Sun, Y.; Yang, X.; Zhao, C.; Wang, X. Enantioselective analysis and degradation of isofenphos-methyl in vegetables by liquid chromatography-tandem mass spectrometry. *Environ. Sci. Pollut. Res.* **2018**, *25*, 18772–18780. [CrossRef]
47. He, R.; Fan, J.; Tan, Q.; Lai, Y.; Chen, X.; Wang, T.; Jiang, Y.; Zhang, Y.; Zhang, W. Enantioselective determination of metconazole in multi matrices by high-performance liquid chromatography. *Talanta* **2018**, *178*, 980–986. [CrossRef] [PubMed]
48. Chen, L.; Capone, D.L.; Jeffery, D.W. Chiral analysis of 3-sulfanylhexan-1-ol and 3-sulfanylhexyl acetate in wine by high-performance liquid chromatography–tandem mass spectrometry. *Anal. Chim. Acta* **2018**, *998*, 83–92. [CrossRef] [PubMed]
49. Zhou, L.; Lou, L.L.; Wang, W.; Lin, B.; Chen, J.N.; Wang, X.B.; Huang, X.X.; Song, S.J. Enantiomeric 8-O-4′ type neolignans from red raspberry as potential inhibitors of β-amyloid aggregation. *J. Funct. Foods* **2017**, *37*, 322–329. [CrossRef]
50. Rimkus, G.G.; Hoffmann, D. Enantioselective analysis of chloramphenicol residues in honey samples by chiral LC-MS/MS and results of a honey survey. *Food Addit. Contam. Part A* **2017**, *34*, 950–961. [CrossRef] [PubMed]
51. Zheng, X.; Qiao, L.; Sun, R.; Luo, X.; Zheng, J.; Xie, Q.; Sun, Y.; Mai, B. Alteration of Diastereoisomeric and Enantiomeric Profiles of Hexabromocyclododecanes (HBCDs) in Adult Chicken Tissues, Eggs, and Hatchling Chickens. *Environ. Sci. Technol.* **2017**, *51*, 5492–5499. [CrossRef]
52. Qi, Y.; Liu, D.; Liu, C.; Liang, Y.; Zhan, J.; Zhou, Z.; Wang, P. Enantioselective behaviour of the herbicide fluazifop-butyl in vegetables and soil. *Food Chem.* **2017**, *221*, 1120–1127. [CrossRef]
53. Nagai, T.; Ishikawa, K.; Yoshinaga, K.; Yoshida, A.; Beppu, F.; Gotoh, N. Homochiral asymmetric triacylglycerol isomers in egg yolk. *J. Oleo Sci.* **2017**, *66*, 1293–1299. [CrossRef]
54. Konya, Y.; Taniguchi, M.; Fukusaki, E. Novel high-throughput and widely-targeted liquid chromatography–time of flight mass spectrometry method for D-amino acids in foods. *J. Biosci. Bioeng.* **2017**, *123*, 126–133. [CrossRef]
55. Taniguchi, M.; Konya, Y.; Nakano, Y.; Fukusaki, E. Investigation of storage time-dependent alterations of enantioselective amino acid profiles in kimchi using liquid chromatography-time of flight mass spectrometry. *J. Biosci. Bioeng.* **2017**, *124*, 414–418. [CrossRef]
56. Nakano, Y.; Konya, Y.; Taniguchi, M.; Fukusaki, E. Development of a liquid chromatography-tandem mass spectrometry method for quantitative analysis of trace D-amino acids. *J. Biosci. Bioeng.* **2017**, *123*, 134–138. [CrossRef]
57. Ye, X.; Ma, S.; Zhang, L.; Zhao, P.; Hou, X.; Zhao, L.; Liang, N. Trace enantioselective determination of triazole fungicides in honey by a sensitive and efficient method. *J. Food Compos. Anal.* **2018**, *74*, 62–70. [CrossRef]
58. Zhao, P.; Wang, Z.; Gao, X.; Guo, X.; Zhao, L. Simultaneous enantioselective determination of 22 chiral pesticides in fruits and vegetables using chiral liquid chromatography coupled with tandem mass spectrometry. *Food Chem.* **2019**, *277*, 298–306. [CrossRef]
59. Zhang, Z.; Zhang, Q.; Gao, B.; Gou, G.; Li, L.; Shi, H.; Wang, M. Simultaneous Enantioselective Determination of the Chiral Fungicide Prothioconazole and Its Major Chiral Metabolite Prothioconazole-Desthio in Food and Environmental Samples by Ultraperformance Liquid Chromatography-Tandem Mass Spectrometry. *J. Agric. Food Chem.* **2017**, *65*, 8241–8247. [CrossRef]
60. Pan, X.; Dong, F.; Liu, N.; Cheng, Y.; Xu, J.; Liu, X.; Wu, X.; Chen, Z.; Zheng, Y. The fate and enantioselective behavior of zoxamide during wine-making process. *Food Chem.* **2018**, *248*, 14–20. [CrossRef]
61. Pan, X.; Dong, F.; Chen, Z.; Xu, J.; Liu, X.; Wu, X.; Zheng, Y. The application of chiral ultra-high-performance

liquid chromatography tandem mass spectrometry to the separation of the zoxamide enantiomers and the study of enantioselective degradation process in agricultural plants. *J. Chromatogr. A* **2017**, *1525*, 87–95. [CrossRef]

62. Aydoğan, C. Chiral separation and determination of amino acid enantiomers in fruit juice by open-tubular nano liquid chromatography. *Chirality* **2018**, 30, 1144–1149. [CrossRef]
63. Ruan, Y.; Zhang, X.; Qiu, J.W.; Leung, K.M.Y.; Lam, J.C.W.; Lam, P.K.S. Stereoisomer-Specific Trophodynamics of the Chiral Brominated Flame Retardants HBCD and TBECH in a Marine Food Web, with Implications for Human Exposure. *Environ. Sci. Technol.* **2018**, *52*, 8183–8193. [CrossRef]
64. Cutillas, A.B.; Carrasco, A.; Martinez-Gutierrez, R.; Tomas, V.; Tudela, J. Thyme essential oils from Spain: Aromatic profile ascertained by GC–MS, and their antioxidant, anti-lipoxygenase and antimicrobial activities. *J. Food Drug Anal.* **2018**, *26*, 529–544. [CrossRef] [PubMed]
65. Yao, G.; Gao, J.; Zhang, C.; Jiang, W.; Wang, P.; Liu, X.; Liu, D.; Zhou, Z. Enantioselective degradation of the chiral alpha-cypermethrin and detection of its metabolites in five plants. *Environ. Sci. Pollut. Res.* **2019**, *26*, 1558–1564. [CrossRef] [PubMed]
66. He, F.; Qian, Y.L.; Qian, M.C. Flavor and chiral stability of lemon-flavored hard tea during storage. *Food Chem.* **2018**, *239*, 622–630. [CrossRef] [PubMed]

Permissions

All chapters in this book were first published by MDPI; hereby published with permission under the Creative Commons Attribution License or equivalent. Every chapter published in this book has been scrutinized by our experts. Their significance has been extensively debated. The topics covered herein carry significant findings which will fuel the growth of the discipline. They may even be implemented as practical applications or may be referred to as a beginning point for another development.

The contributors of this book come from diverse backgrounds, making this book a truly international effort. This book will bring forth new frontiers with its revolutionizing research information and detailed analysis of the nascent developments around the world.

We would like to thank all the contributing authors for lending their expertise to make the book truly unique. They have played a crucial role in the development of this book. Without their invaluable contributions this book wouldn't have been possible. They have made vital efforts to compile up to date information on the varied aspects of this subject to make this book a valuable addition to the collection of many professionals and students.

This book was conceptualized with the vision of imparting up-to-date information and advanced data in this field. To ensure the same, a matchless editorial board was set up. Every individual on the board went through rigorous rounds of assessment to prove their worth. After which they invested a large part of their time researching and compiling the most relevant data for our readers.

The editorial board has been involved in producing this book since its inception. They have spent rigorous hours researching and exploring the diverse topics which have resulted in the successful publishing of this book. They have passed on their knowledge of decades through this book. To expedite this challenging task, the publisher supported the team at every step. A small team of assistant editors was also appointed to further simplify the editing procedure and attain best results for the readers.

Apart from the editorial board, the designing team has also invested a significant amount of their time in understanding the subject and creating the most relevant covers. They scrutinized every image to scout for the most suitable representation of the subject and create an appropriate cover for the book.

The publishing team has been an ardent support to the editorial, designing and production team. Their endless efforts to recruit the best for this project, has resulted in the accomplishment of this book. They are a veteran in the field of academics and their pool of knowledge is as vast as their experience in printing. Their expertise and guidance has proved useful at every step. Their uncompromising quality standards have made this book an exceptional effort. Their encouragement from time to time has been an inspiration for everyone.

The publisher and the editorial board hope that this book will prove to be a valuable piece of knowledge for researchers, students, practitioners and scholars across the globe.

List of Contributors

Tong Liu, Xue-Mei Nie, Feng-Ming Chen and Feng Zhang
Institute of Food Safety, Chinese Academy of Inspection & Quarantine, Beijing 100176, China

Chuan-Sheng Wang
College of Applied Chemistry, Shenyang University of Chemical Technology, Shenyang 110142, China

Jian-Qiao Cheng
Institute of Food Safety, Chinese Academy of Inspection & Quarantine, Beijing 100176, China
College of Applied Chemistry, Shenyang University of Chemical Technology, Shenyang 110142, China

Jingheng Ning, Xin Luo, Min Wang, Jiaojiao Li, Donglin Liu, Hou Rong, Donger Chen and Jianhui Wang
School of Chemistry and Biological Engineering, Changsha University of Science & Technology, Changsha 410110, China

Ana V. González de Peredo, Mercedes Vázquez-Espinosa, Estrella Espada-Bellido, Marta Ferreiro-González, Miguel Palma, Carmelo G. Barroso and Gerardo F. Barbero
Department of Analytical Chemistry, Faculty of Sciences, University of Cadiz, Agrifood Campus of International Excellence (ceiA3), IVAGRO, Puerto Real, 11510 Cadiz, Spain

Ana Jiménez-Cantizano and Antonio Amores-Arrocha
Department of Chemical Engineering and Food Technology, Faculty of Sciences, University of Cadiz, Agrifood Campus of International Excellence (ceiA3), IVAGRO, Puerto Real, 11510 Cadiz, Spain

Junwen Wang
College of Life Science and Technology, Huazhong University of Science and Technology, Wuhan 430074, China

Dan Luo
National Engineering Research Center for Nanomedicine, Wuhan 430075, China

Xiangliang Yang and Wei Liu
College of Life Science and Technology, Huazhong University of Science and Technology, Wuhan 430074, China
National Engineering Research Center for Nanomedicine, Wuhan 430075, China

Ming Liang, Ting Zhang, Xiquan Yin and Ying Zhang
Infinitus (China) Co., Ltd., Guangzhou 510665, China

Jian Jin
Institute of Chinese Materia Medica, Hunan Academy of Chinese Medicine, Changsha 410013, China
School of Chinese Medicine, Hunan University of Chinese Medicine, Changsha 410208, China
2011 Collaboration and Innovation Center for Digital Chinese Medicine in Hunan, Changsha 410208, China

You Qin and Yuhui Qin
Institute of Chinese Materia Medica, Hunan Academy of Chinese Medicine, Changsha 410013, China
School of Chinese Medicine, Hunan University of Chinese Medicine, Changsha 410208, China

Can Zhong, Jing Xie, Hao Liu, Dan Wan and Shuihan Zhang
Institute of Chinese Materia Medica, Hunan Academy of Chinese Medicine, Changsha 410013, China
2011 Collaboration and Innovation Center for Digital Chinese Medicine in Hunan, Changsha 410208, China

Jia Lao and Wei He
Resgreen Group International Inc., Changsha 410329, China

Rongrong Zhou
College of Pharmacy, Changchun University of Chinese Medicine, Changchun 130117, China

Wen Nie, Yu-zhu Li, Shuo Zhang, Yu Wang and Jie Guo
School of Food Science and Engineering, Hefei University of Technology, Hefei 230009, Anhui Province, China

Ke-zhou Cai and Cong-gui Chen
School of Food Science and Engineering, Hefei University of Technology, Hefei 230009, Anhui Province, China
Key Laboratory on Deep Processing of Agricultural Products for Anhui Province, Hefei 230009, Anhui Province, China

Bao-cai Xu
School of Food Science and Engineering, Hefei University of Technology, Hefei 230009, Anhui Province, China
Engineering Research Centre of Bio-Process, Ministry of Education, Hefei 230009, Anhui Province, China

Tianchen Ma and Hui Gao
School of Food Science and Engineering, Northwest University, Xi'an 710069, China

Haoan Zhao and Min Zhu
School of Chemical Engineering, Northwest University, Xi'an 710069, China

Caiyun Liu
Bee Product Research Center of Shaanxi Province, Xi'an 710065, China

Ni Cheng and Wei Cao
School of Food Science and Engineering, Northwest University, Xi'an 710069, China
Bee Product Research Center of Shaanxi Province, Xi'an 710065, China

Xixia Liu, Qi Lu, Sirui Chen, Fang Wang, Jianjun Hou, Chen Meng, Tianyuan Hu and Yaoyao Hou
Hubei Key Laboratory of Edible Wild Plants Conservation and Utilization, Hubei Normal University, Cihu Road, Huangshigang District, Huangshi 435002, China

Zhenlin Xu
Guangdong Provincial Key Laboratory of Food Quality and Safety, South China Agricultural University, Wushan Road, Tianhe District, Guangzhou 510642, China

Julia Keller, Luisa Hantschke and Matthias Koch
Department Analytical Chemistry, Reference Materials, Bundesanstalt für Materialforschung und-prüfung (BAM), Richard-Willstätter-Straße 11, 12489 Berlin, Germany

Hajo Haase
Department of Food Chemistry and Toxicology, Technische Universität Berlin, Gustav-Meyer-Allee 25, 13355 Berlin, Germany

Ivana Mitar, Ivica Ljubenkov, Ivana Anđelić and Nenad Vuletić
Department of Chemistry, Faculty of Science, University of Split, Ruđera Boškovi´ca 33, 21000 Split, Croatia

Nikolina Rohtek
University Department for Forensic Sciences, University of Split, Ruđera Boškovi´ca 33, 21000 Split, Croatia

Ante Prkić
Department of Analytical Chemistry, Faculty of Chemistry and Technology, University of Split, Ruđera Boškovi´ca 35, 21000 Split, Croatia

Yijin Tang and Christine Fields
Applied Food Sciences, Inc., 2500 Crosspark Road, Coralville, IA 52241, USA

Chuan Chai, Xiaobing Cui, Chenxiao Shan, Sheng Yu, Xinzhi Wang and Hongmei Wen
School of Pharmacy, Nanjing University of Chinese Medicine, Nanjing 210029, Jiangsu, China

Qiaoli Xie, Hongbo Zhang, Chunxia Yan, Shuguang Wei, Jianghua Lai, Yunpeng Wang and Bao Zhang
School of Forensic Medicine, Xi'an Jiaotong University, 76 Yanta West Road, Xi'an 710061, China

Fei Yan
School of Energy and Power Engineering, Chongqing University, 174 Shapingba Main Street, Chongqing 400030, China

Ying Zhou, Jian Guan, Weiwei Gao, Shencong Lv and Miaohua Ge
Jiaxing Center for Disease Control and Prevention, Zhejiang 314050, China

Meng-Qi Ling, Han Xie, Si-Yu Li, Yi-Bin Lan, Ruo-Nan Li, Chang-Qing Duan and Ying Shi
Center for Viticulture & Enology, College of Food Science and Nutritional Engineering, China Agricultural University, Beijing 100083, China
Key Laboratory of Viticulture and Enology, Ministry of Agriculture and Rural Affairs, Beijing 100083, China

Yu-Bo Hua
Shandong Taila Winery Co., Ltd., Weihai 264500, China

Jian Cai
College of Biological Resources and Food Engineering, Qujing Normal University, Qujing 655011, China

Chiara Fanali
Department of Medicine, University Campus Bio-Medico of Rome, Via Alvaro del Portillo 21, 00128 Rome, Italy

Giovanni D'Orazio
Istituto per I Sistemi Biologici, Consiglio Nazionale delle Ricerche, Via Salaria km 29, 300-00015 Monterotondo, Italy

Alessandra Gentili
Department of Chemistry, University of Rome "La Sapienza", Piazzale Aldo Moro 5, P.O. Box 34, Posta 62, 00185 Roma, Italy

Salvatore Fanali
Teaching Committee of Ph.D. School in Natural Science and Engineering, University of Verona, 37134 Verona, Italy

Index

www.ingramcontent.com/pod-product-compliance
Lightning Source LLC
LaVergne TN
LVHW081027210926
848485LV00004B/961